D1272895

CHINESE THOUGHT, SOCIETY, AND SCIENCE

CHINESE THOUGHT, SOCIETY, AND SCIENCE

The Intellectual and Social Background of Science and Technology in Pre-modern China

Derk Bodde

University of Hawaii Press

Honolulu

Printed in the United States of America
91 93 94 95 96 97 5 4 3 2 1

Library of Congress Cataloging-in-Publication Data

Bodde, Derk, 1909–
Chinese thought, society, and science : the intellectual and social background of science and technology in pre-modern China / Derk Bodde.
p. cm.
Includes bibliographical references and index.
ISBN 0-8248-1334-0
1. Science—China—History. 2. Science—Social aspects—China—History. 3. Technology—China—History. 4. Technology—Social aspects—China—History. 5. China—Civilization. I. Title.
Q127.C5B63 1991 91-4437
509.51—dc20 CIP

University of Hawaii Press books are printed on acid-free paper and meet the guidelines for permanence and durability of the Council on Library Resources

To Galia

Wife, Friend, Critic

These Six and Fifty Years

and

To My Mentor

J. J. L. Duyvendak (1889–1954)

Professor of Chinese

University of Leiden

CONTENTS

PREFACE

This book, I sometimes like to think, began life as a sailing ship. In 1973, more than fifteen years ago, I visited the British historian of Chinese science, Joseph Needham, at Cambridge University in order to formulate plans with him for the future ship. During the three academic years of 1974–1977, spent by me at Cambridge, her keel was laid, her hull was built, and at the end of my stay I was able to launch and ferry her across the Atlantic to Philadelphia and the University of Pennsylvania. There, after less than two years of further work, she was ready for a second launching and what I optimistically supposed would be a single return voyage to England. To my disappointment, however, that voyage proliferated into a series of back-and-forth crossings, followed by inconclusive headings toward other ports. During these peregrinations, the ship sometimes enjoyed favorable winds and made good progress, sometimes encountered head winds, squalls, or storms, but most often lay idle for long periods because of lack of wind. Repeatedly I repaired or improved her while she was under way. On several occasions I hauled her from the water entirely for major reconstruction.

Finally, in late 1989, a combination of fortunate circumstances enabled me to set her on a course that brought her safely to Honolulu and the University of Hawaii Press. With admirable promptness, the press agreed to convert her from a ship into a book.

Eight organizations and twelve individuals have been of major help at various stages of the building, voyaging, and ultimate conversion of this ship into a book. To all of them I am most grateful.

The organizations are as follows:

The National Endowment for the Humanities, Washington, D.C., whose Senior Fellowship enabled me to spend the academic year of 1974–1975 at Cambridge University.

The University of Pennsylvania, Philadelphia, which granted me sabbatical leave during that same year and then allowed me early retirement so that I could return to Cambridge during the following two years.

The American Philosophical Society, Philadelphia, whose travel grant helped me to return to Cambridge in 1975.

Clare Hall, Cambridge, whose intellectual stimulation and social fellowship greatly enhanced my life during the two years of 1975–1977 when I was a Visiting Fellow.

Gonville and Caius College, Cambridge, whose little house overlooking its playing field provided a cozy haven for my wife and me during the three years that we rented it.

Three libraries and their staffs: Cambridge University Library, especially its Oriental Division (Margaret I. Scott, curator, now retired); the library and librarians of the Faculty of Oriental Studies, Cambridge University; and Van Pelt Library, University of Pennsylvania, especially its East Asian Collection (Nancy Cheng, Chinese Bibliographer, now retired).

And to the following individuals:

Joseph Needham (master of Gonville and Caius College and director of the Needham Research Institute, Cambridge, now retired), whose early volumes of *Science and Civilisation in China* inspired me to write the present book, whose help and friendship meant much to me during my years at Cambridge, and whose library on the history of science in China and the West gave me familiarity with publications about which I would otherwise have been ignorant.

Michael Loewe (Chinese, Cambridge, now retired), who more than once went to unusual lengths to answer my queries on bibliographical and factual matters.

Adele Rickett (Chinese, Maryland, now retired) and W. Allyn Rickett (Chinese, Pennsylvania, also retired), who gave me wise counsel at an important moment.

Erik Zürcher (Chinese, Leiden) and John K. Fairbank (Chinese history, Harvard, now retired), who did all they could.

Victor Mair (Chinese, Pennsylvania), who on his own initiative gave me the idea and impetus to steer the book toward Honolulu and the University of Hawaii Press.

David L. Hall (philosophy, Texas at El Paso) and Roger T. Ames (philosophy, Hawaii), who each went far beyond the call of duty by formulating lengthy and thoughtful comments in the course of reading my manuscript for University of Hawaii Press. I have benefited much from many of these comments and regret that I have been unable to employ them all.

Iris M. Wiley (executive editor, University of Hawaii Press), who has repeatedly shown interest, understanding, and consideration in the course of letters and telephone conversations. Along with her I am very happy to thank Eileen D'Araujo and Cheri Dunn, editors, and Joanne Sandstrom, copy editor, for their extraordinary care while preparing my manuscript for the printed page.

In addition, some eighteen other colleagues, former students, and scholars in various fields have helped me by answering questions, sending their writings, or supplying copies of Chinese and other texts when these were unavailable to me. Although their names are all mentioned in the footnotes, I wish to thank them warmly here as a group.

At the University of Pennsylvania: Roger Allen (Arabic); the late Lloyd W. Daly (classics); Judah Goldin (Hebrew, now retired); Ludo Rocher (Sanskrit).

At Cambridge University: David R. Buxton (formerly Research Fellow, Clare Hall) and Eveleyn A. Silber (former graduate student, Clare Hall).

At other institutions: Susan Blader (Chinese, Dartmouth); Janusz Chmielewski (Chinese, Warsaw); Krzysztof Gawlikowski (compiler of Chinese encyclopedia, Naples); the late L. Carrington Goodrich (Chinese, Columbia); Jean Gordon Lee (curator of East Asian Art, Philadelphia Museum of Art, now retired); James T. C. Liu (Chinese history, Princeton); John Major (Chinese history, Dartmouth; then The Asia Society and the Book-of-the-Month Club, Inc., both New York City); Diane Perushek (Gest Library, Princeton); Renée Pigeaud (Dutch psychotherapist and art historian); Henry Rosemont, Jr. (philosophy, St. Mary's College, Maryland); Alexander C. Soper (East Asian Art, New York University, now retired); Tsien Tsuen-hsuin (professor and librarian of Chinese, Chicago, now retired).

My thanks to all these organizations and individuals, with the hope that this book may seem to them worth the effort.

Philadelphia, Pennsylvania
1 October 1990

CHRONOLOGY OF CHINESE DYNASTIES

(All dates before 841 B.C. are traditional only)

PRE-DYNASTIC

Archaeological Record

A series of stone-age cultures, widely distributed in North and South China and chronologically ranging from early Paleolithic to late Neolithic.

Literary Tradition

A series of legendary sage-rulers and creators of successive stages of civilization, of whom the most important are:

Fu Hsi, Subduer of Animals (2852–2738)
Shen Nung, the Divine Husbandman (2737–2698)
Huang Ti, the Yellow Lord (2697–2598)
Yao (2357–2256)
Shun (2255–2206)

PRE-IMPERIAL DYNASTIES (2205–222 B.C.)

Hsia (2205–1766)

Unconfirmed as yet by archaeology. Founded by Yü the Great, who before becoming ruler saved humankind from a vast flood and made the world fit for sedentary human habitation.

Shang (1765–1123)

Confirmed by archaeology; earliest script and bronze.

Chou (1122–256; actual founding probably ca. a century later)

China's age of "feudalism": many small principalities nominally under Chou suzerainty.

841 B.C. onward: a single standard chronology

Western Chou (1122–771)

Eastern Chou (770–256)

Spring and Autumn *(Ch'un-ch'iu)* period (722–481)

Warring States *(Chan-kuo)* period (403–222)

Increasing warfare between half a dozen large nation-states; rise of iron technology; classical age of Chinese philosophy.

THE EARLY EMPIRES (221 B.C.–A.D. 220)

Ch'in (221–207 B.C.)

Creation of China's first universal empire: a centralized bureaucratic government with unified script, law, weights, and measures; building of the Great Wall.

Han (206 B.C.–A.D. 220)
Age of classical empire; elaboration of Ch'in system of government; expansion into Central Asia and elsewhere through wars and trade; rise of Confucianism to orthodoxy.
Former or Western Han (206 B.C.–A.D. 5)
Wang Mang as Regent and Acting Emperor (A.D. 6–8); then as Emperor of "Hsin" dynasty (A.D. 9–23)
Later or Eastern Han (A.D. 25–220)

PERIOD OF DISUNITY (190/220–589)
Spread of Buddhism; division between north and south; invasion of the north by "barbarian" groups who establish successive non-Chinese states and dynasties.
Three Kingdoms (190/220–280)
Chin dynasty (265–316)
Northern and Southern Dynasties (317–589)

THE MIDDLE EMPIRES (590–1279)
Sui (590–617)
Reunification of north and south into one empire.
T'ang (618–906)
Political grandeur, cultural brilliance, cosmopolitanism, apogee of Buddhism, beginning of printing.
Five Dynasties (907–959)
An interim period of successive brief ruling states.
Sung (960–1279)
A "Renaissance" dynasty: political weakness coupled with great economic development, urbanization, spread of printing, rise of Neo-Confucianism, decline of Buddhism.
Northern Sung (960–1125)
Southern Sung (1126–1279)

THE LATER EMPIRES (1280–1911)
Yuan (Mongol) dynasty (1280–1367)
China's first total rule by non-Chinese; cosmopolitanism; Marco Polo and other Europeans in China.
Ming (1368–1643)
Restoration of Chinese rule; government voyages to South Asia and East Africa (early fifteenth century); then growing autocracy and crystallization of culture.
Ch'ing (Manchu) dynasty (1644–1911)
China's second total rule by non-Chinese, who, however, become strongly Chinese in culture. Political expansion into

Northern and Central Asia. Coming of the Jesuits, ca. 1600 onward; heavy political/economic/cultural impact of the West, early nineteenth century onward.

REPUBLICAN CHINA (1912–)

Republic of China (1912–49)

Sun Yat-sen; China divided by warlord rulers; then reunification (1928) by Kuomintang Party of Chiang Kai-shek; Sino-Japanese War (1937–45); civil war between Kuomintang and Chinese Communists.

People's Republic of China (1949–)

Kuomintang and Chiang Kai-shek expelled to Taiwan (1949) by Chinese Communists and Mao Tse-tung (who heads government until his death in 1976).

I. INTRODUCTION

1. The What and Why of This Book

Some remarkable scholarly achievements have greatly enlarged our understanding of Chinese civilization during the second half of the twentieth century. Two I find particularly outstanding.

The first, still being continued in China by dozens of scholars and hundreds of anonymous workers, is archaeology. Since the founding of the People's Republic of China in 1949, archaeology has unearthed not only such artifacts as the mortuary underground army guarding the tomb of the Ch'in First Emperor but also, more importantly, such hitherto unknown texts as the pre-imperial laws of Ch'in discovered in 1975 in the tomb of a local Ch'in official. How amazing that for centuries and often for millennia these and other artifacts and texts should have been lying underground, as if expectantly waiting to be now uncovered within the space of a few decades.

The second outstanding achievement, still being continued at Cambridge University by a single British historian of science aided by a handful of collaborators and assistants, is Joseph Needham's *Science and Civilisation in China.* Since the first volume of this work was published in 1954, five other volumes and many parts of volumes have appeared, each separately bound. As of early 1990, they have reached fifteen separate tomes containing just over 10,000 pages. The subjects discussed include Chinese mathematics, astronomy, physics, mechanical and civil engineering, paper and printing, alchemy and chemistry, agriculture and botany, and much more. Each topic is treated from its beginnings down to approximately 1600 and, when appropriate, compared with parallel developments in the West and other civilizations. The cutoff date of 1600 was selected because it was then that Matteo Ricci and other Jesuits began going to China, taking with them Western ideas and thus interrupting the hitherto largely separate evolution of Chinese scientific thought.[1]

---

1. In later chapters, the fifteen tomes of this history will be commonly cited under Needham's name, each followed by its own date of publication (e.g., Needham (1954), Needham (1956), etc.). Three of the tomes, however, and portions of others, have been written by other scholars. These will therefore be referred to by the names of their authors as Bray (1984), Huang Hsing-tsung (1986), Kuhn (1988), Sivin (1980), Tsien (1985).

It is this tremendous enterprise that stimulated the writing of the present book. As successive volumes of *Science and Civilisation in China* appeared, my interest increased to the point where I began to wish that I myself could personally participate. The wish resulted in correspondence and then a meeting with Dr. Needham, from which emerged the proposal that I try my hand at writing the greater part of a section planned for the seventh and final volume of *Science and Civilisation.* In it I would deal with the many intellectual factors, either favorable or unfavorable, that I regarded as having significantly influenced China's scientific development down to roughly 1600.

The consequence was three stimulating years in Cambridge (1974–1977) followed by a little over another (1977–1978) in Philadelphia after my return. Most of the present book was written during those four and more years. Only considerably later, when the manuscript began to undergo intensive examination preparatory to publication, did differing opinions become manifest between Dr. Needham and me concerning aspects of Chinese civilization. First among them, but by no means the last, was the question of whether or not the Chinese written language had historically functioned effectively as a medium for the exposition of topics relevant to science. As the next chapter will show, my answer was less sanguine than that of Dr. Needham.

Despite major efforts on both sides, the differences persisted until finally, in 1985, I felt obliged, with great regret, to ask for the return of my manuscript. Since then I have endeavored to broaden the manuscript into a book able to stand firmly on its own feet. Accordingly, I have added several sections that are entirely new and have also tried to update the book by adding comments or references to publications appearing after its gestation in Cambridge. Further, in the hope that the book may interest nonspecialists as well as sinologues, I have added a good many dates, identifications, and explanations for the sake of the former even though the latter will no doubt find many of them superfluous.

Here it should be pointed out that a good many references to *Science and Civilisation in China* can be expected in these pages. Such references are inevitable in almost any book today concerned even peripherally with pre-modern Chinese science. The reason is that *Science and Civilisation* is extraordinarily exhaustive and comprehensive in a field that has hitherto been approached only in a very piecemeal fashion.[2]

Now let us see what the present book tries to say and do. Its subtitle,

---

2. Sivin (1988) provides an excellent survey of pre-1988 scholarship on medicine and science (especially astronomy) in imperial China, but the work appeared too late to be consulted for the present book.

"The Intellectual and Social Background of Science and Technology in Pre-modern China," contains four important paired words. Let us consider the second pair first. What do we mean when we talk about *science* and *technology* in pre-modern China? Of the two, technology is by far the easier to identify. China's Great Wall is obviously a product of technology, as are the Grand Canal, a bridge or a dam, or, reducing the scale, a Chinese mallet irrigation pump, a Chinese wheelbarrow, or even those useful extensions of the fingers that Westerners call chopsticks.

Science, on the other hand, is much more difficult to define because more concerned with theories, procedures, and ideas. In this book's final chapter several definitions of the word will be cited, but they are all modern and Western, and all completely accept the idea of "laws of nature." Thus none could have been formulated in pre-modern China. Nor, for that matter, could they have been formulated in pre-modern Europe. Better for our present purpose, because more general and comprehensive, is that briefest of definitions offered by George Sarton at the beginning of one of his works: "The purpose of this work is to explain briefly . . . the development of science, that is of *systematized positive knowledge*" (Sarton 1927–1948:1.3; emphasis in original). More on the semantic aspects of science in pre-modern China will be noted in the next section.

Before we proceed to the other two words in the subtitle, more must be said about how this book approaches science and technology.

First, we should keep in mind the important differences that have separated science from technology in China. They stem in good part from the fact that what we today term science was primarily the domain of the scholar whereas technology was that of the artisan. What this means will be discussed later and summarized in the final chapter.

Second, we are constantly confronted by the fact that the scientific and technological revolution that transformed Europe from around 1600 onward occurred nowhere else outside the Western world until much later. Why it failed to occur in China, despite China's notable pre-1600 accomplishments in science and technology is, of course, a central problem. In order to address it, our manifold explorations of pre-modern Chinese language, cosmology, religion, institutions, and much else will be accompanied by frequent comparisons with similar intellectual features in Europe both before and after 1600.

Third, the fact that these comparisons seem, scientifically speaking, to be mostly unfavorable to China may lead some readers to suspect that their partial or primary purpose has been to exalt the West at the expense of China. If anyone entertains such a suspicion, I can only deny it in the most emphatic terms. I myself, for example, happen to sympathize strongly with the Confucian opposition to warfare and competition, as against the glorification of both that has too often characterized the West. Yet at the same time I am obliged to recognize that prevailing Western attitudes on these and related matters, in combination with a great many other factors, may have been essential for the rise of a modern science. Or again, faced by the fact that we humans today possess the technological capability to destroy both ourselves and the world, I begin to wonder whether the ancient Taoist distrust of innovative technology may not have had some justification, even though it was no doubt premature and excessive.

However, such moral, aesthetic, or other judgments are quite out of place in this book. Here our sole mission is to determine, as best we can, what may have been the factors in Chinese civilization that either favored or hindered scientific and technological progress. This task I have tried to carry out as objectively and carefully as possible, ever aware of its magnitude, my own limitations, and the inescapable presence of subjectivity in what results.

Now let us turn to the remaining two key words in the book's subtitle. One of them, *intellectual,* is to be understood as often covering a much wider range of thoughts, beliefs, feelings, and other intangibles than come to mind in such expressions as "the intellectual view of life" or "reasoning on an intellectual level." In short, most of the topics presented in this work, ranging from the characteristics of written Chinese to the attitude of Chinese thinkers toward nature, are treated here as broadly "intellectual."

The subtitle's fourth word, *social,* does not signify that a systematic description of Chinese social institutions is to be found in this book. Rather, the book focuses on the attitudes toward these institutions manifested by many Chinese, not upon the structure and functioning of the institutions per se. At this point we should remember the truism that most non-Buddhist educated Chinese (especially but not exclusively Confucians) were deeply concerned with two related problems: how a few individuals could best learn how to become morally responsible rulers or administrators, and how a great many others could best learn how to become cooperative human beings within their several social groups, beginning with the family.

As preparation for later chapters, it may be helpful to see how the intellectual and social factors thought to have influenced pre-modern Chinese science and technology are grouped in this work; how Chinese science and technology may be classified for modern study; and how their development has varied during the course of Chinese history.

First, as to the influencing factors, they and their many ramifications are grouped here under six main topics: language; concepts of time, space, and things; religion; government and society; morals and values; and man's relationship to nature.

Second, as to the classifying of Chinese science and technology, what follows is the scheme used in *Science and Civilisation in China,* which well displays their great diversity. It is here presented with a reduction of the categories to twenty, a slight rearrangement of their sequence, and the reminder that, for the most part, these categories were unknown as such to the pre-modern Chinese themselves. For many of the categories, moreover, I have parenthetically listed a very few of their constituent artifacts or techniques.

1. Mathematics (magic squares, geometry, algebra, etc.)
2. Astronomy (calendars, cosmological theories, astronomical instruments, etc.)
3. Meteorology (climate, precipitation, tides, etc.)
4. Geography and cartography (maps, grids, etc.)
5. Geology, seismology, mineralogy (fossils, seismographs, asbestos, etc.)
6. Physics (light, sound, magnetism including the compass, etc.)
7. Mechanical engineering (animal traction, hydraulic engineering, clockwork, etc.)
8. Civil engineering (roads, walls, bridges, etc.)
9. Nautical technology (ships, navigation, propulsion, etc.)
10. Paper and printing (block printing, movable type, etc.)
11. Alchemy and chemistry (ideas of the afterlife, apparatus, theory, etc.)
12. Military technology (weapons, gunpowder, cavalry, etc.)
13. Agriculture (implements, crops, fertilizers, etc.)
14. Botany and zoology
15. Animal husbandry
16. Medicine
17. Textiles
18. Metallurgy
19. Ceramics
20. Salt industry

Categories 15–16 and 18–20 have not yet been published in *Science and Civilisation* as of this writing; categories 12, 14, and 17 have appeared only in part.

As for the chronological development of Chinese science and technology, it is possible, by telescoping the first two epochs listed in our Chronology of Chinese Dynasties into one, to subsume all Chinese history and prehistory down to the end of the Chinese empire in 1911 under five main epochs. They are presented here primarily for those readers who are unfamiliar with Chinese history. The general comments under each include remarks specifically about science and technology.

1. *Prehistory and early historic* (Neolithic and pre-Neolithic cultures widely scattered across China proper; then three pre-imperial dynasties down to 222 B.C.). Historic China presently begins with the Shang dynasty (trad. 1765–1123 B.C.), which also has the earliest Chinese script and a highly developed bronze technology. Before too long, however, modern archaeology may well push the beginnings of historic China backward by confirming, for example, the historicity of the much later historical accounts concerning the pre-Shang Hsia dynasty (trad. 2205–1766). Someday, too, answers will probably be found to some of the questions that now plague the early history of Chinese science and technology. Two among them are the beginnings of iron technology and of the traction-plow (both, presumably, sometime during the Chou dynasty, trad. 1122–256 B.C.). During the earlier greater part of this long epoch, surviving texts are too sparse, too restricted in subject matter, and too difficult to date to provide more than glimpses of science and technology. Beginning with Confucius (551–479), however, the last centuries of the Chou are marked by a growing social and political maturation of the dozen or more mutually warring Chinese principalities existing at that time in north and central China proper. Written literature also proliferates, culminating during the fourth and third centuries B.C. in the golden age of early Chinese philosophy. The social, political, and cosmological ideas of the late Chou "Hundred Schools" will be repeatedly cited in these pages, but what they concretely tell us about science and technology is relatively slight.

2. *The Early Empires* (Ch'in/Han, 221 B.C.–A.D. 220). A crucial date in Chinese history is the shift from warring states to universal empire in 221 B.C. The subsequent age of classical empire is marked by the elaboration and universalization of the Chinese bureaucratic form of government and a military and cultural expansion of the Chinese *oikoumene* into northern, central, and southern Asia.[3] By the first century B.C. the

---

3. The world of Greek civilization constituted, in Greek eyes, the *oikoumene* or "inhabited world," in contrast to the outlying lands of the "barbarians." Because this dichotomy between civilization and barbarism was also conspicuous in traditional Chinese thinking, the word *oikoumene* will be occasionally used in these pages to refer not merely to the world of China as politically defined, but also, and more importantly, as pervaded by Chinese moral and social values.

empire of Han has become larger than its contemporary Rome at its greatest extent. Most of the Chou "Hundred Schools" of thought disappear and are replaced by a government-fostered eclectic Confucianism, but literature grows into many genres. A notable result is Ssu-ma Ch'ien's *Shih chi* (Historical Records), which is China's first "universal history" (from the legendary beginnings down to ca. 100 B.C.). Science and technology flourish and produce such inventions as the world's first paper and first seismograph.

3. *The Period of Disunity* (A.D. 190/220–589). The disintegration of classical China, nominally in 220 but actually in 190, is followed during the next four "medieval" centuries by a continuing political fragmentation. The major division is between a series of "barbarian" dynasties in the north, ruled by successive non-Chinese tribal groups, and a parallel series of Chinese-ruled dynasties in the south (actually in central China). Buddhism, which in Han times (beginning of the first century A.D.) had entered China from India via Central Asia, becomes the major intellectual/religious force during the Period of Disunity. Institutional or "religious" Taoism also becomes important, whereas Confucianism is weak and makes little intellectual progress. Scientific development is intermittent and scattered.

4. *The Middle Empires* (Sui/T'ang/Sung, 590–1279). The renewal of unified empire in 590 is followed by political grandeur, geographical expansion, and cosmopolitanism. Under the T'ang, Buddhism reaches its apogee in the tenth century but then slowly declines. Under the Sung, Confucianism revives (eleventh century onward) in the form known to Westerners as Neo-Confucianism. It eventually becomes the state orthodoxy, a status it is to hold until the twentieth century. Printing begins during the T'ang and spreads greatly during the Sung, which is an age of major socioeconomic and cultural importance despite its political weakness. Science probably reaches its height during the Sung.

5. *The Later Empires* (Yuan/Ming/Ch'ing, 1280–1911). Under the Yuan all of China is for the first time ruled by non-Chinese (the Mongols); under the Ch'ing this phenomenon is repeated (the Manchus). Both dynasties are for the most part politically powerful, whereas the intervening Chinese dynasty of Ming starts powerfully but then gradually declines. These later empires are marked by significant cultural advances in some fields (rise of fiction and drama, new norms of humanistic critical scholarship, government-sponsored compilations of enormous encyclopedias and the like) but a growing crystallization of

culture in others. By and large, science also suffers from stagnation, so that by around 1600, when the Jesuits begin to enter China, the creativity of most Chinese science, save for a few notable exceptions, has largely ceased.

A bird's-eye view of these five epochs reveals that two contrasting phases dominate the history of Chinese science and technology: one a phase mostly of vigor and originality, extending from pre-imperial times through the Sung dynasty; the other a phase of post-Sung gradual decline and stagnation, sharply contrasting, toward the end, with the rise of the "new science" in Europe from about 1600 onward.

For technology, these phases are confirmed by a table in Needham (1954:242) in which thirty-six Chinese mechanical or other techniques are listed under twenty-six lettered groups. They include items like the piston-bellows, efficient harness for draught animals, crossbow, kite, gunpowder, magnetic compass, paper, printing, porcelain, and so on. For each item the approximate lag in centuries is indicated between its earliest known appearance in China and subsequent appearance in Europe. The time lags (from one to seventeen centuries) sometimes represent demonstrable transmissions of the techniques from East to West, sometimes unproven but probable transmissions, and sometimes the seemingly independent appearance of the techniques at differing times on the two sides of Eurasia.

What concerns us here, however, is that no fewer than sixteen of these thirty-six techniques go back to Han or pre-Han times, whereas only two postdate the Sung. More will be said in our final chapter about the two phases here demonstrated for technology, as well as a simpler "layman's" way of grouping China's scientific achievements. Other than this, no systematic attempt will be made there or elsewhere to describe the rise and fall or other characteristics of particular scientific theories or techniques.

In conclusion, it should be noted that a few small parts of the book have already been published elsewhere, either unchanged or with only minor modifications:

Some sixteen passages are quoted verbatim in the study of social evolution by Hallpike (1986), especially his Chapter 6, Section 2, discussing what he calls the "core principles" of pre-modern Chinese society.

The present book's Chapter VII, Section 2a, "Did 'Laws of Nature' Exist in China?," is largely though not entirely the same as Bodde 1979.

Chapter VI, Section 2, "The Question of Sex," has been published as Bodde 1985.

Chapter V, Section 3, "Social Classes," and the Appendix, "The

Four Social Classes," have been published, in combined and shortened form, as Bodde 1990.

Chapter II, Section 6, "Punctuation," presently constitutes Bodde (forthcoming).

2. Problems

A book such as this inevitably raises various problems. Perhaps most immediately evident is that of the book's wide subject matter. How, one may ask, can any single individual do justice to the variety of subjects that this book is obliged to cover? A possible solution might be to have the writing done by a group of specialists. Sometimes the product of such collective effort is very good, but sometimes, too, it is either too bland and impersonal or too inconsistent and divided. On the whole, I prefer the more personal and perhaps more forceful or controversial one-man or one-woman approach, despite its admitted risks.

Another problem is that this work has to do with Chinese science and yet is written by someone who is neither a scientist nor a historian of science but a humanist. The problem is a real one, yet it may not be as serious as seems at first sight. This is because the major focus of the book is not upon science and technology per se but upon the intellectual and social factors that may have influenced them; these factors are overwhelmingly humanistic. Take, for example, the second chapter on language, where well-known passages from the classics, early thinkers, and histories are cited to illustrate certain linguistic features. It might be argued that had lesser-known scientific texts been cited instead, the conclusions might have been different. I myself, however, doubt this presumption, except marginally, because the classical and other texts that were actually cited were, to a far greater extent than any of the scientific texts, the cultural heritage of virtually all pre-modern educated Chinese. They pervasively influenced the thinking of all of them, including those who were scientifically minded.

This same second chapter, although it discusses in detail the characteristics of the classical written language known as Literary Chinese, says nothing about the suitability or unsuitability of this language for purposes of formal logic. I regret this omission yet have made no effort to repair it. The reason is simply that my training in linguistics and logic is inadequate for the task.

A number of scholars have approached the same subject in recent years, including Chmielewski (1962–1969), Hansen (1983), and Bloom (1981). Using symbolic logic, Chmielewski concludes that valid examples of both explicit and implicit logic are to be found in early Chinese texts. His conclusions are criticized by Hansen (1983:5), who then elaborates a theory of his own about what he calls "mass nouns" in ancient Chinese prose. According to this theory, the noun *ma,* "horse," for example, supposedly actually signified to the ancient Chinese something like "the-horse-kind-of-stuff." This idea and a good deal else in Hansen's book has in turn been strongly criticized by Harbsmeier (1983–1985). Meanwhile Bloom (1981), by asking Chinese and American students to answer questionnaires respectively prepared in Chinese and English, concludes that Chinese speakers have much more trouble than their American counterparts in handling counterfactual statements or theoretical statements formulated in an abstract manner. This conclusion, based primarily on the written form of modern spoken Chinese, is shown to be inapplicable to Literary Chinese by Garrett (1983–1985), who questions its modern validity as well.

The linguistic-logical debate has been lively, ongoing, stimulating, and often inconclusive. Its topics include, among many others, the question of whether or not the noun is functionally dominant in Literary Chinese, the comparability of Aristotle's categories with those of the Later Mohists, and the "rather surprising conclusion that the Chinese language is a well organized system which requires only minimal corrections when put into new, logico-philosophical, surroundings" (Reding 1986b:55; see also 1986a). Its debaters include, besides those already named, such scholars as Fang Wan-chuan (1984), Bao Zhiming (1985), Gernet (1985), Wu Kuang-ming (1987), Hall and Ames (1987:esp. 253–255, 261–268, 290–304), and Graham (1989:390–428).

Confronted by this assemblage, I find it quite impossible to do them justice within any reasonable space. Instead, I can only say that this book's next chapter was written completely independently from any of them and is less theoretical and more pragmatic in its approach to the central question of whether Literary Chinese has historically constituted an effective medium for "scientific" communication.

As to Chinese logic, the important thing to keep in mind is that very few ancient or later Chinese thinkers ever showed much interest in logical demonstration and kindred topics. Many of the writings of these few thinkers subsequently disappeared or became textually corrupt owing to the indifference or misunderstanding of most of their transmitters before the present century. However, the school of thought known as the Mohists was exceptional in early China because of its deep interest in logical-semantic matters combined with its general sociological importance. Mohist ideas will often be discussed in later chapters.

From this excursion let us turn now to other problems confronting this book. Among them is the intangible quality of so much of what it discusses. How, for example, is one to measure with any certainty the impact of language, social institutions, or religion on scientific development? Obviously, in the absence of quantifiable data, what purports to be solid judgment can often only amount to personal opinion. This is why one sometimes wonders whether the social sciences, especially when applied to pre-modern societies, really deserve the name of "science."

This last word introduces a further problem with respect to China: the fact that before that country experienced extensive contacts with the modern West, its people had no single conception or generic term equivalent to the word "science":

> The sciences were not integrated [in China] under the dominion of philosophy, as schools and universities integrated them in Europe and Islam. Chinese had sciences but no science, no single conception or word for the overarching sum of all of them. Words for the level of generalization above that of the individual science were much too broad. They referred to everything that people could learn through study, whether of Nature or human affairs. (Sivin 1982:49)

Thus whenever Chinese science is mentioned in this work, we must remember that the term represents an abstraction unknown as such to the people being talked about. The fact that culture in pre-modern China was diffused, that is to say, that philosophy, art, cosmology, science, and other creations of the mind were not recognized as autonomous disciplines, increases the difficulty of isolating any of them, including science, for individual analysis.

Throughout this book, also, we should heed the hint made earlier not to think of "science" solely in terms of the "new science" that arose in Europe from around 1600 onward. Many pre-1600 ideas may perhaps be regarded by some scientists today as too magical, superstitious, or imaginative to be called "scientific." Historians of science, however, will view them differently, because they will know their historical contexts and understand that within these contexts the ideas represent systematic attempts to explain natural phenomena coherently. Accordingly, these historians will recognize them as belonging to a given scientific tradition.

This is true, for example, of Chinese theories about the Five Elements or the yin and yang principles. These theories did not originate because people at an early time deliberately carried out systematic observations and experiments. Nevertheless, they belong to the Chinese tradition of science because, when they were formulated, they represented systematic attempts, often devoid of supernaturalism, to arrange and classify the data of the natural and human worlds into meaningful interrelationships. Only later, when the theories froze into barriers effectively blocking any subsequent potential attempts to approach closer to reality, did they lose their scientific significance and become outmoded dogma.

In an article discussing why the scientific revolution did not occur in China, Sivin (1982) warns against several fallacies. Among them is "the fallacious assumption that one can make sense of the evolution of science by looking at intellectual factors alone, or socio-economic factors alone, according to preference" (Sivin 1982:58). This fallacy was very much in my mind when I wrote this book, which obviously by itself can provide only a partial picture. It should be complemented by whatever other studies may be relevant, such as, at present, those by Ray Huang (1986 and 1988).

Another fallacy noted by Sivin is one

> often adduced to explain the failure of China to beat Europe to the Scientific Revolution despite an early head start, namely the predominance of a scholar-bureaucrat class immersed in books, faced toward the past, and oriented toward human institutions rather than toward Nature as the matrix of the well-lived life. But in Europe at the onset of the Scientific Revolution we are faced with the predominance of the Schoolmen and dons, immersed in books, faced toward the past, and oriented toward human institutions rather than toward Nature. They did not prevent the great changes that swept over Europe. (Sivin 1982:57)

This warning, although well taken, has not, I fear, always been heeded in this work. On the contrary, "the predominance of a scholar-bureaucrat class" has here been mentioned more than once as a probable inhibitor of Chinese science. This is because I believe that Sivin's East-West parallel is probably not as absolute as he assumes. Thus the so-called predominance of the Schoolmen and dons did not mean that they always surpassed everyone else in power and prestige. On the contrary, they had to share these qualities with royalty, the nobility, the military and a rising mercantile and entrepreneurial class—groups whose thinking often differed from that of the Schoolmen and dons and whose power often exceeded theirs. That is why the dons and Schoolmen did not prevent the great changes that swept over Europe.

In China, on the other hand, the power and prestige of the scholar-bureaucrats usually surpassed that of anyone else save the emperor and his close intimates. During a large part of Chinese imperial history, the emperor, at least outwardly, shared with the scholar-bureaucrats their veneration for books, their facing toward the past, and their orientation toward human institutions. The nobility usually enjoyed little political power; the military were subordinate to the civil bureaucrats except in times of dynastic crisis; and the merchants were traditionally discriminated against even though their status gradually improved in later centuries. All this will be discussed at length in this book. It explains why the scholar-bureaucrats of China, unlike the dons and Schoolmen of Europe, retained their predominance until the beginning of the present century.

3. Technicalities

3a. Chronology

Dates of Chinese dynasties and reigns accord with the chronological tables found in Tchang 1905. For convenience, a Chronology of Chinese Dynasties appears near the beginning of this book. The acronyms A.D. and B.C. are often omitted when the context or sequence of dates sufficiently indicate that either an A.D. or B.C. chronology is involved. Examples are 590–617 (dates of the Sui dynasty, sequentially increasing and therefore obviously A.D.) or 551–479 (dates of Confucius, sequentially diminishing and therefore obviously B.C.).

3b. Romanization

Names of well-known present-day Chinese cities and provinces are spelled according to the romanization system formerly used by the Chinese Post Office (Peking, Sian, Shansi, Fukien, etc.). Names of Chinese authors, when they write in Western languages, are spelled according to the preferences shown by the authors themselves (Fung Yu-lan instead of Feng Yu-lan, Wing-tsit Chan instead of Ch'en Jung-chieh, etc.). All other Chinese names and terms, past and present, are romanized according to the Wade-Giles system, with the single exception that Wade-Giles *i* becomes *yi*.

3c. Chinese Characters

Characters for all the Chinese-language authors cited in this work and for the Chinese titles of their books or articles will be found in the Bibliography. Characters for almost all other Chinese persons, place names, or terms will be found in the Chinese Glossary. The only exceptions are a few names (those of dynasties or rulers, places like Shanghai or Hupei, persons like Mao Tse-tung) that are so familiar that they do not need to be included in the Glossary at all. Occasionally, for convenience of the reader, the characters for certain Chinese terms or passages occur in the main text of this book on the same pages on which the terms or passages themselves are discussed. Normally, however, for the sake of the printer, such pages contain only romanizations, whose characters must therefore be sought either in the Bibliography or Glossary.

3d. Bibliographical Citations

Pre-1900 Chinese works: These are listed in Section A (Chinese Original Sources) of the Bibliography. Such pre-1900 works, when referred to in the general text of this book, are sometimes cited only by chapter number (*Mo-tzu* 11), sometimes by chapter and page number (*Huai-nan-tzu* 2/12a = p. 12a of ch. 2), and sometimes by chapter number followed, within parentheses, by *chüan* and page number (*Hsün-tzu* 23 (17/4a) = *Hsün-tzu* ch. 23, p. 4a of *chüan* 17). In the case of those pre-1900 Chinese works referred to in this book only in a general way by title, the Bibliography does not indicate any specific edition. Nor does it do so for such well-known classics as the *Lun yü, Li chi,* and the like, for which Western translations are widely known and available.

Post-1900 Chinese works and works of all periods in other languages: These are listed in Section B (Secondary Works) of the Bibliography. Here again particular editions are not specified for those secondary works that are referred to only in a general way, without citation of page or chapter.

3e. Western Translations

References to original Chinese sources by chapter, page, and so on, are commonly accompanied by references to translations into Western languages, if available and not obsolete. For a few major Chinese texts, two rather than one such translation may be cited, if both are of good quality and especially if one includes the original Chinese text absent from the other. For example, the entry "*Li chi* 17; tr. Legge 1885:28.124–125 and Couvreur 1913:2.102," signifies that this passage from *Li chi* chap. 17 occurs at pages 124–125 of Legge's 1885 English translation, constituting volume 28 of the *Sacred Books of the East* series (where, however, the Chinese text is not included). It further signifies that the passage also occurs in volume 2, page 102 of the 1913 edition of Couvreur's French edition (which does include the Chinese text). However, in order to avoid overly long insertions in the main text, references to a few well-known early Chinese texts—ones like the *Lun yü, Meng-tzu,* and *Lao-tzu,* whose section divisions are both short and easily identifiable—are often not followed by references to translations at all. Nor, for the same reason, are more than two translations normally cited even in the case of texts for which multiple translations may be available.

Important Note

The primary purpose of citing Western translations in this book is to facilitate identification of the particular Chinese passages that are being mentioned or quoted. Rarely, however, do such quotations consist simply of word-for-word reproductions of the cited Western translations. Much more commonly they are modifications that sometimes depart only slightly from the cited translations but sometimes are radically different.

As to quotations in English in the present work from books or articles originally written in French, German, and so on, these are my own renditions. The only exceptions are those books and articles that have already been published in English translation before being quoted here.

II. THE DYNAMICS OF WRITTEN CHINESE

1. Basics of the Chinese Language

The following summary of the basic features of Chinese may prove helpful to persons who know little about the language; those who are better acquainted would do well to turn directly to Section 2. Ramsey (1987) provides a good general survey of the various Chinese dialects, including their history and grammar, but says very little about the written language known as Literary Chinese, which is the main topic of this chapter.

1a. Chinese as a Spoken Language

• Both geographically and demographically, Chinese holds a predominant place among the Sino-Tibetan languages—a linguistic group that covers China, Tibet, Vietnam, Laos, Thailand, most of Burma, Taiwan, and certain cities such as Hong Kong and Singapore.

• What for the sake of convenience is called *Chinese* consists of some half dozen or more major dialects, with numerous lesser variants. Sometimes their linguistic differences are as great as those between French and German, so that some scholars would prefer to term them separate languages. In these pages, nevertheless, I shall continue to call them dialects. Most of the dialects occupy a belt of coastal southeastern China running southward from the lower Yangtze River all the way to the Vietnamese border but rarely extending more than six hundred miles inland at any point. The rest of China (including Manchuria) is covered by the dialect traditionally known as Mandarin (discussed in the next sect.). With many local variants, this dialect is the native tongue of well over two-thirds, and possibly even three-quarters or more, of the total Chinese population.

• The common characteristic of all Chinese dialects is that they lack inflection. Words undergo no morphological change, regardless of whether used in the singular or plural, masculine or feminine, past, present, or future. Many words (of course not all) can, depending on their position in a sentence, function as nouns, verbs, or adjectives without undergoing any change. This means that context and word order are of central importance.

• A great many spoken Chinese words consist of single syllables—a phenomenon facilitated by the absence of inflection and made conspicuous in writing by the fact that each spoken single syllable is recorded by a single written symbol. However, it is wrong to believe, as has often been asserted, that not only the classical written language, but modern spoken Chinese as well, is monosyllabic. This is because the spoken language, besides its many one-syllable words, also includes many bisyllabic (and less often trisyllabic) wordlike compounds. The components of these compounds can sometimes be taken apart for separate use as one-syllable words in spoken Chinese. An example is *kung chi* 公鷄, "rooster," literally "male chicken." The term's components, *kung,* "male," and *chi,* "chicken," can both be used separately in speech. Other compounds, however, cannot thus be broken apart when speaking (though this is usually possible in the classical written language). An example is *chih-chu* 蜘蛛, "spider," whose components, *chih* and *chu,* are each separately defined as "spider" in the dictionaries, yet have to be joined with one another when one refers to a spider in speech. More will be said about this later.

• The paucity of separate one-syllable vocables in modern spoken Chinese (only a little over 400), and the fact that many words are monosyllabic, means that *homonyms* are extremely common (analogous to "to," "too," "two" in English). However, the resulting semantic confusion is helped by the just-mentioned frequency of compounds. It is also alleviated by *tones,* which are a feature of all Chinese dialects. In other words, each vocable acquires a particular meaning (or usually a group of meanings) according to the particular tonal inflection given to it in speech. Four such tones exist in the Mandarin dialect, but in Cantonese there are no fewer than eight. An example is the vocable *hsing,* of which *hsing*¹ 星 (first tone, upper level) means "star"; *hsing*² 行 (second tone, lower level) means "to walk, move"; *hsing*³ 醒 (third tone, rising) means "to awaken"; and *hsing*⁴ 姓 (fourth tone, departing) means "surname". Note that each of these words is graphically distinct as soon as it is put into writing. However, even with the added differentiation provided by the tones, a great many homonyms remain. This too will be discussed later.

1b. Chinese Writing

• China's writing system goes back to the Shang dynasty (trad. 1765–1123) and possibly earlier. It began with pictographs and ideographs and has never completely turned its back on such written forms. In other words, Chinese is unique among the scripts surviving from antiquity in that it neither began with an alphabet or syllabary nor did it develop them later. Instead, each of the graphic symbols that we call Chinese characters represents, with a few exceptions, a single monosyllabic word, as used in the traditional written language.

• *Graphically,* the characters have evolved greatly from their early complexity and concreteness toward increasing simplification and abstraction. *Structurally,* however, the principles used to form the characters have changed but little. *Numerically,* the characters have proliferated from the somewhat more than 3,000 forms that are separately identifiable on the Shang inscriptions to the almost 50,000 listed in modern unabridged dictionaries. In such dictionaries, however, many graphs are obsolete, rare, technical, or deviants of more standard characters. In China today, it is perhaps in a general way possible to read newspapers or comparable materials with a knowledge of some 3,000 characters, if one has access to a dictionary.

• Although conventionalized pictographs and ideographs are still often recognizable in modern characters, the proliferation of new characters from Shang times onward was primarily made possible by another principle of character-making. This lay in combining a graphic symbol, to be used in the new character for its semantic value, with another graphic element, to be used as a rebus for its phonetic value only. Thus the resulting new compound character consists of a graphic element that has semantic value plus another graphic element that serves as phonetic. As an example, let us take the character *yang* 羊, "sheep" (originally a pictograph showing a sheep's body surmounted by horns). In combination, this character loses its semantic value and functions simply as a phonetic, *yang,* as in the following examples: (1) *yang* 洋, "ocean" (*yang,* "sheep," used as a phonetic, combined with 氵, an abbreviated pictograph of water ripples; this left-hand element appears in almost all characters having to do with water); (2) *yang* 恙, "to be sad, disquieted" (again *yang,* "sheep," used as a phonetic, but this time placed above 心, pictograph of a heart; this pictograph appears in most characters having to do with thoughts and emotions). Most characters today have been formed in this manner. Had this principle been applied consistently and universally, it might conceivably have led to a completely phonetic system of writing. Why this did not happen is too complicated and technical to be discussed here.

• Traditionally, the Chinese characters have often been described as "pictographic" or "ideographic," but this terminology is today under scholarly attack, as by DeFrancis (1989:220–230 and passim), who stresses the above-mentioned fact that a phonetic compound is present in most characters. The controversy results from differing emphases given by different people to the same body of data. Because even the phonetically functioning components present in characters are commonly traceable to early pictographs or ideographs, "pictographers" prefer to refer to Chinese writing as basically pictographic/ideographic. But because these same components contribute their sound rather than their meaning to the characters, "phoneticists" prefer to refer to Chinese writing as basically phonetic (see Bodde 1991). In these pages we shall try to avoid this controversy by referring to the characters only as graphs or characters except when making specific mention of their pictographic/ideographic origin.

• The characters have always retained their meanings, irrespective of their varying pronunciations in different dialects or at different historical epochs. This persistence of meaning has continued even when the characters have been taken over as script for the very different languages of Korea and Japan. The cultural importance of this phenomenon is discussed in section 3 below.

• Traditionally, the continuity of the characters since ancient times has often given them great value as instruments for revealing aspects of early Chinese civilization. Today, however, in the People's Republic of China, this value has been diminished or obliterated in the case of many common and important characters. This is because, beginning in 1956, the Chinese People's Republic has systematically replaced such characters with new abbreviated forms in which, very often, the graphic elements formerly present have been altered beyond recognition. An example is *li,* "ritual, ceremonial, principles of civilized behavior," formerly written as 禮. In this important Confucian written word, the left-hand element is present in most characters having to do with supernatural matters. To its right stands a conventionalized pictograph that portrays a sacrificial vessel standing on a table or altar. Today, however, the new abbreviated form of this same character has been reduced to 礼 . Obviously the single line replacing the former vessel/altar pictograph is far easier to write and memorize, but equally obviously the cultural significance of the old pictograph, and with it of the entire *li* character, is lost.

2. Speech and Writing

The differences between Chinese as a spoken and a written medium, only briefly touched on in the preceding pages, are all-important, so that they need more extended discussion here. The common designation for spoken or colloquial Chinese is *pai hua,* "white [i.e., undecorated] speech," a term theoretically applicable to any of China's spoken dialects, but in practice commonly equated with the most widespread of them, that known in imperial times as *kuan hua,* "Mandarin" (lit. "official speech"). This dialect, called such because it was the commonest medium of spoken communication between Mandarin officials who might be natives of differing dialect areas, was and is spoken (with numerous local variants) throughout northern and most of southwestern China. Traditionally, however, its "purest" form was considered to be the speech of Peking, long China's capital. After the downfall of the empire in 1911, the term gradually became obsolete, being first replaced by the term *kuo yü* ("national language") and then, under the People's Republic of China, by *p'u t'ung hua* ("ordinary" or "standard" speech). All three terms, however, apply to essentially the same linguistic area.

In contrast to the colloquial, the overwhelming bulk of Chinese written literature, throughout virtually all of China's literary history, was recorded in a medium possessing its own distinctive vocabulary and syntactic rules, known as "Literary Chinese" (*wen yen,* lit. "written words"; another common name for this medium is "Classical Chinese"). Before the present century, almost the only kinds of written literature recorded in a written equivalent of *pai hua* rather than in Literary Chinese were short stories, novels, and plays.[1] Because of the enormous prestige enjoyed by the literary medium, however, these written examples of the vernacular were traditionally regarded by most scholars as too frivolous to deserve serious attention. Only in 1916–1917, as part of a broad movement for cultural, social, and political reform, did the *pai hua* movement begin—the movement, that is, to replace *wen yen* by written *pai hua* as the general vehicle for scholarly, creative and popular writing.[2] The success of the movement during the following decades was great but far from total. Since the establishment of the People's Republic in 1949, however, written *kuo yü* (renamed *p'u t'ung hua*) has become well-nigh universal in mainland China, where *wen yen* has almost disappeared; the latter survives, but only to a limited extent, in such overseas areas as Taiwan, Hong Kong, and Singapore.

The spoken language and Literary Chinese share a considerable common vocabulary, written of course with identical characters whenever the spoken language is reduced to writing. A great many other words, however, are peculiar either to spoken or Literary Chinese or, if

---

1. There are a very few exceptions to this general statement. During the Sung dynasty, for example, the conversations of Chu Hsi (1130–1200) and other Neo-Confucianists were sometimes recorded in a semivernacular style, as were (both then and earlier) the conversations of some Buddhist monks, especially those of the Ch'an (i.e., Zen) school. In Sung times, too, the vernacular was sometimes used even for the writing of certain government documents, but this development did not continue after the dynasty.

2. A good account by a major protagonist is that in Hu Shih (1934).

common to both, are differently used. Likewise, major syntactical and stylistic differences separate the two media. In general, the literary language is laconic, the spoken language relatively more verbose. That is to say, the literary language is basically monosyllabic, so that each word (i.e., each separate meaningful lexical unit) consists of a single syllable written with a single character. *Wen yen* does, however, also employ a relatively small number of polysyllabic (almost always dissyllabic) compound units. By contrast, the spoken language—at least that known for the last millennium and more—is basically polysyllabic (usually dissyllabic), though at the same time it contains a considerable number of meaningful monosyllabic units.

This vernacular/literary dichotomy rests on two basic phenomena: (1) the phonetic poverty of spoken Chinese of the last fifteen hundred years and more, that is, the existence in it of numerous monosyllabic homophones to which multiple meanings are attached by the dictionaries, but which are indistinguishable to the ear (compare our *to, too, two*); (2) the ideographic rather than phonetic origin of Chinese writing, which means that each graph is visually distinct from every other graph even though a great many share the same pronunciation.

As an example, early Chinese had two words meaning "clothing," today respectively pronounced *yi* 衣 and *fu* 服 (compare our use of two separate words, "clothing" and "garment"). Because the written graphs for *yi* and *fu* are visually distinct, it was possible for them to continue as separate words in Literary Chinese down to the present.

In spoken Chinese, on the contrary, neither *yi* nor *fu* can be used alone to mean "clothing" because so many other *yi*'s and *fu*'s exist. Many of these are pronounced identically, including the same tone, and many others differ only in tone, but all bear different dictionary meanings. Examples for the vocable *yi* are words meaning "one," "easy," "explain," "different," and so on. The solution, therefore, has been to combine *yi* and *fu* into the auditorily distinct compound *yi-fu,* which serves as the spoken word for "clothing." Such, in the opinion of many scholars, is possibly the commonest way in which polysyllabic words have come into being in spoken Chinese, although some scholars deny this (see n. 3 below). The sequence of syllables is almost always fixed by convention. Thus one could not, either in speech or in writing, reverse *yi-fu* into *fu-yi* and still be understood.

Rather than go into further detail, let us illustrate in a general way the difference between literary and spoken Chinese by comparing the original text of the famous opening sentences of the *Meng-tzu* (1a.1; tr. Legge 1960:2.125) with a modern (1960) *pai hua* translation. The original sentences read in a non-literal English translation: "Mencius [ca. 372–289] had an interview with King Hui of Liang. King Hui said: 'Venerable Sir, you do not find a thousand *li* [here simply meaning a long distance] far to come. Do you also [in addition to taking the trouble to come] have something that will profit my country?"

The following is the original text with literal translation:

孟 子 見 梁 惠 王 惠 王 曰 叟 不
Meng tzu chien liang hui wang hui wang yüeh sou pu
Meng Tzu saw Liang Hui King. Hui King said: "Venerable-Sir not
(Mencius)

遠 千 里 而 來 亦 將 有 以 利
yüan ch'ien li erh lai yi chiang yu yi li
find-distant thousand *li* and come. Also going to-have whereby to-profit

吾 國 乎
wu kuo hu
my country eh?"

Here is the same passage translated into *pai hua* (anon. 1960:1):

孟 子 謁見 梁 惠 王 惠 王 說 老 先生
Meng tzu yeh-chien liang hui wang hui wang shuo lao hsien-sheng
Meng Tzu visited-saw Liang Hui King. Hui King said: "Old First-born.
(Sir, Mr.)

您 不 辭 千 里 長 途 的 辛勞 前
nin pu tz'u ch'ien li ch'ang t'u ti hsin-lao ch'ien
You not decline thousand *li* long road 's bitterness-laboriousness forward
(discomfort, trouble)

來 那 對 我 的 國家 會 有 很 大
lai na tui wo ti kuo-chia hui yu hen ta
come, that toward I 's country-family could have very great
(my) (nation)

利益 吧
li-yi pa
profit-advantage eh?"

Three major points of difference are evident: (1) The vernacular version is more than one-third longer than the original text, requiring thirty-eight characters to say what the latter says in twenty-four. (2) The original version is truly monosyllabic, each character standing for a separate meaningful word, whereas the vernacular version includes five dissyllabic compounds: *yeh-chien* (visit-see), *hsien-sheng* (First-born), *hsin-lao* (bitterness-laboriousness), *kuo-chia* (country-family), *li-yi* (profit-advantage). (3) The two versions differ so sharply in vocabulary that they share only thirteen characters in common. Five of these are proper names (Meng Tzu, Liang Hui Wang), three are originally monosyllabic words that the vernacular expands into compounds *(chien* becomes *yeh-chien, kuo* becomes *kuo-chia, li* becomes *li-yi),* and only five are common words that remain as monosyllables in the vernacular (*pu,* "not"; *ch'ien* + *li,* "thousand *li*"; *lai,* "come"; *yu,* "have"). It is not necessary to go into detailed grammatical analysis to see how greatly the spoken and literary languages differ.

How did these differences arise? A common answer is that the spoken language in the course of its long history underwent phonetic changes that led what had once been phonetically distinct monosyllabic words to lose their final consonants or otherwise merge phonetically until they became indistinguishable homophones. It was this growing homophonization of monosyllabic units, according to this theory, that moved the spoken language in the direction of dissyllabic compounds and thus away from the written language. The latter meanwhile changed differently and much more slowly, so that it retained its original monosyllabism. Its relative slowness of change was reinforced by the tendency of later writers to imitate the style of the earlier ones.[3]

Acceptance of this theory (one that I myself accept only in part) suggests the further conclusion that there must once have been a time—politically it would correspond roughly to the Chou dynasty (trad. 1122–256 B.C.); linguistically it is known as the age of Archaic Chinese—when little distinction between spoken and written Chinese existed; both were basically monosyllabic, and the latter was essentially a written reproduction of the former. Such indeed is the conclusion reached by several distinguished scholars, among them Bernhard Karlgren, who pioneered

3. What probably happened, according to this theory, has been described very clearly in a series of articles by Janusz Chmielewski (1949, 1957a, 1957b, 1964). It should be noted, however, that not all scholars accept this thesis. See, for example, Kennedy (1951), Yoshikawa Kōjirō (1955), and Serruys (1959). Kennedy writes (1951: 164) that "the concept of a people doing incessant battle with its language is wholly irrational." Yoshikawa conjectures (1955:129) that "the Chinese always tended to speak in what, for practical purposes, we may term polysyllables; but . . . when they began to write, the ideographic nature of their script made it unnecessary to represent by two syllables what could be conveyed to the eye by one." And Serruys takes this idea further by suggesting (1959:163) that pre-Han Chinese contained, in addition to monosyllabic words, those that were bisyllabic and even trisyllabic, all of which, however, were equally written by single graphs only. The subject is complex and cannot be discussed here other than to say that in my opinion the viewpoints just expressed either seem inherently improbable or are unconvincing because of lack of evidence or because they leave too many factors unexplained. My own view is expressed at the end of the present section.

in the reconstruction of the phonetic evolution of Chinese. He writes (Karlgren 1929:177–178):

> It ought to be clear to any unbiased reader that the dialogues of *Lun yü, Meng-tsï* and *Chuang-tsï,* the dramatically narrated episodes of the *Tso chuan* etc., are the purest possible reproduction of a spoken language. We can positively hear the speakers, with all their little curious turnings, anacoluthic sentences, exclamations etc. I go so far as to say that I believe even in the Han period the written language was not very far removed from the colloquial. There are often passages that are clearly intended to reproduce verbally what has been uttered, and they are nevertheless exactly what we call 'literary Chinese.'

To prove his contention for the Han dynasty (206 B.C.–A.D. 220), Karlgren goes on to cite how the *Shih chi* (Historical Records), in its biography of the statesman Chou Ch'ang (d. 191 B.C.), reproduces his stuttering as he violently criticizes the founder of the Han dynasty: "Your servant de- . . . definitely knows it won't do."[4]

Chmielewski accepts Karlgren's thesis and believes that the time when the written language ceased to be a reproduction of the spoken came "about the beginning of our era."[5] Likewise Dobson (1959:xvii) writes with reference to what he calls Late Archaic Chinese (fourth–third century B.C.): "The extent to which it departs from the vernacular of the period is uncertain, and cannot be assumed always to have done so simply because the literary language and the vernaculars of later periods diverge." Then he continues more positively in a footnote (ibid., n. 6): "While certain of the material bears plainly the marks of artifice (rhyme, metre, parallelism, the formal archival entry, and so on), the maxims of the sophists used in open debate . . . at the very least must be examples of the spoken language."

I am ready to believe that quite a few of the Chou texts were probably fairly close to the speech of their time in vocabulary and syntax. However, I would also suggest that *stylistically* speaking, a very considerable difference between the two even then often existed. This I believe to be at least true for most of the literary works not expressly written in dialogue form, which means the greater part of surviving Chou literature. Thus only the *Lun yü* or *Analects* and *Meng-tzu* or *Mencius* (fifth and third centuries B.C.) consist predominantly of free dialogue, whereas in virtu-

---

4. *Ch'en ch'i . . . ch'i chih ch'i pu k'o* 臣期期知其不可. See *Shih chi* 96; tr. Watson (1961:1.260). Chou Ch'ang's whole statement reads: "Your servant's mouth is inept at speaking. Nevertheless, your servant de- . . . definitely *(ch'i ch'i)* knows it won't do. Even should Your Majesty wish to discard the crown prince, your servant de- . . . definitely *(ch'i ch'i)* would not accept the command."

5. Chmielewski (1949:413). See also Chmielewski (1957a:74).

ally all the other texts, exposition is varyingly mixed with dialogue. In these mixed texts we may expect to find, to varying degrees, Dobson's "marks of artifice (rhyme, metre, parallelism, the formal archival entry, and so on)." Thus it is obvious enough that, except very occasionally, people could not have spoken in rhyme in the age of the *Shih ching* (Songs Classic, ninth–fifth centuries B.C.) any more than they would have used blank verse in Shakespeare's day. Likewise, when we turn to prose literature, it is impossible to suppose that the many highly cadenced passages often found in Chou as in later writings could, stylistically, have borne any close resemblance to the speech of their time. Even in some of the dialogue texts, in fact, literary devices such as formulaic repetition, parallelism, and antithesis occur, including instances in the *Lun yü*[6] and *Meng-tzu*.[7] When this happens, it can only mean that, as in virtually all dialogues found in later Literary Chinese, they are only stylized redactions of vernacular speech. Finally, as to the stutter cited by Karlgren as evidence for the recording of the spoken language in the *Shih chi,* this occurrence by no means guarantees that most or even many of the other speeches in that history represent the vernacular of their time. It would seem much more likely that the *Shih chi*'s author, Ssu-ma Ch'ien, singled out Chou Ch'ang's stutter for special record precisely because it was such a distinctive feature of the man.

In short, the conclusion seems quite plausible that even in Chou times only a lesser part of the written literature was stylistically very

---

6. See the highly formulaic chain syllogism in *Lun yü* 13.3; tr. Legge (1960: 1.263–264):

> If terms are not correct, speech will not go smoothly.
> If speech does not go smoothly, things will not succeed.
> If things do not succeed, rites and music will not flourish.
> If rites and music do not flourish, corporal and other punishments will not hit the mark.
> If corporal and other punishments do not hit the mark, the people will have no place to set hand or foot.

7. *Meng-tzu* 1b.7; tr. Legge (1960:2.204–205), in which Mencius advises King Hsüan of Ch'i how to select or reject officials: "When those about you all say he is good, this won't yet do. When your great officers all say he is good, this won't yet do. When the people of the country all say he is good, then look into it and if you see goodness in him, then employ him." Mencius then continues with two variant formulations of the same sentences, the first of which leads to dismissal of the official under consideration and the second to his execution: (1) "When those about you all say he won't do, don't listen . . . ," etc.; (2) "When those about you all say he may be executed, don't listen . . . ," etc. The original Chinese is even more cadenced and formulaic than is the translation.

close to the spoken language—this notwithstanding the fact that in vocabulary and syntax the similarity was considerably greater, and certainly a great deal greater than in post-Han times.[8] In the following sections, whenever I discuss features of the Chinese language, it will usually be with Literary Chinese primarily in mind.

3. Significance of the Written Language

Throughout its history, China has been overwhelmingly a scribal rather than a verbal civilization. Already during the early stages of script, it would seem that whenever significant verbal utterances were made, such as the divination formulas addressed to the Shang royal ancestors, or the exhortations, speeches, and homilies in the *Shu ching* (Documents Classic), or again the songs in the *Shih ching* (Songs Classic), these were all conscientiously reduced to writing.[9]

The process must have been slow and laborious, owing to the elaborateness of the characters (much more pictographic and many-stroked in Shang–Chou times than later) and the awkwardness of the early writing materials and implements (bones, tortoise shell, bronze vessels, wooden or bamboo strips and tablets, as well as silk scrolls, on which the characters were variously incised with a knife, cast in metal, or written with a stylus or sometimes a brush).[10] Indeed, I would like to suggest that the ever-present emphasis on terseness in Literary Chinese may in part have stemmed from the physical difficulty of recording the characters, especially before the invention of paper (first century B.C. or possibly earlier). If so, this would support the theory that the dichotomy between spoken and Literary Chinese began much earlier than often supposed, for it would point to a two-phase process: first, a movement of the literary medium away from the spoken, beginning already in the early stages of Chinese writing, and then, much later, the well-known contrary movement of the spoken medium away from the written, induced by the phonetic evolution of the spoken language. The extreme terseness of the earliest written records strengthens such a hypothesis.

8. Of course, if one accepts the hypothesis (see n. 3 above) that spoken Chinese was always polysyllabic, its difference from Literary Chinese would have been absolute from the beginning.

9. This fact supports, rather than contradicts, the well-known fact that some of the most notable Chinese philosophical texts, beginning with the *Lun yü* or *Analects* of Confucius, are written redactions of what were originally conversations. A sympathetic account of this conversational tradition is given by Holzman (1956).

10. See the accounts in Tsien Tsuen-hsuin (1962 and 1985).

In all this there is a striking difference from India, where for centuries the Vedas were transmitted orally before being committed to writing. In India it was the oral recitation of a composition that made it fully meaningful and effective. In China it was above all its reproduction in written or printed form. Famous calligraphers have been legion in China, famous orators very rare. This is in no way to deny the frequency of oral memorization and transmission of texts in China. On the contrary, scholars memorized incredible amounts of classical literature, and future officials did the same in order to get through the civil service examinations. At a very early age, schoolboys were expected to *pei shu,* "turn their backs on the text," and then recite it from memory to the teacher. This procedure, however, was only a mnemonic adjunct to the reading and writing of the texts. In no way was it intended to substitute for such readings and writings.

So great was the mystique of the written word that in imperial times papers bearing writing could not be indiscriminately discarded in the streets. This was not so much because they polluted the streets as that such an act would show disrespect to the written word. The author, as a student in Peking in the 1930s, remembers still occasionally seeing trash receptacles in public places inscribed with the traditional exhortation: *Ching hsi tzu chih,* "Respect and care for written paper." It is perhaps no accident that three of China's greatest inventions—paper, block printing, and movable type—are all closely associated with writing.

Unquestionably, a major reason for the cult of the written word in China has been the pictographic-ideographic origin of the script, and the numerous resulting opportunities for calligraphic variety. As everyone knows, calligraphy has been a major art in China, where scrolls or sometimes gigantic individual characters are hung in private houses and public places alike, their message strengthened by the graphic forms in which it is written. A very considerable part of Chinese humanistic scholarship has been devoted to study of the evolution of the written characters, much less to the phonological evolution of the spoken language.[11]

Superficially, it may seem strange that writing was so valued in a civilization where never, until recent decades, was more than a tiny portion of the population literate. Yet surely the explanation lies in the very rarity of literacy, the difficulty of acquiring it in a literary medium sharply different from spoken Chinese and requiring a knowledge of several thousand characters, the aesthetic qualities of the script, and, very

---

11. It should be noted, however, that phonology became an important component of the new humanistic scholarship, more scientific in its methodology, that developed in China from the seventeenth century onward. See Elman (1982 and 1984:212–221).

importantly, the avenue that literacy provided to prestige and power. The written word, as a result, was imbued with an almost magical potency that was felt by illiterates as well as scholars.

This potency, almost surely, manifested itself in the political sphere as well. Inseparable from bureaucracy, whether Chinese or any other, are of course letter writing, records keeping, files and archives, and, in general, an enormous amount of paperwork, often in multiple copies. Thus it seems evident that the development of the Chinese bureaucratic state must have encouraged an emphasis on the written word. I would also suggest, however, that the cult of the written word, evidenced already long before the beginning of empire in the tendency to put everything deemed important into writing, must reciprocally have encouraged the growth of the bureaucratic ethos.[12] And this in turn, more narrowly speaking, must have fostered the maintenance of detailed records about natural phenomena and thereby the accumulation of scientific knowledge.

On the broader plane, as has often been said before but deserves reiteration, the written language has contributed in two major ways to the continuity, spread, and homogeneity of Chinese civilization. In the first place, its linguistic stability, ensured by separation from the living spoken language, makes it possible for a modern Chinese conversant with Literary Chinese to read—with effort, of course—classics written two and more millennia ago as well as contemporary documents. This contrasts with the situation confronting the English-speaking scholar who approaches *Beowulf* or even Chaucer. Secondly, the fact that the written characters retain their meaning irrespective of the phonetic values attached to them (just as 1, 2, 3 always signify the same numbers regardless of whether pronounced one, two, three or *un, deux, trois*) means that all literate Chinese—and especially, in old China, government officials—can and could, irrespective of their own particular spoken dialect, communicate with one another through the characters.

In short, the script and its accompanying literary language gave China a cultural *continuity in time* and *unity in space* so powerful that without it one may seriously wonder whether the Ch'in–Han creation of the bureaucratic state could ever have happened—or, perhaps more impor-

---

12. Although Chinese bureaucracy was first made universal only during the Ch'in and Han empires, its beginnings probably go back much earlier—some scholars even maintain as early as the Shang dynasty (trad. 1765–1123). Keightley (1978:221), for example, asserts very positively that "Chinese administration may be regarded as already proto-bureaucratic in Shang times." To which he adds: "Written documents [then] certainly played a major role in the organization of the state."

tantly, whether it could have been repeated, as it was under Sui and T'ang, after the centuries of disunity that followed the fall of Han.

The Western analogue is of course Latin, but the role of Latin as a politically unifying factor was weakened after Roman times by the fact that its primary users were churchmen whose highest allegiance commonly was to the Kingdom of Heaven rather than to a secular state. A more subtle, but perhaps in the long run more significant, circumstance was that the phonetic symbols of the Latin alphabet lacked the ideological cohesiveness of the lapidary Chinese symbols. Besides being used to represent the Latin words, they could equally well be reassembled in endless new combinations to spell out the verbal changes in the developing spoken languages of Western Europe. Thus, in concert with a great many other political, economic, and intellectual factors, they facilitated the rise of localized speech forms to the status of national written languages—a step achieved in China only in the present century with the replacement of *wen yen* by *kuo yü*.

The Chinese characters, besides serving to unify the Chinese world internally, were also potent transmitters of Chinese civilization to neighboring peoples, including not only those, like the Vietnamese, closely related to the Chinese linguistically, but also others, like the Koreans and Japanese, whose spoken languages belong to totally different linguistic families. Because each character carries from its cultural past its own distinct connotations, the acceptance of Chinese writing by other peoples has meant, to a considerable extent, their acceptance of Chinese cultural and moral values as well.

Conversely, the Chinese script has long proved a major deterrent to the free entry of foreign ideas and values into Chinese culture, because it meant that these ideas and values, when put into writing and translated into Chinese, could reach the Chinese consciousness only through the filter of the characters.[13] Resulting difficulties and failures in communication were as well known to Buddhist missionaries a millennium or fifteen hundred years ago as to the Jesuits and others of the seventeenth century onward. Wright (1953b:287) has vividly summarized some of the translation problems connected with the Chinese language and its manner of writing:

---

13. The converse, of course, is also sometimes true, despite what has been said above. Needham remarks at one place (1954:4): "The very difficulty of the characters has proved an almost insuperable barrier to understanding [by outsiders]." And yet, as he points out later (1954:218), in the fourteenth century the Chinese characters were regarded by Rashīd al-Dīn al-Hamdānī as superior to alphabetic systems for science because the meaning of a character is independent of its pronunciation.

> Thus the monks from medieval India, the Jesuits from Renaissance Europe, emissaries of modern scientific thought such as Bertrand Russell, and representatives of the Comintern all spoke inflected polysyllabic languages. . . . Structurally Chinese was a most unsuitable medium for the expression of their ideas, for it was deficient in the notations of number, tense, gender, and relationships, which notations were often necessary for the communication of a foreign idea. . . . Moreover, Chinese characters as individual symbols had a wide range of allusive meanings derived from their use in a richly developed literary tradition. . . . Further, the Chinese was relatively poor in resources for expressing abstractions and general classes or qualities. Such a notion as "Truth" tended to develop into "something that is true." "Man" tended to be understood as "the people"—general but not abstract. . . . These characteristics of the Chinese language reduced many proponents of foreign ideas to despair. . . . Kumarajiva (344–413), devoted Buddhist and stouthearted missionary, . . . was moved to sigh: "But when one translates the Indian [Buddhist texts] into Chinese, they lose their literary elegance. Though one may understand the general idea, he entirely misses the style. It is as if one chewed rice and gave it to another; not only would it be tasteless, but it might also make him spit it out."

Whereas it is the *style* of Chinese translation that Kumārajīva laments, Wright's central focus is on the possibility that the *meaning* may be distorted as a result of the particular nature of Literary Chinese and its graphic elements.[14] Obviously, both style and meaning are vitally important and closely interrelated, but modern scholars—those, for example, who work with Chinese Buddhist texts—will probably attach relatively greater weight to the question of whether or not the translation is semantically and psychologically true to the original. Some of the linguistic features remarked on by Wright, such as absence of inflection, will be discussed later.[15]

Here we may summarize the essential points in this and the preceding section by saying that the distinction between the literary and the spoken language may, stylistically at least, have begun considerably earlier than is often assumed; that the emphasis on the written word was probably a significant factor in the creation and perpetuation of the bureaucratic state; and that the strong urge to record all sorts of matters

---

14. For example, Wright says (1953b:289): "While it was strategically advantageous to equate an Indian term of this sort [*dharma*] with such a venerable and revered Chinese word as *tao,* the very power and authority of the indigenous term and concept tended to make it absorb the Indian notion into its already ramifying galaxy of meanings."

15. The reader's attention is also drawn to the way in which the Chinese language was adapted for modern science, especially chemistry, beginning soon after the mid-nineteenth century. See Needham (1976:252–262).

in writing, observable already very early, would seem to have been conducive to scientific development. In short, the script and the literary language that went with it were immeasurably powerful agents for the preservation and diffusion of Chinese civilization over a very wide span of time and space.

By the same token, however, the script and the literary language were significant barriers to change. For example, a nineteenth-century statesman memorializing the emperor on how to handle the Western barbarians would be using essentially the same medium of communication—and perhaps even some of the same metaphors, allusions, and illustrations—as would a Han statesman dealing with the Hsiung-nu tribes two thousand years earlier. The script and literary language were also barriers to the proper understanding—let alone adoption—of foreign ideas. Furthermore, if it be true that science prospers in an atmosphere of intellectual diversity, one might argue that the cultural uniformity that was encouraged, especially on the level of the ruling elite, by use of a common literary language, could well have been unfavorable to original scientific thought. Of course, this literary language was only one of several factors conducive to cultural uniformity. Also important was the age-old drive toward political and intellectual unity that was so apparent in Chinese history (see V.1 below).

In sum, the script and the written language, vital as they were for China's cultural and political continuity, may conceivably—though of course no measurement is possible—have been more harmful than beneficial to scientific development. This suggestion is made without reference to other linguistic features to which we shall now turn.

4. Morphology, Grammar, Meaning

So far we have been concerned with what may be termed the externals of Literary Chinese: its relationship to the spoken language and its general significance in Chinese history and society. Now we are ready to examine some of its internal features: structure, semantics, and the like. Before doing so, however, two matters require emphasis.

• Our concern is *not* the aesthetic qualities of Literary Chinese viewed as an instrument for writing beautiful or moving prose or poetry. Rather, it is the question of whether Literary Chinese constitutes an effective medium of communication from the point of view of clarity, precision, and directness. This is why, for example, the practice of sometimes inserting repetitive synonyms into a sentence of Literary Chinese is negatively regarded below, despite its possible stylistic acceptability when judged purely from a literary point of view.

• When evaluating the passages from ancient Chinese texts to be cited later, we should alert ourselves to two possible kinds of ambiguity: (a) ambiguities seemingly inherent in the original wording and structure of a given cited text; (b) ambiguities arising because the particular meanings originally held by key words in a given text and known when that text was first written have possibly become subsequently altered or forgotten during the long centuries of transmission down to our own time. This topic is touched on below.

The topics to be discussed in this and the next section, though widely known, require such discussion because they are central to our inquiry: to what extent has the Chinese language, and especially Literary Chinese, encouraged or inhibited the accurate and clear communication of facts and ideas, and thereby the development of scientific ways of thinking? A good many persons, both scholarly and otherwise, have expressed themselves on this subject, sometimes vehemently and emotionally. Some will be referred to in due course, though the weakness of too many is a tendency toward dogmatic generalization inadequately supported by concrete facts. In what follows, we shall try to support each point with actual examples.

The first point centers on what Karlgren (1949) calls "the most important characteristics of current Chinese." This is its "absence of inflection in the true sense of the word, the absence of word derivation, i.e., the formation of new words from a common stem, the absence of grammatical distinctions such as different forms for different word classes—all this which we call its 'isolating nature.' "[16]

This situation, Karlgren maintains further, did not necessarily always exist. Basing himself on his much earlier research, he finds traces of case inflections for the personal pronouns to be present in the *Lun yü* and *Meng-tzu;* in these texts the preferred nominative/genitive form for the first person (I, my) is *wu* 吾 (Archaic pronunciation *ngo*), whereas for the dative/accusative (me) it is *wo* 我 (Archaic *ngâ*). Similarly, for the second person (thou, you), the preferred nominative/genitive form is *ju* 汝 (Archaic *ńio*), whereas for the dative/accusative it is *erh* 爾 (Archaic *ńia*). Besides the above, Karlgren also detects "a few weak traces of tense inflection in certain verbs," as well as a very few other seeming indications of inflectional differences between noun and verb, adjective and verb, and so on. "These interesting features," he writes, "point to the fact that the character of Proto-Chinese was much more like that of our Western languages in essential points. Like the Indo-European tongues, it must have possessed its system of inflections

16. Karlgren (1949:98 and, for the facts and quotations in the next four paragraphs, 72–76 and 98–100).

and of word derivation, its formal word classes, in short, a more or less rich morphology."

These distinctions, however, even if they once existed, disappeared very early, making their modern rediscovery extremely difficult. (The task would have been impossible without Karlgren's reconstruction of the phonology of Archaic Chinese.) Even in the *Lun yü* and *Meng-tzu*, for example, the case differences for personal pronouns are far from absolute, and by Han times they were forgotten. In thus losing whatever inflection it may once have had, "Chinese is analogous to English, although it has proceeded much further along the road towards simplification and ossification." For the reader of Literary Chinese, the consequences are profound. To quote Karlgren once more:

> The lack of pointers in the shape of inflections and derivations that we are accustomed to in our Western languages, apply in actual practice just as much to the oldest texts of nearly three thousand years ago as to those of the later literature. The vestiges of an earlier and grammatically clearer stage of the language which we have just described are much too few to be of any appreciable help in the analysis and, worst of all, these vestiges are completely hidden by the screen created by the modern pronunciation of the archaic words.

In short, study of the ancient morphology is of historical and linguistic interest only. For the ordinary reader it has little or no practical value in helping him to handle his text, whether old or new.[17]

Our second topic is that of syntax and grammar. Words in Literary Chinese can, according to circumstances, commonly function in at least two, and more often several, grammatical roles (i.e., as nouns, verbs, adjectives, adverbs, etc.). An effective example of this phenomenon is the frequent conversion of what are basically nouns or stative verbs into transitive verbs. We saw this happen, for example, in the *Meng-tzu* quotation of the preceding section, where *yüan,* ordinarily a stative verb meaning "to be far, distant," was made to function as a transitive verb, "to distance" (a thousand *li*). Because of the lack of any English equivalent, we had to render this very weakly as "to find-distant." Of course the same phenomenon is widespread in other languages (cf. the verbalization of "man" in "to man the boats"), but in Chinese it is an especially common and effective stylistic device. And when it occurs in inflected languages—even one as little inflected as English—the verbalization is usually clearly indicated (e.g., "black" becomes "blacken,"

17. It must be added that a very few scholars disagree with Karlgren's theory that traces of morphological change can be found in very early Chinese. It is severely criticized, for example, by Ping-ti Ho (1975:261–264). This or similar criticisms, however, regardless of whether justified or not, do not invalidate our own conclusion as just expressed in this two-sentence paragraph.

but see also "man" above). In noninflected Chinese, on the other hand, this and other grammatical changes are enormously harder to detect.

This is why syntax becomes of such overriding importance. Contrary to popular belief, the wide range of grammatical functions open to Chinese words is by no means arbitrary. These functions depend in each instance on rules of word order which, in major respects, resemble those governing English: subject precedes verb which precedes object; adjective and adverb respectively precede the noun and verb they modify, and so on. On the whole, these syntactical rules are more fixed for Literary Chinese than they are for English.

Even with such strictness, however, absence of inflection makes it all too possible to arrive at more than one interpretation—each grammatically plausible—for the same phrase or sentence. And this in turn may sometimes disastrously affect one's understanding of a considerable passage. At this point an adequate context—cultural as well as grammatical—becomes vital. Its absence makes the reading of fragmentary Chinese inscriptions often virtually impossible, and, as we shall see, the interpretation of some of the most important utterances in the Confucian *Lun yü* so uncertain. Although similar ambiguities can of course sometimes be found in the texts of Indo-European or other languages, they are rarer and probably easier as a rule to overcome, owing to the existence of inflection.

Another grammatical feature causing numerous difficulties is what Dobson (1959) calls "the rule of economy." By this he means that although certain important grammatical devices (notably the particles) do exist in Literary Chinese to indicate verb mood, noun number, nominalization of words, and the like, their use is by no means always mandatory and depends very much on the whim of the writer:

> A lexic [a word] used as a verb may be invested environmentally [i.e., contextually] with mood, voice and aspect, but the verb may equally occur uncommitted, or neutral in any of these regards. Similarly, a lexic in nominal usage may be invested with number, or occur uncommitted to number. The language is not governed by rules which require the introduction of such indications as a matter of prescribed necessity.[18]

Much the same point has been expressed by Victor Purcell in language that is less scholarly, more intemperate, but for that reason more arresting:

> The rule is, if you can possibly omit, do so. The result may be that the meaning is quite hidden, but the reader is supposed not only to have an encyclo-

---

18. Dobson (1959:xxvi). His statement is made with specific reference to the Chinese of the fourth–third centuries B.C. (Late Archaic Chinese), but it is generally applicable to almost all Literary Chinese (but see next note).

> paedic knowledge to assist him in his guess work, but to have unlimited time for filling in ellipses. This does not mean that the language has no words to fill in the ellipses, or that there are no words to convey tense, number, or mood. It merely means that the spirit of the language is against their use.[19]

A very common ellipse occurs when the author fails to indicate explicitly either the subject or object of his or her sentence, leaving it to the reader to deduce them from the context. Or the object is indicated merely by using the ambiguous third personal pronoun *chih* 之 (variously translatable as "it, him, her, them"), for which the text may have two or more possible antecedents.[20] Or yet again, Literary Chinese (until its later stages) lacked a copula, "to be"; instead it placed an interjective word, *yeh* 也 (comparable to "oh," "ah," or "indeed") *after* the two nouns or nominal phrases it wished to equate. Thus the standard formula for equating A with B is "A, B *yeh*" ("A, B indeed"), that is, A is (or is equivalent to) B. In this usage, *yeh* normally comes at the end of a sentence. However, not infrequently the word may be omitted entirely, either because the writer thinks the meaning is clear without it, or for reasons of euphony, or, very likely, simply because of the "rule of economy."[21]

To illustrate some of what has been said, let us see how, within the successive clauses of a single sentence that are marked by absolute

19. Purcell (1936:93). During the centuries between Han and T'ang there was a temporary tendency for Literary Chinese to become somewhat more wordy through free use of grammatical adjuncts and other words that, strictly speaking, were not absolutely necessary for comprehension. Yoshikawa Kōjirō (1955:126–130) has demonstrated this phenomenon by taking a passage of seventy-five characters from the *Shih-shuo hsin-yü* (New Account of Tales of the World), by Liu Yi-ch'ing (403–444), and showing that even if thirty-two of the seventy-five were deleted, the essential meaning of the passage would still be conveyed.

20. Kenneth Robinson, of the Needham Research Institute, Cambridge University, has pointed out to me that ambiguity of this sort, when properly controlled, may sometimes be quite useful. An ambiguous pronoun like *one* in English or *on* in French, for example, enables a writer or speaker to formulate a generalization that is equally applicable to reader, author, and possible third parties without the need to include further data required for greater specificity but irrelevant to the generalization being made. What Robinson says is of course true. However, I very much doubt that many Chinese writers were deliberately using "controlled ambiquity" as a device for formulating generalizations on those many occasions when they failed to name the subject or object in a sentence. The sort of vagueness they practised is well exemplified by the passage from the *Hsiao ching* discussed below.

21. A further complication is that *yeh,* when it comes at the end of a sentence, sometimes does not indicate an equation at all but is placed there merely to mark the conclusion of the sentence. In this function, too, it can be readily omitted.

grammatical uniformity, common sense requires us to make repeated mental shifts in determining their subjects and objects, these not being explicitly stated in the text. The passage comes from the tenth chapter of the *Hsiao ching* or *Classic of Filial Piety* (late Chou or early Han), which begins with the words: "In the filial son's service to his parents" (tr. Legge 1899a:480). Five parallel clauses then follow (in the accompanying literal translation I indicate alternative possibilities for certain words):

居 則 致 其 敬
chü tse chih ch'i ching
not active, then brings-forth his/their respect,

養 則 致 其 樂
yang tse chih ch'i lo
caring-for, then brings-forth his/their pleasure,

病 則 致 其 憂
ping tse chih ch'i yu
sick, then brings-forth his/their concern,

喪 則 致 其 哀
sang tse chih ch'i ai
mourning/die, then brings-forth his/their grief,

祭 則 致 其 嚴
chi tse chih ch'i yen
sacrificing, then brings-forth his/their reverence.

In the first clause, initial *chü,* "not active," literally means "to dwell (quietly)," and the whole clause describes the filial son: "When he is not active, he brings forth [i.e., manifests] his respect [toward the parents]." In the second clause, the parallel first word, *yang,* "caring for," refers once more to the son, but this time when he is actively engaged in physically caring for the parents (providing them with food and the like). However, the final word *lo,* "pleasure," probably no longer refers to him but to the parents only: "When he cares for them, he brings forth [i.e., generates in them through his solicitude] *their* pleasure." In the third clause, on the other hand, the situation is exactly reversed: it is the parents who are *ping,* "sick," and the son who then manifests "his" concern for them. In the fourth clause, the initial *sang* can mean either "to mourn," in which case it is the son who mourns, or "to die," in which case it obviously refers to the parents. Commentators generally support the former interpretation, but either way, it has to be the son who thereupon manifests his grief. The fifth clause presents no problem and refers to the son throughout.

Thus, with the addition of the introductory statement, we arrive at the following translation: "In the filial son's service to his parents, when he is not active, he manifests *his* respect; when he physically cares for them, he engenders *their* pleasure; when *they* are sick, he manifests *his* concern; when *he* mourns for them [at death], he manifests *his* grief; and when he sacrifices to them [after death], he manifests *his* reverence."

This passage is not cited to demonstrate that Literary Chinese is hopelessly ambiguous. On the contrary, with the use of common sense, its shifting sequence of subjects and objects can be traced, though with some effort. What the passage does illustrate, however, is the unnecessary burden of time and energy such imprecise writing imposes on the reader by obliging him or her to focus on the text's externals rather than its inner content. Other similar instances could be found, having meanings that are less readily apparent.

Before going further, it is important to note that the ambiguities and ellipses already discussed (as well as the allusions and other features to be discussed later) are all particularly common in prose of a literary or hortatory nature, for which style is of prime importance. In more prosaic and factual writings, such as those having to do with science or technology, the language used may often be simpler, more direct, and less ambiguous. This tendency was of course desirable for Chinese science, yet its presence in any given piece of scientific writing can by no means always be taken for granted. This is because classical literary and hortatory texts, such as are often cited in these pages, constituted the humanistic heritage of virtually all Chinese intellectuals before quite recent times, which meant that they influenced even those individuals who explored the realms of science or technology.

We come now to the final topic, that of meaning: the fact that one and the same character can often embrace a considerable range of different meanings, many of them obviously semantically related, but others sometimes seemingly completely arbitrary. In view of the pictographic/ideographic origin of the characters, it is not surprising that their meanings should often begin with the physical and concrete and proceed from these through a series of increasingly metaphorical abstractions.

A good deal has been written about the influence this phenomenon may have had upon the Chinese psyche, such as the alleged Chinese tendency to see the abstract and general only in terms of the concrete and particular.[22] There may be some truth in this idea, and the topic of

22. See, among many scholars and amateurs, Fenollosa (1934), Duyvendak (1923: 286–287), Asycough (1929–1934:1.9–16), Richards (1932:58–59), Granet (1934:chs. 1–2 but esp. summary on p. 82), and Pound (1952), as well as Pound's commentary on Fenollosa (1934).

particularity is one to which we shall return in a different context (see sect. 9 below). As far as the written characters are concerned, however, I prefer to speak of their possibly psychological influence only in passing, because of what I feel is the great difficulty of dealing satisfactorily with such an intangible topic.

At present, then, our interest is simply in the concrete fact that many or most characters have more than one meaning, and in the practical difficulties often caused by this fact. The following are three examples of words having multiple meanings as defined in Karlgren's "Grammata Serica Recensa":[23]

Luan 亂 : "disorder, confusion, rebellion; end of a piece of music; as loan for *ssu* 𤔲 = *ssu* 司 : regulate, bring into order."

Chien 間 : "crevice, interstice; interval, space between; middle, in, among; interval in time; find a crevice in, find fault with; to separate, alienate, differences; favourable moment, occasion; intermeddle; replace, supersede; alternate; to spy on (look in a crevice); insert"; also, when pronounced *hsien* (interstice in time): "leisure; peace."

I [*Yi*] 易 : "change, exchange; easy; negligent; at ease; well cultivated (meaning a field)."

Of course, multiple meanings are not restricted to Literary Chinese; other languages have them as well. And again, of course, in Literary Chinese, as in other languages, context may solve what might otherwise be uncertain. Thus it is unlikely that many readers of Literary Chinese will confuse *luan,* "end of a piece of music," with the more usual *luan,* "disorder," any more than English readers will confuse *loom,* "machine for weaving yarn or thread into fabric," with *loom,* "to appear indistinctly."

Unfortunately, however, Literary Chinese also has quite a few multiple meanings that present considerably more difficulty. An example is the "disorder/regulate" dichotomy that, as shown above, pertains to the same single written graph, *luan.* The result is the curious phenomenon that the single term *luan ch'en* 亂臣 can, in different contexts, mean either "order-bringing ministers" or "rebellious ministers." Thus in the *Shu ching* (Documents Classic), when King Wu of Chou is about to overthrow the preceding Shang ruler (trad. in 1122 B.C.), he states to his followers: "Of order-bringing ministers *(luan ch'en),* I have ten men."[24] By contrast, Mencius, after asserting that Confucius compiled the short

23. Karlgren (1957:180c, 191a, and 850a, respectively).

24. *Shu ching* 5.1b; tr. Legge (1960:3.292).

text known as the *Ch'un ch'iu* (Spring and Autumn Annals), states the consequence of this act of composition to have been that "rebellious ministers *(luan ch'en)* and villainous sons were struck with terror."[25]

In Literary Chinese, the problem of multiple meanings is exacerbated by the fairly frequent occurrence of loan characters and the further fact that these loans are often unrecognized as such by readers. An example is the foregoing *luan,* "disorder," when it is used, as indicated by Karlgren, as a loan for *ssu* and therefore comes to mean "regulate, bring into order." The resulting confusion can scarcely be said to be helped by the Chinese lack of inflection. On the contrary, it seems a reasonable conclusion that multiple meanings are easier to identify in languages whose words undergo morphological change than in Literary Chinese, whose every written word, irrespective of usage (as a noun, verb, etc.), remains eternally the same.

In sum, I believe that multiple meanings are probably more frequent and difficult to identify in Literary Chinese than in English or other European languages. This, admittedly, is a subjective opinion, reached after half a century of experience with Literary Chinese, but unsupported by any attempts to measure the frequency of multiple meanings either in Literary Chinese or in other languages.

Let us see now how the ambiguity of a single Chinese word can sometimes create prolonged confusion among writers concerned with some particular scientific or technological subject. In the example that follows, the subject is the manufacture of porcelain and the ambiguous word is *ch'ing* 青, a designation for color that sometimes means "green," sometimes "blue," and sometimes points to other colors. In the *Ching-te-chen t'ao-lu* (Report on the Potteries of Ching-te-chen), a treatise published in 1815, Cheng T'ing-kuei writes about the word:[26]

> Among porcelains all equally [said to be] *ch'ing,* such *ch'ing,* when mentioned for the Ch'ai and Ju ceramics, nevertheless comes close to being a light blue

---

25. *Meng-tzu* 3b.9; tr. Legge (1960:2.283). For a much longer discussion of the dual meanings of *luan* and the possible relationship of the word to myth, see Girardot (1983:114–133).

26. 10/8b; tr. Bushell (1910:xvi) and Sayer (1951:115). Not included in Julien (1856), who translates only *chüan* 1–7 and, despite the passage here quoted, always renders *ch'ing* as "bleu." The proper names that follow (unessential for our purpose) are those of ceramic types of the Sung dynasty. More important are Cheng's terms for their colors, which are difficult, so that my renditions (and those of Bushell and Sayer) are only approximations. The *Ching-te-chen t'ao-lu* originated with materials gathered by Lan P'u, but was greatly enlarged and put into its present form by Cheng T'ing-kuei. I am indebted to Professor James T. C. Liu of Princeton for providing me with a copy of this passage.

> *(ch'ien lan)* color; for the Kuan, Nei, Ko, Tung and Hsiang ceramics, it comes close to being a light jade-green *(ch'ien pi)* color; for the Chang ceramics of Lung-ch'üan, it comes close to being a kingfisher *(tsui)* color; and for the Yüeh and Yo ceramics, it comes close to being a pale blue *(p'iao)* color. Yet the ancients, when they talked about ceramics, simply called all of these *ch'ing*.

Such comment on the loose use of *ch'ing* is exceptional in pre-modern Chinese writings. At any rate, I have not encountered it elsewhere. Yet the ambiguity of the word is already to be found in classical literature.[27] Of the several Western scholars who have discussed the word,[28] all seem to be interested only in determining the word's several color possibilities. None mention what is our central concern: the fact that the ambiguity of *ch'ing* exemplifies a not uncommon phenomenon in Literary Chinese.

From this ambiguity of a single word let us turn now to that of a single sentence. Our example will be the famous passage in *Lun yü* 2.16 (tr. Legge 1960:1.150), in which Confucius criticizes unorthodoxy. Its eight characters all represent everyday words, but major difficulty arises because two of them, and especially the first, can be understood in more than one way. The alternative possibilities are indicated in the literal English renditions below:

攻	乎	異	端	斯	害	也	已
Kung	*hu*	*yi*	*tuan*	*ssu*	*hai*	*yeh*	*yi*
Attack	on	strange	shoots	this	harmful	is	indeed.
Study	of	strange	shoots	these	harmful	are	indeed.

First of all, there is the problem of the complete absence of context: we have no idea when or why Confucius made this utterance. Hence it is difficult to guess just what he meant by *yi tuan*, "strange shoots," a term that occurs here for the first time. *Yi* is often a pejorative, meaning "strange" in the sense of "irregular," "abnormal," "monstrous." *Tuan* is a neutral word, signifying the tip or end of something, and hence something that begins to sprout. Mencius (*Meng-tzu* 2a.6) uses it to denote the four "beginnings" or "shoots" of goodness (we would call

---

27. In the *Tso chuan*, Hsi 26 (634 B.C.; tr. Legge [1960:5.198]), the statement appears: "In your fields there is no green grass *(ch'ing ts'ao)*." Here *ch'ing* can only mean "green." In *Chuang-tzu* 1 (tr. Watson [1968:30] and Graham [1981:43]), the mythical P'eng bird is said to "mount on the back of the wind and shoulder the blue sky *(ch'ing t'ien)*." Here *ch'ing* can only mean "blue."

28. See von Strauss (1879), Hirth (1887:7–9), Boodberg (1979:178–179), and Schafer (1982).

them instincts) which, he maintains, are present in all men at birth. We are dependent on later commentators for the assertion that *yi tuan,* as a compound, means "unorthodox, heretical, or uncanonical doctrines" (strange or monstrous sprouts). On the strength of this single reference by Confucius, the term has crept into all the surviving law codes of imperial times (beginning with the T'ang Code of 653), where it appears in the statute proscribing the (in Confucian eyes) dangerously subversive politico-religious doctrines that pertain to various Taoist or Buddhist underground sects and secret societies. We have little alternative but to project the conventional interpretation, "heterodox doctrines," back to the *Lun yü,* even though we do not know what Confucius really meant when he coined the term.

"Strange shoots," however, is only the beginning of our problem. The first word of the sentence, *kung,* is basically a verb, "to attack" (or, followed by the preposition *hu,* "to make attack on"). Used metaphorically but quite commonly, however, *kung* also means "to deal with, to devote oneself to" (as we say "to attack a problem"); in an ideological situation, as here, it can very well mean "to study" (or, followed by *hu,* "to make study of").

The other ambiguous word, the demonstrative pronoun *ssu,* means either "this" or "these," depending on what it refers back to. Unfortunately, here as in many other texts, the antecedent is uncertain. Is it the act of attacking (or studying) the heterodox doctrines, or is it the doctrines themselves? These uncertainties, coupled with the fact that *kung* might just as well be in the imperative as in any other mood, bring us face to face with four different possibilities for our seemingly simple eight words:

1. "To attack heterodox doctrines: this is harmful indeed!"
2. "Attack heterodox doctrines [because] these are harmful indeed!"
3. "To study heterodox doctrines: this is harmful indeed!"
4. "Study heterodox doctrines [because] these are harmful indeed!"

All four interpretations seem equally valid grammatically. Hence we have to make our choice on purely ideological grounds. The first interpretation can be readily rejected, because its civil libertarian flavor, while eminently suitable for Voltaire, would be quite unlikely for Confucius.[29] The fourth possibility also seems unlikely in a Confucian context (whereas if it came from a military writer, it would agree well with the famous adage: "Know yourself and know the enemy, and in one

29. Nevertheless, Lyall (1925:5) has adopted this very interpretation: "To attack strange doctrines does harm."

hundred battles you will have one hundred victories"[30]). This leaves us with the second and third versions. Most commentators and translators have accepted the third ("To study heterodox doctrines . . ."), but I find myself equally drawn to the second ("Attack heterodox doctrines . . ."). In either case, the opposition to heterodoxy is evident and, irrespective of what Confucius may himself have meant, has strongly colored all later Confucian thinking on the subject.[31]

The linguistic features we have been discussing—lack of inflection, flexibility of grammatical function coupled with syntactical strictness, the "rule of economy," and multiple meanings—have enabled the Chinese to do things with their written language that would be impossible in any other. A striking example is the *Ch'ien-tzu wen* or *Thousand Characters Text,* attributed to Chou Hsing-ssu (d. A.D. 521). The work consists of exactly one thousand characters, none repeated, arranged in rhyming couplets of four characters each, making a total of 250 lines. Despite these extraordinary requirements, the text reads intelligibly and covers a wide range of history, morals, and cosmology.[32]

5. Stylistic Balance: Parallelism and Antithesis

The features we have been discussing, with the sole exception of the "rule of economy," are all inherent in the Chinese language and its written medium. Thus any resulting difficulties are inescapable for readers and writers alike, who, at least before modern times, were no doubt scarcely aware of their existence. A very important stylistic feature *does* exist, however, to help the writer achieve not only literary beauty but grammatical and semantic clarity as well. This feature,

---

30. This saying derives from two almost identical utterances in the *Sun-tzu ping-fa* (Master Sun's Art of War) (fourth century B.C.) 4 and 10; tr. L. Giles (1910:24–25 and 112).

31. Forke (1924), in a special study of this passage, supports the interpretation of *kung* as "to study" (rather than "to attack"), but after reading his arguments, I am still convinced that both interpretations are possible. Another very different interpretation, not included above because it seems so improbable, is that of Waley (1938:91), who often proposes something very original: "He who sets to work upon a different strand destroys the whole fabric." The Hung-wu Emperor, founder of the Ming dynasty (r. 1368–1398), is said to have supported the interpretation of *kung* as "to attack" rather than "to devote oneself to," which was the interpretation espoused by Chu Hsi. See Ku Chieh-kang (1938:300).

32. There are several Western translations of the *Ch'ien-tzu wen.* See Cordier (1904–1924:2.1439–1440). Probably the best known is that of Julien (1864).

which we shall call *stylistic balance,* involves the writing of clauses, sentences, and larger prose units in cadenced patterns. The basic building blocks consist of successive clause or sentence units (groups of syntactically related words), all or most of equal length. These we shall hereafter refer to as cadence-units. Those of four characters are the most common, but other longer or shorter cadence-units may be introduced at any time to meet the exigencies of exposition or to vary the cadence. Prepositions and some other unstressed words are quite often ignored in measuring the basic length of cadence-units.

The reading aloud of such patterned prose produces a strongly rhythmic effect, sometimes close to poetry, despite the absence of rhyme (customary in Chinese poetry). That is one reason why, as we have seen, oral memorization and recitation were important aspects of traditional education, even though the listener could not possibly understand the classical texts thus recited unless he was already familiar with them (as of course, if he were the teacher, he commonly was).

Whereas good written style in English normally avoids frequent repetition of the same word, except very occasionally for deliberate emphasis, the balanced structure of Literary Chinese quite often requires repetition not only of separate words but sometimes of phrases or even clauses as well. Yet as suggested above, the stylistic requirements are by no means as absolute and mechanical as this necessarily bald account might suggest. In Literary Chinese, as in any other medium, the skilled writer knows how to vary rhythm and pattern to retain freshness and vigor.

The two basic aspects of Chinese stylistic balance are *parallelism* (the more common of the two) and *antithesis.*[33] Parallelism may be either quantitative, qualitative (for which another term might be *functional*), or both. Quantitative parallelism is one in which an equal number of monosyllables (i.e., characters) occurs in each of the two or more groups of words that are juxtaposed in parallel construction. Qualitative (or functional) parallelism is one in which there is syntactic and semantic congruence between the juxtaposed word groups. The combination of both quantitative and qualitative parallelism results in strict or absolute parallelism. An example in English would be the saying: "Spare the rod and spoil the child." Here, on the two sides of the axial conjunction "and," the parallelism between the two juxtaposed clauses is both quantitative (three monosyllabic words in each clause) and qualitative (each embodies a verb/object construction so that there is close syntactic similarity between the two).

---

33. For the following discussion of parallelism and antithesis, I am indebted for valuable suggestions to Professor Janusz Chmielewski, who in conversation with me stressed the importance of the topic. See also Chmielewski (1979:70 and n. 10).

Parallelism can also be described as internal, external, or a combination of both. Internal parallelism means parallelism within a given sentence. It can occur between various kinds of lexical components, such as clauses ("spare the rod" and "spoil the child"), phrases ("green grass" and "blue sky"), or individual words ("grass" and "sky" used alone). External parallelism means parallelism between rather than within separate sentences. A crude example would be: "The sheep graze on the green grass. The birds fly in the blue sky." In Chinese, external parallelism can sometimes lead to very complex linguistic structures.

Antithesis may be, and often is, included in parallel structures. In such instances, the prime point of difference from an ordinary nonantithetical parallelism is that, in an antithetical statement that is at the same time parallel, the relationship between its juxtaposed parallel word groups is one of antonymy or near antonymy rather than synonymy or similarity. The tendency for Chinese antithetical statements to be expressed in parallel structure is very strong, and when this happens, it is common for their basic feature of semantic antonymy to be combined structurally with quantitative parallelism. An example in English would be the saying "Easy come, easy go," in which the antonymy between the key words "come" and "go" is coupled with quantitative parallelism between the two juxtaposed parallel phrases.

At first sight it is tempting to believe that if, in Chinese, absolute parallelism is readily possible for nonantithetical parallel statements, then something like it should also be possible for those parallel statements that are at the same time antithetical. In other words, the structure of such parallelistic antithetical statements should display not only quantitative and qualitative balance but also something which, for purposes of this discussion, might be called "antithetical balance." That is to say, antonymy or near antonymy should exist not only between their key terms, but also between many or all of the other mutually corresponding words belonging to their juxtaposed parallel word groups. A moment's thought, however, is enough to show the virtual impossibility of such an achievement, whether in Chinese or any other language. Probably as close as we can come in English is the saying "Penny wise and pound foolish." For here, in addition to the qualitative parallelism existing between the two juxtaposed phrases and the perfect antonymy between "wise" and "foolish," there is antonymy to some degree between "penny" and "pound," inasmuch as these are respectively the smallest and largest units (i.e., the opposing poles) on a common monetary scale.

Both antithesis and parallelism—the latter often in absolute or near absolute form—are far more common and easy in Literary Chinese than in Indo-European languages. This fact is obviously attributable to the distinctive linguistic features of Literary Chinese: its monosyllabicity, lack of morphological change, and determination of grammatical function by syntactic position. As a result, what might almost be called the "linguistic polyphony" of a passage of Literary Chinese can sometimes become very elaborate: a network of parallelisms and/or antitheses that interweave, both internally (within sentences) and externally (between sentences), with one another. Such a balanced complexity is plainly unattainable in any Indo-European language.

Parallelism and antithesis of a much simpler sort, moreover, continue to appear in modern colloquial Chinese literature, as well as in the set phrases and proverbial sayings that, to a considerable measure, are heritages in modern Chinese from the old language. In both media, too, antonymal binomes have long been used to represent abstract concepts connected with quantitative measurement. Examples are *ta* 大 (large) and *hsiao* 小 (small), which combine to form a compound, *ta-hsiao,* "size"; *ch'ang* 長 (long) + *tuan* 短 (short) = *ch'ang-tuan,* "length"; *yüan* 遠 (far) + *chin* 近 (near) = *yüan-chin,* "distance," and so forth.[34]

No doubt parallelism and antithesis had their start as stylistic features, but so ubiquitous did they become in Literary Chinese that they have acquired grammatical significance as well. In other words, the skilled writer uses them not only for stylistic effect but also, quite deliberately, to give syntactic and semantic clarity to what he is saying. Thus they provide the reader with an indispensable tool for "parsing" the text, which he or she can ignore only at his peril. Supposing, for example, that there is a pair of parallel statements, of which one is clear in structure and meaning but the other is obscure. If, say, the third word in the clear statement is a verb, the reader will do well to assume that the corresponding word in the unclear statement is also a verb, even though in its case such verbal use may be quite exceptional.

But the importance of parallelism and antithesis transcends style and grammar and extends to thought as well. Worth considering, for example, is the possible relevance Chinese stylistic parallelism may have for the widespread form of Chinese reasoning sometimes referred to as "analogical" or "correlative" thinking. "Essentially," it has been said, "the Chinese method [of argumentation] was analogical—like causes bring like effects, as it was then so it is now, and so it will be for ever.

---

34. See Rygaloff (1958:371). On p. 375, Rygaloff also lists examples of compounds that, not being strict antonyms, are used to express abstract terms of a nonquantitative sort. They include *shan* 山 (mountains) + *shui* 水 (water) = *shan-shui,* "landscape," and *shui* (water) + *t'u* 土 (soil) = *shui-t'u,* "climate."

This faith was profound. Hence the great dominance of history (and all its ancillary sciences) throughout Chinese history."[35]

It is my belief that the linguistically generated principle of stylistic balance, as exemplified in parallelism, served to stimulate, or at least reinforce, the Chinese tendency to look at the world in terms of symmetrically paired analogies.[36] And I also believe that there is a further link between this and the broader Chinese search for cosmic order and harmony—especially as exemplified in the concept of the "harmony of man and nature." Let us look at a concrete example. Stylistic parallelism appears very early in Chinese literature, and when it does, it is significant that it is already conjoined with the just-mentioned view of man and nature. For example, the very first poem in the *Shih ching* (Songs Classic) says in its opening stanza (four lines of four characters each):[37]

> Watchful, watchful, the osprey,
> On the islet in the river.
> Comely, goodly, the maiden,
> A fit mate for our lord.

The traditional interpretation makes the first line read quite differently: " *'Kuan! Kuan!',* [so cries or cry] the osprey[s]." In the new interpretation of Steininger (1967), here adopted, the doubled first word *kuan* 關 (lit. "to close," which obviously makes no sense) is no longer taken as imitative of the call of the osprey, but instead as being a loan word for homophonous *kuan* 觀, "to observe, to watch"; this results in the rendition "Watchful, watchful *(kuan kuan),* the osprey."[38]

Once we accept the new interpretation, the poem's inner meaning emerges: a female osprey—symbolic in the poem of wifely devotion—perches or sits on an islet in the river (quite conceivably she sits on a nest), where, with undivided watchful attention *(kuan kuan),* she awaits the return of her mate. In similar fashion, the beautiful and good maiden of the poem's third line is expectantly awaiting the arrival of the lord who is to take her in marriage. There is obvious parallelism here, both quantitatively and qualitatively, between the doubled *kuan-kuan* of the first line, descriptive of the osprey, and the rhyming binome of the

---

35. Needham (1965b:16). The use of analogy will be touched on again (see II.8).

36. See III.2 for a detailed discussion of Chinese concepts of symmetry and centrality.

37. Tr. Waley (1937:81) and Karlgren (1950a:2), but modified, especially in the first line, in the manner to be described below.

38. See Steininger (1967), whose persuasive but rather technical arguments are, for the sake of simplicity, omitted here.

third line, *yao-tiao* 窈窕, descriptive of the maiden—which I have inadequately rendered as "comely, goodly."[39]

Much more important than such linguistic parallelism, however, is what might be called the *topical parallelism* of the poem: the balance it establishes between the little vignette of conjugal devotion drawn from nature in the first two lines and the scene of human rejoicing over marital felicity suggested in the third and fourth. Here, in embryo, is the all-important "harmony of man and nature." Similar juxtapositions occur frequently not only in the *Shih ching* but in all later Chinese poetry. And just as, stylistically, antithesis is the obverse of parallelism, so, topically, antithesis is quite as possible as parallelism. In topical antithesis, the effect depends on ironically contrasting, rather than comparing, the natural and the human scene. A good example comes once more from the *Shih:*

> Oh, the flowers of the bignonia,
> Gorgeous is their yellow!
> The sorrows of my heart,
> How they stab![40]

Just as it was earlier suggested that the principle of linguistic parallelism may have its broader ideological ramifications, so I would now sug-

---

39. Descriptive binomes such as these are common in Literary Chinese and fall into two main categories: (1) *shuang sheng,* "doubled [initial] sounds" (i.e., alliterative binomes) and (2) *tieh yün,* "piled-up rhymes" (i.e., rhyming binomes; *yao-tiao* belongs to this group). A third category, deriving from these two, is that in which a monosyllable is simply doubled (as in the case of *kuan-kuan*). Since all three kinds of binomes function as describers of appearance, manner, activity, and the like, binomes in any one of the categories are felt by the Chinese to be not only quantitatively but also qualitatively parallel to those in either of the other two. On the development and significance of such binomes, see Chmielewski (1957a:72 n. 14). Even if *kuan-kuan* were to retain its traditional interpretation as representing the call of the osprey, it would still, as a binome descriptive of a sound, be regarded as parallel to *yao-tiao.* In the new interpretation, however, the parallelism between the two is much more pronounced—thus strengthening the plausibility of the interpretation.

40. *Shih ching* no. 233; tr. Waley (1937:324). See also Karlgren (1950a:184). Waley (1937:15) writes concerning topical antithesis:

> Again and again it is as contrast and not as comparison that the things of nature figure in early Chinese songs. The lotus is in the pool and the pine-tree on the hill. Nature goes its accustomed way, only man changes; such is the burden of many songs. The birds and beast[s] come back to the farm at dusk; but the warrior knows no rest. . . . The kite soars into the clouds, the fish hides in the depths of its pool; but for man there is no escape.

gest that there may be similar links leading from linguistic antithesis to topical antithesis, and from both to Chinese dialectical thinking. By the latter I mean the kind of thinking that views the universe in terms of a never-ending flux of interacting processes, rather than an array of unchanging substances and qualities. "One Yin, [then] one Yang: such is the Tao."[41] How and why the kind of Chinese thinking connected with parallelism and antithesis differs from the thinking of speakers of inflected Indo-European languages will be briefly suggested at the end of this section.

Meanwhile, let us turn to the specifics of stylistic balance and see with concrete examples how it operated. A popular literary medium serving to perpetuate the ideal of cosmic harmony into the present century is that of the ubiquitous pairs of hanging wall scrolls, with complementary lines of calligraphy, known as *tui-tzu* (lit. "juxtaposed things"). As a young student in Peking, I had the privilege, around 1933, of accompanying my father on a visit to Hsü Shih-ch'ang (1855–1939), once one of the warlord presidents of China (1918–1922) but, at the time of the visit, a scholar in his late seventies living in retirement in Tientsin. The old man, though not a remarkable president, had been an active proponent of scholarship and was still famed for his calligraphy.[42]

President Hsü had expected my father's visit and so had prepared a *tui-tzu* (pair of scrolls) in advance to give him. However, my own coming had been quite unanticipated. Yet after talking politely for a few minutes and learning what I had been studying, the old man left the room and in a very short time returned with an additional *tui-tzu* whose ink was barely dry. These scrolls, since treasured by me for the personal circumstances of their composition as well as their calligraphy, read as follows:

客 子 光 陰 詩 卷 裏
K'o tzu kuang yin shih chüan li
Guest('s) son light darkness poetry scrolls within.
(day night)

春 花 消 且 雨 聲 中
Ch'un hua hsiao chü yü sheng chung
Spring flowers melt oh rain sounds amidst.

---

41. *Yi ching,* Great Appendix, sect. 1; tr. Legge (1899b:355) and R. Wilhelm (1950: 2.319).

42. The large compilation known as the *Ch'ing ju hsüeh-an* (Schools of Confucian Philosophers during the Ch'ing), containing biographies of about 1,690 scholars, was prepared under Hsü Shih-ch'ang's editorship and published in 1938 shortly before his death.

"Guest" refers of course to my father, and "guest('s) son" to me. The "poetry scrolls" were the *Shih ching,* which I had said I was studying. The time was spring, so that the reference to spring flowers melting away amidst the downpour of rain was from a literary point of view entirely appropriate. (Realistically, it was less so, because rain rarely falls in force during the dry North China spring; the words apply much better to Central or South China. On the use of literary clichés, something will be said in sect. 8 below.) Here is how the lines read in a less literal translation:

> My guest's son day and night is buried within the books of poetry.
> The spring flowers melt away, oh, amidst the sounds of rain.

Here we find the same balance between man and nature as appeared a few pages back in the opening poem of the *Shih,* although with the natural and human scenes put in reverse order. Quantitatively and qualitatively there is good balance between "guest('s) son" and "spring flowers," as well as between "poetry scrolls within" and "rain sounds amidst." The parallelism, however, is not strictly maintained in the intervening words of the two lines. It should be noted that the Chinese first line contains no verb (an omission made possible by the "rule of economy"; in the English version the verb has to be added), whereas in the second line the verb has been followed by an expletive, "oh," so as to fill out the requisite seven words. A brilliant performance, one would say, from an old Chinese scholar composing and writing for a young Westerner whom he had seen for the first time barely half an hour earlier.

Stylistic symmetry, however, is after all just what one would expect in formalized compositions such as poetry and *tui-tzu* scrolls. Let us look further, therefore, to see how the same principle operates in ordinary prose. *Lun yü* 15.38 provides the simplest possible example:

有	教	無	類
Yu	*chiao*	*wu*	*lei*
Have	teaching,	not-have	category.
			categories.

Yu, "to have," and *wu,* "not to have," are antonyms. Hence the balance is antithetical rather than parallel, thereby causing the second situation to be conditional upon the first: "[If or when] there is teaching, [then] there is/are no category/categories."

Despite the everyday words and simple structure, the meaning is far from certain. Aside from the question of just what Confucius meant by *chiao,* "teaching," the basic semantic problem concerns *lei.* Ordinarily this word means "kind, sort, category," but early commentators give it a special twist here to mean "[social] class." Legge (1960:1.305) translates accordingly: "There being instruction, there will be no distinction of classes." On the basis of these four words, scholars have drawn sweeping conclusions about Confucius the democrat who abhorred all class distinctions. The danger of such speculation is shown by the fact that, other than in this passage, *lei* apparently is not used in the sense of "social distinction"; the usual Confucian word for this concept, popularized by Hsün Tzu (ca. 298–ca. 238), is *fen* (lit. "distinction, division, section, allotment").[43] Chu Hsi (1130–1200), as a Neo-Confucian moralist, does not interpret *lei* as referring to social classes at all, but to the categories of morally good and bad men.[44]

It might be argued that the ambiguity of *lei* is not inherent here but results from the failure of later generations to understand properly a shade of meaning perhaps perfectly clear to Confucius's contemporaries. Even if this were so (and in view of what has just been said, such an assumption is far from certain), I believe the real cause of ambiguity lies less in the uncertain meaning of *lei* itself than in the fact that *lei* is embedded here in an antithetically balanced sentence consisting of only four words. It is most unlikely that Confucius himself ever spoke in such a precisely balanced and highly concise manner. Rather, the sentence represents the effort of a redactor straining to reduce the probably considerably more prolix verbal utterance of the Master into a written statement that would be pithy and balanced. Stylistically, the result is dramatic. Semantically, it is probably much more ambiguous than Confucius's original utterance.[45]

---

43. For Hsün Tzu's use of *fen*, see V. 2.

44. Chu Hsi's commentary is translated by Legge (1960:1.305 n.), where he rejects it. Couvreur, however, who regularly follows Chu Hsi, translates accordingly (1895: 248): "Le sage admet à son école tous les hommes sans distinction [de bons ou de méchant, d'intelligence ou de peu perspicace, afin que tous cultivent la vertu]." Quite a mouthful to derive from four words, but closer to actuality than Waley (1938:201), who, citing a passage in *Meng-tzu* 2a.2, translates the *Lun yü* passage as: "There is a difference [between us and the sages] in instruction but none in kind." Still another interpretation of *lei* is that of Miyakawa, who thinks that the word means "race" in this passage and comments: "This important remark means, to my mind, that in principle Confucianism may be propagated to the foreign tribes as well as to the lower classes in Chinese society" (Miyakawa 1960:26).

45. In the final paragraph of II.2 above, the conclusion reached concerning the extent of similarity between spoken and written Chinese in Chou times was that "the conclusion seems quite plausible that even in Chou times only a lesser part of the written literature was stylistically very close to the spoken language."

A far more complex example of balance is found in Hsün Tzu's chapter on the evilness of human nature.[46] In the following presentation, the sentence is schematically broken down into its five component sections:

凡 古 今 天 下 之
1. *Fan* *ku* *chin* *t'ien* *hsia* *chih*
Everybody past-present Heaven-below 's

所 謂 善 者
2. *so* *wei* *shan* *che*
that-which called good-ones:

正 理 平 治 也
3. *cheng* *li* *p'ing* *chih* *yeh*
alignment pattern equableness good-order are;

所 謂 惡 者
4. *so* *wei* *ô* *che*
that-which called evil- ones:

偏 險 悖 亂 也
5. *p'ien* *hsien* *pei* *luan* *yeh*
one-sidedness irregularity turbulence disorder are.

The first and last words of the first line (for reasons given at the beginning of this section) do not count in determining the length of the basic cadence-unit, which thus consists of the intervening four words. These in turn are subdivided into two words of duration (*ku-chin,* "past-present") and two of space (*t'ien-hsia,* "Heaven-below," i.e., the entire known world under Heaven). Their combination in a single unit conveys the image, so congenial to Chinese thinking, of a single time-space continuum.

In lines 3 and 5, again, the final copulative *yeh* (lit. "oh" or "indeed") is in each case ignored for word count. Thus the overall structure is that of a single sentence consisting of five cadence-units, each basically four words long. Among these units, the first line is introductory; the two following pairs of lines (2, 3 and 4, 5) are antithetical in their total message; and their individual lines exhibit an alternating pattern of balance and nonbalance. This last statement means that the relationship between lines 2 and 4 and between lines 3 and 5 is one of antithesis that is at the same time structurally parallel, whereas there is no balance at all between lines 2 and 3 or lines 4 and 5.

This high degree of structure removes all grammatical doubts from the sentence, leaving the major translation problem that of finding ade-

---

46. *Hsün-tzu* 23 (17/4a); tr. Dubs (1928:307) and Watson (1963:162).

quate renditions for the four antonymous terms in each of lines 3 and 5. All eight words have several dictionary meanings, among which one must find those that are diametric opposites within each pair, yet that keep within the psychological frame of reference being discussed by Hsün Tzu.

In particular, the difficulty lies within the pairs *li* and *hsien* and *p'ing* and *pei*. *Li* is the famous "pattern or principle of organization" of Neo-Confucianism, whereas the most common meaning of its opposite, *hsien*, is "danger, dangerous," which obviously does not fit. However, "danger" is actually only a secondary abstraction derived from *hsien*'s basic concrete meaning of "steep or rugged terrain"; this, abstracted in turn into "irregularity," seems an adequate antonym to "pattern." In the case of *p'ing* and *pei*, *p'ing* is basically a concrete word, meaning "level or even," but it can also, in a more extended psychological sense, mean "equable" (one thinks of the calm evenness of still water). *Pei*, on the other hand, is basically psychological (it contains the "heart" significative), and has to do with perverse and impassioned thoughts and impulses. "Turbulence" is a word that covers both physical and psychological connotations, and thus seems a suitable antonym to "equable," having a comparable semantic range.

Thus we reach the following nonliteral translation of the entire sentence: "For everyone, past and present, within the entire world, what are called goodness are alignment, pattern, equableness, and good order; what are called evil are one-sidedness, irregularity, turbulence, and disorder." Dubs (1928:307), translating the same passage, completely destroys the antithesis by telescoping the four words in line 3 into two binomes while retaining their fifth line counterparts as four separate values: "In whatever age or place on earth, in ancient times or in the present, men have meant by goodness true principles and just government. They have meant by evil partiality, a course bent on evil, rebellion, and disorder." Watson's translation (1963:162) succeeds better in retaining the antithesis, but does so, curiously, by reducing each set of terms from four to three: "All men in the world, past and present, agree in defining goodness as that which is upright, reasonable, and orderly, and evil as that which is prejudiced, irresponsible, and chaotic."

Obviously, tremendous skill is needed to maintain ongoing rhythm, pattern, and variety in Literary Chinese. It is not at all surprising, therefore, that even great writers should on occasion resort to literary devices that, viewed by the outsider, seem to diminish the quality of their exposition. (It is quite unlikely that the writers themselves would have shared this unfavorable opinion or even been aware that any problem of form versus content arises in Literary Chinese.)

The more common of these devices is that of adding synonymous words or phrases to a cadence-unit that otherwise would be too short to maintain the rhythm and balance of the surrounding text. This device has the curious effect of bringing redundancy into what in other respects is probably the world's most concise language.[47] Its results are immediately apparent in Western translation, where redundancy is regarded as a defect; in the original Chinese they are hardly noticeable because of the constant emphasis on balance. In the following example, Mencius describes what the primeval world was like before the culture-hero Yü drained the land of its flood waters and made it fit for human habitation (*Meng-tzu* 3a.4; tr. Legge 1960:2.250):

洪	水	橫	流
Hung	*shui*	*heng*	*liu*
Flood	waters	horizontally (in all directions)	flowed,

氾	濫	於	天	下
fan	*lan*	*yü*	*t'ien*	*hsia*
inundated	poured	over	Heaven-below. (the world)	

草	木	暢	茂
ts'ao	*mu*	*ch'ang*	*mao*
Grasses	trees	luxuriated	flourished,

禽	獸	繁	殖
ch'in	*shou*	*hsi*	*chih*
birds	beasts	proliferated	multiplied.

The preposition *yü* in the second line is not counted in determining the four-word cadence-unit. In this and the two lines following, the doubling of each verb to fill out the word pattern is semantically redundant: "The flood waters flowed in all directions, they *inundated* and *poured* over the world. Grasses and trees *luxuriated* and *flourished,* birds and beasts *proliferated* and *multiplied.*"

More serious from the point of view of content are instances in which the author, wishing to express his thought in a stylistically compelling manner, allows its written formulation to differ somewhat from what he actually has in his mind. In *Lun yü* 14.11 we read:

---

47. Here and elsewhere, our emphasis is on Literary Chinese as an instrument for clear and simple communication, not on its stylistic qualities as judged from a literary point of view. See II.4, second paragraph.

貧 而 無 怨 難 富 而 無 驕 易
P'in erh wu yüan nan fu erh wu chiao yi
Poor and not-have resentment difficult, rich and not-have pride easy.

In this pungent apothegm, the wording of the second half is obviously determined less by considerations of accuracy (for to be rich and yet lack pride is in fact often *not* easy) than by insistence on retaining an unbroken stylistic antithesis to the first half. Legge (1960:1.279) translates faithfully: "To be poor without murmuring is difficult, to be rich without being proud is easy." He adds in a note: "This sentiment may be controverted." It is evident, however, that Confucius (or rather the disciple who recorded the Master's utterance) used the pithy style deliberately to capture the reader's attention, without expecting everything to be understood literally. Waley (1938:182) translates accordingly: "To be poor and not resent it is far harder than to be rich, yet not presumptuous." This surely captures the intent of the original but in so doing destroys its brevity and appositeness.

By now, it is hoped, the reader has become convinced of the aesthetic qualities, creative difficulties, and semantic problems of Literary Chinese. Stylistic patterns other than those illustrated here could have been shown if space permitted. More important than style or even language, however, are the ideological ramifications that quite possibly stem from the dominant role of stylistic balance in Chinese.

It has been suggested earlier that stylistic and topical parallelism may both be connected with Chinese analogical thinking, as well as with the Chinese search for order and harmony, especially as expressed in the concept of the harmony of mankind and nature. Likewise it has been suggested that there may be a relationship between antithesis (both stylistic and topical) and the Chinese dialectical thinking which sees the universe in terms of ever-shifting phenomenal change rather than unchanging categories. These several aspects of the Chinese world view contrast sharply with the tendency of Indo-European language speakers (the ancient Greeks, the Hindus, and others) to classify universal phenomena under relatively unchanging categories of substance, qualities, activities and so on.

The final suggestion, therefore, is that, just as Chinese thinking has, I believe, been strongly influenced by a linguistically determined tendency toward parallelism and antithesis, so, just conceivably, the "categorical" (i.e., classificatory) thinking of the speakers of Indo-European languages may have been influenced by the linguistic structure of their inflected languages. For in these languages, the fact that the different parts of speech are so readily distinguishable owing to their morphological shapings might well have encouraged a mental outlook that, accustomed to the existence of categories in language, would, perhaps unconsciously, seek for such categories in the natural world as well. Of course, this can only be proposed as an unprovable hypothesis.[48]

6. Punctuation

Aside from stylistic balance, there is another device, punctuation, which, had it been perfected and popularized, would have substantially lessened the difficulties of Literary Chinese.[49] As we shall see, a crude proto-punctuation goes back at least as early as the third century B.C., yet until recent times punctuation remained underdeveloped, was used only sporadically, and was often looked down upon as a vulgar device unworthy of the true scholar. That traces of the same attitude sometimes appear even now is suggested by what an excellent student of comparative literature, Achilles Fang, has said in his witty article, "Some Reflections on the Difficulty of Translation":

> Most Chinese texts can be readily punctuated. . . . If a translator cannot correctly put dots and circles in the body of his text, obviously he is not ready for translating; he will have to wait some more years. . . . The so-called punctuation marks in Chinese texts, which any school child of ten can put down, represent nothing much beyond breathing pauses. They are neither grammatical nor logical. (Fang 1953:273)

Perhaps they are neither grammatical nor logical from the point of view of an Indo-European grammarian, yet their importance for textual understanding has, I hope, by now been demonstrated. If Chinese punctuation is really as simple as Mr. Fang asserts, one wonders why famed commentators and modern scholars have so often differed in their understanding of Chinese classical texts, syntactically as well as semantically. The following example, selected because of its brevity, comes from the *Lun yü* 9.1:

子 罕 言 利與命 與仁
Tzu han yen li yü ming yü jen

---

48. In formulating this statement of the contrast between Chinese and Indo-European languages and its possible consequences, I wish again to express indebtedness to Professor Janusz Chmielewski.

49. With slight changes, this section on punctuation is to be separately published as Bodde (forthcoming).

The passage is traditionally regarded as a single sentence and hence rendered as: "The Master rarely spoke about profit and fate and humanity" (see, e.g., tr. Legge 1960:1.216). However, two major difficulties confront this rendition: intellectually, the fact that Confucius did in fact speak very often about humanity *(jen)* in the *Lun yü,* and grammatically, the fact that the use of *yü* twice here in the same sentence as a conjunction (". . . and [*yü*] . . . and [*yü*] . . .") is quite exceptional. Suppose, however, that we divide this allegedly single sentence by inserting a full stop in the middle. The passage now reads: *Tzu han yen li. Yü ming yü jen.* With this division into two sentences, the two *yü* words can no longer remain conjunctions and have to function as verbs: "to permit, allow, give forth, share." At once the passage becomes both intellectually and grammatically acceptable: "The Master rarely spoke about profit. He gave forth [his ideas] on fate [and] gave forth [his ideas] on humanity." The change in meaning is made possible merely by inserting a single full stop.[50]

When Mr. Fang asserted that any ten-year-old school child can put down the Chinese dots and circles, he probably did not think of the marking of proper names as part of the punctuation process. Yet modern Chinese punctuation, as compensation for the fact that the characters cannot be capitalized, commonly marks off proper names by inserting a straight line under or beside them. (For book titles, the line is a wavy one.) Of course, there is little gain in so doing when the names are well known, but when, as often happens, they are obscure (as in many historical texts) or when they have common meanings of their own that can confuse the reader, the failure to indicate them can sometimes cause real trouble.

In the *Shih chi* (Historical Records), for example, near the end of its biography of the Lord of Meng-ch'ang (famous statesman of the state of Ch'i, fourth–third century B.C.), there is an episode (*Shih chi* 75/3a) in which the lord has fallen into disfavor with the king of Ch'i. His faithful follower speaks on his behalf to the king, telling him that if the Lord of

---

50. For further details, see Bodde (1933). It would be unfair to pretend at this point that no other elucidations of *Lun yü* 9.1 have been attempted. Several scholars, both Eastern and Western, have in fact made them, of which the most recent and comprehensive is that by Boltz (1983). His conclusion is that the key word *yü* here means "together with" (rather than merely "and"), thus leading to the suggested rendition: "The Master rarely spoke about profit together with *(yü)* fate or together with *(yü)* humanity." This is grammatically and contextually possible, but seems to me platitudinous, it being self-evident that Confucius in all likelihood would not often speak about profit in conjunction with either fate or humanity. More meaningful, at least to me, as well as grammatically equally acceptable, is the rendition reached by dividing the passage into two sentences.

Meng-ch'ang is cast off, then "Ch'in will become the cock and Ch'i will become the hen." In other words, Ch'in will triumph in the long struggle between these two Warring States principalities. After saying this, the envoy adds the following cryptic words:

雌	則	臨	淄	即	墨	危	矣
Tz'u	*tse*	*lin*	*tzu*	*chi*	*mo*	*wei*	*yi*

Hen, then overlook Tzu thereupon ink in-danger indeed!

Still today, Tzu remains the name of a river in modern Shantung province (where Ch'i was situated) and therefore should constitute one of the knowns in this dubious equation. Let us suppose that one of Mr. Fang's precocious schoolchildren (in all probability a boy, not a girl) is handling the text and has also studied sufficient English to attempt a translation into that language. He, armed with the indubitable fact that Tzu is the name of a river, will no doubt also know that *chi,* "thereupon," is customarily followed by a verb. Accordingly, he will quite likely clutch at the idea that *mo,* "ink," when used as a verb, can conceivably mean "to become black, to darken." These suppositions will thus result in the sentence: "Being the hen, it [Ch'i] will then overlook the Tzu [river] and thereupon will indeed become dark and in danger!" This at least translates all the words and conveys a sense of impending doom, even though it leaves unexplained why and how the state should be overlooking the Tzu and—a strange grammatical feature—why "then" should be followed by "thereupon" in the same clause.

Let us suppose now that the schoolchild has followed Mr. Fang's advice to wait a few years before translating the sentence. During the interim the child will have learned that Lin-tzu, "Overlooking-the-Tzu," is actually the name of the Ch'i capital, so named because it stood on the bank of the Tzu River. He will accordingly confidently translate: "[If Ch'i] is the hen, then [its capital] Lin-tzu will thereupon indeed become dark and in danger!" So happy will he be to have made such good sense that he too will probably not be unduly bothered by the redundancy of "then" followed by "thereupon."

Finally, let us suppose that the schoolchild has become a mature scholar. By now he knows that words in Literary Chinese are not always what they seem, and he also notices, while examining the sentence, that its surface meaning fails to provide the kind of stylistic balance one normally looks for. Realizing that Lin-tzu is a place name, he will consult a Chinese historical geographical dictionary, and there, with pleased surprise, will make the discovery that *chi mo,* "thereupon ink," is actually the name of a city in the state of Ch'i. Accordingly, he will triumphantly produce a translation that yields excellent sense as well as perfect stylistic balance: "[If Ch'i] is the hen, then [its cities] Lin-tzu and Chi-mo will indeed be in danger!" How simple it all would have been if Ssu-ma Ch'ien (ca. 145–ca. 86), author of the *Shih chi,* had only taken the trouble to mark his proper names to start with! But also how much less challenging to the reader!

Let us look now at some examples of punctuation in pre-modern China. Primitive division-markers very occasionally occur in texts recovered from archaeological sites of the Han dynasty (206 B.C.–A.D. 220).[51] From its latter years come the surviving fragments of seven Confucian classics that were inscribed on stone tablets in A.D. 175. On the largest of them, a small portion of the *Kung-yang chuan* (Kung-yang Commentary) on the *Ch'un ch'iu* (Spring and Autumn Annals), dots are used to divide each yearly entry from its neighbors.[52] Analogous dots (or sometimes horizontal bars) also appear on somewhat earlier Han administrative documents, notably those recovered from the Chü-yen military outposts in Inner Mongolia (approx. 100 B.C.–A.D. 100).[53]

It would be possible to cite still earlier Han examples, but since 1975 this is no longer necessary because of the discovery in that year of the Ch'in legal texts unearthed at Yün-meng in Central China. All of these predate the Ch'in unification of China in 221 B.C. and thus belong to the preceding three quarters of the third century B.C. Punctuation dots occur on a very few of them, where their primary purpose is to indicate topical rather than syntactical divisions (which, however, necessarily coincide with the ends and beginnings of sentences). Of the 383 Yün-meng items that Professor Hulsewé has recently numbered and translated into English, only 46, according to my count, contain punctuation dots—a ratio of 1 out of every 8.32 items.

---

51. For information on Han punctuation, I owe much to Dr. Michael Loewe, now lecturer emeritus in Chinese at Cambridge University, who took considerable trouble to list more references than can conveniently be indicated here.

52. See Tsien (1962:76 and 78, Plate XIIIA), where, for example, a dot can be seen in the second column from the right immediately before the entry beginning: "In the fifteenth year . . ."

53. Examples occur in Loewe (1967:nos. 14.II, TD I and 7, MD 6, 8, and 9). Others could be cited, but in toto they would only be an infinitesimal fraction of the more than 10,000 inscriptions and inscription fragments recovered from Chü-yen alone. The purpose of the dots is evidently to indicate the beginning of an inscription. In these inscriptions, however, this convention has little practical value because, even without a dot, the beginning of the inscription is adequately indicated either by starting the inscription on a new wooden strip or, when on the same strip, by separating it by a blank space from the preceding inscription.

In the 46 items, moreover, the dots frequently appear only very irregularly and incompletely.[54]

The feature common to all these Ch'in and Han examples, therefore, is that the dots (or horizontal bars), when they appear at all, serve primarily to indicate topical divisions. Moreover, because of spacing or using of a new wooden strip to begin a new section, most such divisions would in any case be fairly self-evident without the dots. Thus the primary purpose of the dots is *not* to indicate syntactic pauses, either between or within sentences (although, as noted above, divisions in subject matter necessarily coincide with divisions between sentences). In other words, the dividing dots are basically *topical* rather than *syntactic,* and as such more analogous to paragraph indentations than to true punctuation marks. At most, therefore, they represent only a crude proto-punctuation and not a punctuation system per se.

Five hundred and more years after the Han, punctuation had apparently made some but still only limited progress, to judge from several of the Tun-huang manuscripts. Without going into detail, we may mention two, the texts of songs *(ch'ü)* that were brought back early in this century by Paul Pelliot to Paris. The song on one manuscript consists of two seven-character lines followed by six lines of five characters each; a black punctuation dot marks the end of each line. The other manuscript contains three songs with lines of varying length, each punctuated in the same way save that the dots are written in red ink.[55] In all these cases, although the dots represent a progressive development, their value is somewhat limited because even without them the rhythm of the songs would be strong enough to reveal their basic pattern.

Only after printing became common, and then only rarely, were Chinese books produced with dots or small circles systematically inserted to indicate successive cadence-units. Examples of such punctuation are found in printed books of the Southern Sung (1126–1279), but it is possible that the practice goes back somewhat earlier.[56] From a later period

---

54. See Hulsewé (1985), in which 3 of the 110 items in sect. A have dots; none of the 29 items in B; 12 of the 26 items in C; 22 of the 190 items in D; 9 of the 25 items in E; and none in sects. F and G, respectively containing 2 and 1 items.

55. The songs (catalogued as Pelliot chinois 3251 and 3911) are reproduced in figs. 44–48 of Plates XXII–XXIV in Jao and Demiéville (1971). For description of the songs, including their punctuation, see Jao's introduction in Chinese, pp. 92–95. The dating of the entire collection ranges from the K'ai-yüan period (713–741) to 979.

56. Yeh Te-hui (1973:33–34) lists examples of Southern Sung punctuation. However, Professor Tsuen-hsuin Tsien (University of Chicago), in a personal communication

(early Ming) a particularly handsome example of a punctuated work (although in manuscript only) is the largest encyclopedia ever compiled by man, the *Yung-lo ta-tien* (Yung-lo Great Encyclopedia), whose 11,095 manuscript volumes were completed in 1408 but never printed because of the enormous expense. In the 349 volumes that have survived a series of disasters culminating in the Boxer uprising of 1900, the punctuation markers, consisting of small circles in red ink, are clearly visible.[57] Although these markers fail to distinguish between what in Western punctuation would be marked by commas, full stops, question marks, and so on, and make no attempt to indicate proper names, their presence is still extremely helpful. Thus it is cause for real regret that the example of the Yung-lo encyclopedia was not widely followed, so that as late as the nineteenth century, works punctuated in this way constituted only a small fraction of the total printed output. The popularization of punctuation had to await the present century, when it was inspired by Western models.

Today practically all Chinese books (except photographic reproductions of old texts) are printed with modern (Western) punctuation which, however, does not always completely utilize the signs available. Curiously, the publications least fully punctuated tend to be newspapers, in which proper names are often unmarked even though the constant inflow of new ones from all parts of the world would seem to make such marking particularly desirable. The reason, perhaps, is the extra typographical burden the practice would impose on a medium necessarily pressed for time. Many texts from the past, however, have in recent years been reprinted in new punctuated editions both in the People's Republic of China and in Taiwan.

How does the Chinese experience compare with that in other civilizations? In India, before the introduction of modern (Westernized) printing, Sanskrit manuscripts and inscriptions maintained no divisions between words. Furthermore, the tendency of the highly inflected language was to build up words by combining more and more lexical units, until a single compound "word" could have as many as twenty separate meaningful components and a single "sentence" could cover two or three pages. The only punctuation mark was a vertical bar used fairly often in inscriptions to indicate the end of an epigraph, less often the end of a section or a sentence. However, there was no consistency in this regard. No device was used to indicate proper names—a fact that, because most such names also have a meaning, could, as in China, cre-

---

to me, expresses the opinion, after referring to Yeh's book, that printed punctuation probably predates the Southern Sung.

57. For details on the encyclopedia and its history, see Kuo Po-kung (1938).

ate much confusion. In short, punctuation was less developed and common in pre-modern India than it eventually became in pre-modern China.[58]

In the Islamic world, before the present century, the situation was virtually the same. No punctuation was used, whether it be full stops, commas, or paragraph divisions; individual words were separated in some texts but not all; and no specific device served to indicate proper names—again, as in Sanskrit and Chinese, a cause of much difficulty.[59]

In Hebrew the situation has long been somewhat similar. Yet the greatest problem has lain not in the absence of punctuation in the ordinary sense but rather in that the letters of the ancient Hebrew script represented only consonants and not the intervening vowels (though a few consonants did at times also serve as vowels). This meant that uncertainty and ambiguity could often arise for any written word that could be vocalized in more than one way and thus yield more than one meaning. For a long time, traditions of pronunciation were transmitted orally, but then, beginning around the seventh century A.D., a system of points or accents was devised to indicate the vowel values following the consonantal letters to which the accents were attached. These marks thus ended the possibility of ambiguity for any later texts that used them. However, they did not end uncertainty about the correct vocalization of texts written before the system had been devised. Moreover, in later centuries the vocalization signs were by no means universally used. Even today, for example, the Torah scroll (Pentateuch) that is read in synagogues as part of the service must be completely unpointed and unvocalized in any way.[60]

In the Western world the situation has been both different and more

---

58. Personal communication from Ludo Rocher, professor of Sanskrit, University of Pennsylvania, whose help is gratefully acknowledged, supplemented by Basham (1959:389–390, 396), and by Sircar (1965:5).

59. Personal communication from Roger M. A. Allen, professor of Arabic, University of Pennsylvania, to whom I am likewise much indebted, the more so as this subject has not been systematically studied.

60. For this and further information I am grateful to Judah Goldin, professor emeritus of biblical studies, University of Pennsylvania. A good example of ambiguity resulting from the absence of vocalization is the sentence in Isaiah 54:13: "When all your sons are instructed by the Lord, through your sons peace will increase." A *midrash* (exegesis of scripture) suggests that the word here vocalized as *banayikh,* "your sons," may just as properly (for homiletical purposes) be vocalized as *bonayikh,* "your teachers," resulting in the sentence "When all your teachers are instructed by the Lord, through your teachers peace will increase." See Goldin 1983:159.

complicated.[61] In ancient Greece, inscriptions were for the most part written continuously save for a few instances from before the fifth century B.C., in which phrases were separated by a vertical row of two or three points. Likewise, in the oldest literary texts (papyri of the fourth century B.C.), the beginning of a line introducing a new topic might be underlined by a horizontal bar. Such is the only punctuation mentioned by Aristotle. The first real punctuation system, in which prototypes of the full stop, comma, and colon were used, as well as marks of quantity, accents, and so on, is attributed to Aristophanes of Byzantium, librarian at Alexandria, ca. 200 B.C. For centuries, this system was seldom employed in its full form; it did, however, become the basis for modern Greek punctuation beginning in the ninth century.

In the Roman world, punctuation was commonly sporadic and unsystematic. Although Latin inscriptions from the earliest examples onward often used points to separate words, books were for centuries written with continuous script. Significantly, the first text to be really well punctuated was the Vulgate Bible, for which a system indicating phrases was devised by its translator, Saint Jerome (d. 419/420).

In medieval times, the major developments consisted of improvements in the punctuation of biblical and liturgical manuscripts that were devised under Charlemagne by his Anglo-Saxon adviser Alcuin (director of the palace school at Aachen, 782–798). This system spread through Europe together with the new Caroline minuscule script and reached its perfection in the twelfth century. By the fifteenth century, the early printed Bibles were all carefully punctuated.

Thus punctuation took a course in Europe very different from that followed in the non-Western world. Although the system invented around 200 B.C. was often ignored for long periods thereafter, it left a seed that was ready to spring up when conditions became propitious. Religion would seem to have been a major impelling factor, especially when printing enabled the Bible to be brought to larger numbers of people than ever before.

In the non-Western civilizations, one may strongly suspect that the differing situation stemmed, at least in part, from a differing conception of the purpose of writing. In China, writing was regarded as an instrument of prestige and power, to be exercised by a small elite primarily for communication among its own members rather than for mass commu-

61. For European punctuation, I am indebted to a personal communication from the late Lloyd W. Daly, emeritus professor of classical studies, University of Pennsylvania, supplemented by Brown (1974). The achievements of Aristophanes of Byzantium are discussed in greater detail in Pfeiffer (1968:178–180). See also, for developments in the fifteenth century, Hirsch (1967:136–137).

nication. Without this proprietary attitude, there would seem no reason why in a country where printing had been invented and where a rudimentary punctuation had been known for a considerable time, the latter could not have been eventually elaborated and universalized just as it was in the West. The failure of this to happen until the twentieth century, together with the parallel failure of written *pai hua* to replace the esoteric literary language until the same century, are equally indicative of deeply rooted traditional attitudes toward the popularization of knowledge. In the Western world, where a similarly proprietary attitude had also of course long been prevalent, the fact that it ultimately broke down may in part be due to ideological factors such as religion. Probably, however, it derived even more from a wide range of socioeconomic forces that significantly differentiated Western Europe from other societies.

Among the nonusers of punctuation, China at least enjoyed the advantage that its written symbols were readily distinguishable from one another, in contrast to the *scriptura continua* too often found in other civilizations. Moreover, in contrast to European and other languages, the vocabulary of Literary Chinese includes a goodly number of final particles and introductory words that fairly often—although far from invariably—help to determine the beginnings and ends of sentences or clauses even without explicit punctuation. Some scholars would therefore argue that Literary Chinese had less pressing need to develop a punctuation system than, say, did English.[62]

To this other scholars (including myself) would reply that such advantages are more than offset in Literary Chinese by several syntactic and stylistic features, among which the absence of morphological change is only one. To take Arabic as a single contrasting example, its usual lack of word separation was "not as desperate a situation as may at first appear, since the morphology of Arabic is such that it is usually possible to work out each word as a discrete unit on morphological grounds without the necessity of referring to the graphological symbol on the page."[63]

In Literary Chinese, as we have seen, failure to punctuate could sometimes create serious ambiguity. Probably more important, however, is that this and other difficulties, even when successfully resolved

---

62. This is the thesis of Dr. Joseph Needham of Cambridge University, personal communication. Some of the common final particles in Literary Chinese are *yeh* 也, *yi* 矣, *yen* 焉, *hu* 乎, and *tsai* 哉, but there are others as well. An excellent discussion of Chinese syntactic matters is that of Harbsmeier (1981).

63. Remark by Roger M. A. Allen, in personal communication referred to in n. 59 above.

in individual instances, placed the reader under a constant (though perhaps often unconscious) pressure to focus his attention on the externals of the text—grammar and stylistics—at the expense of its inner content. The same criticism, *pari passu,* no doubt applies to the creator of the text as well as its reader. All of this, we must suppose, reinforced a tendency among many writers and readers to direct more attention than would otherwise have been necessary toward literary and philological matters, and correspondingly less toward broader topics outside the world of language and literature.

Perhaps most important of all was the already mentioned proprietary attitude toward literacy and book knowledge, of which nonpunctuation was by no means the only example. On all these counts, the factors we have been considering must have impeded efforts to turn Literary Chinese away from its literary conventions and direct it more positively toward becoming a truly simple, unadorned, and practical instrument for precise communication.

7. Classification

As the daily users of alphabetic scripts, we Westerners are scarcely aware, because they are so ubiquitous, of the enormous benefits that are ours because of the principle of *alphabetization.* In dictionaries, encyclopedias, library catalogues, indexes, telephone books, and countless other media, this technique for classifying words has so transformed our lives that it is impossible to visualize our civilization without it. Yet the technique was by no means always exploited in the alphabetic civilizations, nor of course is it available in nonalphabetic China today. Before describing some of the alternatives employed by the pre-modern Chinese, let us look briefly at the development of alphabetization in some other civilizations. Generally speaking, the story is remarkably parallel to that of punctuation.

In India, the famed classical dictionaries or *kośa* (perhaps ca. sixth–twelfth centuries) are of two kinds: those in which synonymous or nearly synonymous words are grouped under subject headings (such as sky, earth, plants, man, etc.) and those consisting of words having more than one meaning. Only in the second category are words grouped phonetically rather than topically. The main systems, some with variants, include:

1. Words arranged according to the length of the verses written to explain them (a whole verse, half verse, quarter verse)
2. Words arranged according to their own length (dissyllabic words,

followed by trisyllables and others of greater length); grouped within each category according to gender and then arranged in alphabetic sequence according to their initial letter
3. Words arranged alphabetically according to their final consonant
4. Words first arranged according to the number of their syllables and then under a double alphabetic system: first according to the final consonant and then according to the initial letter

There are further variations, for example, the number of syllables, then the final consonant, and then the suffix. Since none of these systems was simple, completely alphabetic, or accepted as the only standard, the difficulty of finding words in dictionaries of this sort can be imagined. Even so, however, they are easier to use than the other major group, that of topically arranged synonyms.[64]

In the Islamic world, greater use was made of alphabetization alone, although still awkwardly. The major classical dictionaries, ranging in date from the eighth to the fifteenth centuries, display the following systems:

1. Words not listed alphabetically but according to sounds, beginning with guttural and ending with labial letters. Under each category, two-letter (biliteral) words are grouped first, followed by those derived from triliteral roots. This is the system used in the first major dictionary, *Kitāb al-'ayn,* by al-Khalīl Ibn Ahmad (d. 786).
2. Words arranged alphabetically according to their last, then middle, and then first consonants (most Arabic words being triliteral). This arrangement reminds us of some of the Indian dictionaries.
3. Words arranged as in Western alphabetic order, that is, by first, middle, and last consonants. This system, today generally adopted for Arabic dictionaries, is first found in the *Asās al-balāghah,* by al-Zamakhsharī (d. 1143).[65]

In Europe, the history of alphabetization, as of punctuation, is quite complex.[66] The Greeks had used the Phoenician alphabet for some five centuries before making any attempt at alphabetization. The earliest example comes from the libraries founded at Alexandria by Ptolemy

64. Personal communication from Ludo Rocher (see n. 58), supplemented by Zachariae (1897:14) and Renou and Filliozat (1947–1953:2.100–101).

65. Personal communication from Roger M. A. Allen (see n. 59), supplemented by Chejne (1969:44–45).

66. The information below is all taken from Daly (1967), supplemented by Professor Daly's personal communication.

Philadelphus (r. 285–246), where Callimachus (d. ca. 270 B.C.), in his *Pinakes,* catalogued 120 books according to categories (epics, tragic poetry, etc.) and then, according to what is said by a later writer, apparently listed the authors under each category according to the first letters in their names. This beginning of alphabetization is particularly striking because it parallels so closely the creation of the first real punctuation system by the librarian Aristophanes of Byzantium at Alexandria around 200 B.C. In the words of Daly (1967:25): "It is a reasonable and attractive hypothesis that the principle [of alphabetization] was first put to effective use by the scholars of Alexandria in response to the problems they faced as a result of the accumulation under their supervision of such an unprecedented mountain of literature and information."

Thereafter alphabetization continued to be particularly associated with scholars and writers, even though it was also practiced in the elaborate bureaucracy of Ptolemaic Egypt. For a long time it was little used elsewhere, being almost unknown in Athens, for example. In Roman literature, however, instances occur in Plautus (ca. 254–184), Varro (116–27 B.C.), and Pliny the Elder (A.D. 23–79), who tends to be apologetic for "what he apparently felt to be a rather unimaginative procedure" (Daly 1967:36). Curiously, alphabetization is scarcely known in Roman bureaucratic procedure: out of several hundred thousand recovered Latin inscriptions (seventh century B.C.–fifth century A.D.), only four are indubitably alphabetic. The first clear examples of *absolute* alphabetization (alphabetization by every letter of every word of a given entry instead of merely the first letter of its first word) dates from the late second century A.D.[67]

During the Middle Ages, alphabetization fell into decline except for continued use in Latin glossaries, but it was brought back in improved form during the revival of learning starting in the twelfth century. During the thirteenth century, the ledgers used in French countinghouses were alphabetical, as were the financial records of the Papacy from the

---

67. Only much later did absolute alphabetization become common. Conceivably, lack of motivation may have been partially responsible, but this is hard to prove. Much more important, in the opinion of Daly (1967:85–90), is that the use of separate slips for preparing alphabetic entries was prohibitively expensive before paper became common in Europe. Until then, therefore, the making of alphabetical lists involved preparation of sheets with spaces left blank under each letter, and then the filling in of these spaces by entries as they occurred. While this procedure was good enough for alphabetization by first letter only, it made absolute alphabetization extremely difficult. Rather ironically, China, which for many centuries had had an abundance of paper, was prevented by its script from using the paper for alphabetization.

mid-fourteenth century onward (perhaps as a result of contact with France during the Avignon period). A very important event, with which we can end this account, is the creation of the first biblical concordance (based on the Paris Vulgate Bible) under the editorship of Hugh of Saint-Cher (first half of thirteenth century). In this work—the beginning, one might say, of modern indexes—each word is listed, without context, in alphabetical order, followed by complete references to all chapters where it occurs (with letters *a* to *g* indicating locations within each chapter). The religious motivation here manifested reminds us once more of the similar motivation evidenced in the punctuating of the fifth-century Vulgate Bible and of the early printed editions of the Bible.

Word classification in China is of course something utterly different.[68] The earliest dictionary, the *Erh ya* (late Chou or early Han), is really only a glossary of words arranged under such categories as heaven, birds, implements, relationship terms, and so on. Not until the most important of all Chinese dictionaries, Hsü Shen's *Shuo-wen chieh-tzu* or *Script Explained and Characters Elucidated* (completed ca. A.D. 100), are written words for the first time systematically grouped under the graphically meaningful components known as "radicals" or "keys." There are 540 such radicals, under which the dictionary, as presently arranged, lists and defines 9,353 separated characters plus 1,279 variants. This system, although an incalculable improvement over the topical groupings of the *Erh ya,* still requires great effort to locate any given word, inasmuch as the sequence of the 540 radicals is determined solely by rough similarities of shape, and that of the subsumed characters seems quite arbitrary.

The number of radicals was reduced in later dictionaries until it reached its lowest figure of 214 early in the seventeenth century, which thereafter remained the norm as, for example, in the *K'ang-hsi tzu-tien* or *K'ang-hsi Dictionary* of 1716. Another improvement in the *K'ang-hsi Dictionary* is that the radicals are systematically arranged according to the number of strokes required to write them, from 1 to 17 strokes. Under them, 49,030 characters are listed and defined. This has been the model for all radical dictionaries since, save for some now appearing in China, in which account is taken of the simplified writings of many characters inaugurated in the 1950s.

Besides its reform of the radicals, the *K'ang-hsi Dictionary* has instituted another significant improvement by placing the characters that

68. The facts that follow are for the most part common knowledge, but two convenient reference works, among many, are Teng and Biggerstaff (1971) and Kennedy (1953).

fall under each radical into further subgroups according to their number of strokes. Even so, the time required to find a given character under one of the "populated" radicals is often considerable. Under the "hand" radical (no. 64), for example, no fewer than 85 characters belong to the five-stroke subcategory (that is, require five strokes for their writing in addition to the strokes needed to write the "hand" radical present in each of them). The sequence of the 85 is furthermore arbitrary, as it is for all the other number-of-strokes groupings. Before leaving the radicals, we should note that their numbering is a device used only in Chinese-Western dictionaries. The Chinese themselves learn the 214 radicals by unnumbered sequence only, just as we learn the letters of the alphabet.

Besides arrangement of words by radicals, the other major system for classifying words is that found in the rhyming dictionaries. These are dictionaries whose entries are arranged phonetically (by rhyme) under a fixed sequence of vocables consisting of well-known characters selected for their phonetic values rather than their meanings. For example, under the first vocable, pronounced *tung,* are grouped all words pronounced *tung,* followed by other rhyming groups of words pronounced *t'ung, chung, ch'ung, kung, hung,* and so on. The major rhyming dictionary of 1011, the *Kuang yün* (Expanded Rhyming Dictionary), contains 206 such basic vocables, known as "rhymes" *(yün),* which were subsequently reduced to 106. The rhyming dictionaries are invaluable for reconstructing Chinese historical phonology, but extremely difficult for practical use by anyone not a linguistic specialist. Nor is the situation helped by the fact that some of the words grouped under a given rhyme no longer rhyme today, owing to phonetic changes occurring since the rhyming system became fixed by the dictionary. In the old days, nonetheless, every educated Chinese was expected to know the 106 standard rhymes by heart. A curious byproduct of this situation is that the rhymes were often "made to serve as a key for serial numbering. The first thirty graphs, for example, were used in telegrams to express the days of the Chinese months."[69]

Other devices were also used for serial numbering. By far the most awkward is that found in the *Tao tsang* or *Taoist Canon,* whose 1,464 titles are arranged in 512 groups under the first 512 characters of the *Ch'ien-tzu wen* or *Thousand Characters Text* (on which see above, II.4 end). Obviously, for anyone who has not memorized the initial half of this text, such classification is useless. Much more practical, because much shorter, are the sixty pairs of characters (*chia-tzu, yi-ch'ou,* etc.) compris-

69. Kennedy (1953:49). Telegraphy, of course, is relatively recent, but before its introduction, the rhymes were used for other serial purposes.

ing the sexagenary cycle. For still shorter sequences, use is often made of either of the two components of the sexagenary cycle, namely the ten "heavenly stems" or (more commonly) the twelve "earthly branches." In the *K'ang-hsi Dictionary,* for example, the radicals and their entries are distributed under twelve parts or books, each designated by one of the "branches" and further broken down into subdivisions labeled "upper," "middle," and "lower." Pagination is not continuous, but begins anew for each of the three subdivisions under each of the twelve "branches."

Mention of pagination brings us face to face with a major inconvenience whose rectification could seemingly have been achieved at any time but had to await the introduction of Western-style printing in the twentieth century. This was the absence of continuous pagination for any book consisting of more than a single *chüan* (part or division). The reason, of course, is that Chinese books originally consisted either of successive rolls of silk, known as *chüan,* or successive units of bound wooden strips, called *p'ien.* Much later, when bound paper books with numbered pages came into existence, the words *chüan* and *p'ien* were retained as respective designations for the larger and lesser divisions of a book, that is, its "parts" *(chüan)* and "chapters" *(p'ien).* Pagination was made to start over again with each *chüan,* and the several *chüan* of a book were consecutively numbered but not titled unless, as sometimes happened, the book consisted solely of *chüan* without any lesser *p'ien* divisions. In that case the *chüan* bore the titles that the *p'ien,* if they had existed, would have had.[70]

Until recent times (and quite often even today), the absence of continuous pagination has meant that scholars never cited texts by page numbers but only by *p'ien* (or *chüan*) title or, more rarely, by *chüan* number. This has been beneficial to the extent that it has enabled scholars to cite texts irrespective of the particular editions available either to themselves or to their readers. But it has also been harmful inasmuch as:

1. Citation in this way is a good deal more awkward than by page number.
2. It gives trouble to any reader who wishes to look up the citation in

70. Strictly speaking, the terms *page* and *pagination* are anomalous when used, as in this paragraph, to refer to traditional Chinese books; *folio* and *foliation* would be more accurate because traditional Chinese books do not consist of single-sheet pages but of folded double-pages (folios), the recto and verso sides of which bear identical "page" numbering (distinguished, when cited in Western scholarship on China, as p. 1a, p. 1b, etc.). For the sake of convenience, however, I prefer to use the more familiar terminology.

the original text, especially if the cited chapter happens to be a long one.

3. Citation by chapter title rather than chapter number imposes the further task on the reader of first locating the title in the cited work's table of contents.
4. Because the procedure tends to downgrade the importance of specific editions, the result may be a failure to notice variant readings found in different editions.

At this point one may wonder why ordinary numerals, despite their obvious convenience, were so little used for numbering more than relatively short sequences. As we have just seen, continuous pagination did not extend beyond each individual *chüan* in a book, nor, as noted earlier, were the 214 radicals numbered in the dictionaries. Traditionally, numbers were used to designate the days of the month, months of the year, and years of a reign (or "reign period"), but no attempt was made to apply them to dynasties or to still longer epochs comparable to those demarcated in other civilizations by events like the Hegira or the birth of Christ.[71]

The reasons for this situation are obscure, the more so as the Chinese place-numerals already used in Shang times were much more suitable for serial numbering than, say, the cumulative numerals of the classical West (Needham 1959:5–17). The persistent Chinese tendency, despite this superiority, to use nonnumerals rather than numerals for serial purposes is clearly shown by the continued dominance of the sexagenary cycle for chronology until long after the Shang dynasty. Apart from cultural inertia, this persistence may conceivably have been indicative of the Chinese fondness for *symmetry:* the fact that nonnumerical symbols all equally consist either of one or (in the case of the sexagenary cycle) two characters, and therefore are aesthetically preferable to numerical symbols which, depending on their magnitude, variously require one or several characters for their writing.

Here, just possibly, is the reason why, in the tables of "general rules" *(fan li)* that commonly preface large dictionaries, encyclopedias, and other scholarly compilations, the individual rules are each equally introduced by the numeral "1," placed at the top of a new column, instead of being consecutively numbered. This means that no matter how many rules a given table may contain, each rule is uniformly prefaced by a single digit. The resulting typographical regularity is roughly suggested by the following English equivalent:

---

71. Very lengthy time spans, based on the movements of the heavenly bodies (the longest was 23,639,040 years), were in fact recognized in Chinese astronomy, but they did not form part of ordinary chronology. See III.3, beginning.

1. Such-and-such a rule
1. Another such-and-such rule
1. Still another such-and-such rule
   (And so on for the remaining rules, of which there happen to be eighteen in the "general rules" section of the *K'ang-hsi Dictionary.*)

If this explanation is valid, we have here another ramification of the insistence on stylistic balance, which was discussed in Section 5 above.

If a pre-modern Chinese scholar could have been asked why all these cumbrous procedures had not been simplified long ago, he no doubt would have answered that the real scholar would never dream of avoiding them because, by definition, he would have a photographic memory, encyclopedic knowledge, and ample time at his disposal. A scarcely satisfactory reply, one fears, for the modern scholar who, deficient in all these respects compared with his predecessors, and obligated by his research to scan lengthy Chinese texts in search of some particular topic, name, or quotation, finds himself confronted by solid blocks of unpunctuated text running from page to page without any paragraphing or other indications of shifts in time or topic. Not all texts, of course, are as bad as this. The basic annals in the dynastic histories, for example, usually start each year's entry with a new paragraph. How helpful it would have been if all texts had adopted topical and chronological paragraphing!

Still more helpful but less common is the insertion of subtitles into the text to indicate new topics. A variant practice—less efficient but still helpful—is to place subtitles at the beginning of a chapter after the main heading, but not at the same time into the text itself. This is done, for example, in the "Treatise on Ritual" *(Li yi chih)* in the *Hou Han shu* (History of the Later Han Dynasty), whose three chapters on the subject (14–16) are each headed by such subtitles as "Spring's Beginning," "The Five Offerings," and so on (thirty subtitles in all). By comparison, the corresponding "Treatises on Ritual" in subsequent dynastic histories from Han to T'ang show a great decline in this respect: no subtitles at all and, at most, infrequent paragraph divisions according to chronology or topic.[72]

A further example of how a little typographical thoughtfulness could have saved untold drudgery for generations of users is provided by the way the *K'ang-hsi Dictionary* arranges its entries. These, to be sure, are printed in larger boldface type than the definitions that follow them, but they are also scattered irregularly through the text in such a way that the reader often has to scan several pages of packed print (from top to

---

72. Such treatises occur in *Chin shu* 19–21, *Sung shu* 14–18, *Nan Ch'i shu* 9–10, *Wei shu* 108A–D, and *Sui shu* 6–12.

bottom and side to side) in order to locate the particular entry he wants. It would have saved enormous trouble if each entry had been clearly demarcated by being placed at the top of a new column, and above all, if marginal listings (catchwords) of the same entries had also been printed along the tops or sides of pages. These are precisely the typographical improvements made in more modern dictionaries such as the *Tz'u yüan* (1915) and *Tz'u hai* (1938), which otherwise, until rather recently, have followed the *K'ang-hsi Dictionary* in their lack of continuous pagination and division of their material into the twelve compartments provided by the duodenary "stems."[73]

Chinese bibliography is another field in which—lacking alphabetization or (until the present century) indexes and card files—a good memory is needed. Some three centuries after Callimachus had produced the world's first alphabetical book catalogue (120 items) at Alexandria, Liu Hsin (ca. 53 B.C.–A.D. 23), imperial librarian under Wang Mang (r. A.D. 9–23), prepared his catalogue of 677 works belonging to the imperial collection.[74] His classification scheme embraced six major categories ("Classics," "Philosophers," "Poetry," and so on). These categories were in turn divided into subcategories (nine such under "Classics," for example). Finally, under each subcategory, titles of relevant texts were listed in roughly chronological sequence. The catalogue is an outstanding monument of analytical ability, despite weaknesses inevitable in a pioneer effort. One of these is the listing of all histories (later to become a huge category) under the *Ch'un ch'iu* or *Spring and Autumn Annals* (a brief chronicle covering the years 722–481), which itself is merely one of the nine subcategories under "Classics." Beginning in the mid-third century, this situation was remedied by instituting a new "four-branch" classification plan consisting of Classics, Histories, Philosophers, and Belles Lettres. This system, with many refinements and elaborations, has come down to modern times. For example, it provides the classification for Alexander Wylie's *Notes on Chinese Literature* (1867) and—with further modifications—for notable contemporary Chinese collections like the Harvard-Yenching Library at Harvard.

By now the difficulties posed by a nonalphabetic script for the ordering of knowledge should be fully apparent. In the Chinese experience,

---

73. Marginal or top-of-page entries are in fact not totally unknown in pre-modern Chinese books, though their use is excessively rare. See, for example, the eleventh-century historical text pictured in Hartwell (1971:711). To provide easy reference, key phrases from its text are printed as catchwords at the top of the page above the text proper.

74. It now constitutes the thirtieth chapter of the *Ch'ien Han shu* (History of the Former Han Dynasty).

these difficulties were compounded by the partial or total neglect of such simple typographical devices as continuous pagination, paragraphing, subtitles, catchwords, and the like. As in the case of punctuation, one may suspect that this neglect stemmed, at least in good part, from the proprietary attitude of the traditional Chinese scholar toward his professional skills.

During the present century such an attitude has of course disappeared entirely except among a few traditionalists. If indexes still remain exceedingly rare in works of contemporary Chinese scholarship, many of highest value have nevertheless been compiled for China's past literary monuments. Beginning in the 1920s, two rival systems were devised for listing characters under numerical designations. One was the *kuei-hsieh* ("pigeon-holing") system used by the Harvard-Yenching Institute for its indexes, the other the *ssu-chiao hao-ma* ("four-corner numbering") system used by the Commercial Press, Shanghai, for dictionaries and other publications. The essence of each was the numbering of characters according to the kinds of strokes (horizontal line, vertical line, right angle, etc.) required to write them. In other words, each character was designated by a sequence of several digits representing the graphic structure of the character. I have the impression, perhaps erroneous, that relatively few scholars ever bothered to master either system sufficiently well to use it freely. Many, I believe, were content instead to rely on the cross-listings by radicals that usually accompanied both systems, or, in Harvard-Yenching publications, the cross-listings according to Wade-Giles romanization.

A serious defect of both systems was their joint inability, owing to the graphic similarity of many characters, always to distinguish separate characters by separate numbers. In the *kuei-hsieh* system, for example, the number III 33081 covers three semantically quite different but graphically similar characters: *lo* 落, *p'u* 菩, and *jung* 蓉. In the four-corner system, likewise, 7722_0 is the numerical designation for no fewer than six characters: *tan* 丹, *yung* 用, *t'ung* 同, *chou* 周, *t'ao* 陶, and *chiao* 脚.

From the early nineteenth century onward, many attempts have been made to replace the characters by romanization.[75] All, however, have foundered on two stubborn facts: the inability of *any* extended passage in Literary Chinese to remain intelligible when romanized and the difficulty of *most* people, even when using *pai hua,* to write *exactly* as they would speak and thus to avoid occasional ambiguity. Of course, abandonment of the script after more than three thousand years would entail major psychological consequences as well. No doubt these are among

75. The standard account is DeFrancis (1950).

the reasons why the People's Republic of China, after apparently considering the possibility of complete romanization in the 1950s, has gone no further than simplifying the writing of the more common or complex ideographs (see end of sect. 1 above). While this step is undoubtedly helpful for learning and writing purposes, it in no way solves the other problems arising from the use of a nonalphabetic script.[76]

Important among these problems, in addition to those already discussed, is the pressing need, in every modern society, to record and duplicate written materials rapidly, easily, and inexpensively. Heretofore, so far as I know, it has not proved possible effectively to adapt the traditional typewriter, linotype machine, and the like to the needs of the Chinese symbols. (Chinese typewriters, while long known, have been little used because they are so cumbersome, slow, and difficult to learn.)

The arrival of the computer age, one would suppose, could solve this situation before too long. DeFrancis (1989:267), however, is not optimistic. He writes that in comparison with "alphanumeric" symbols, "there is not, and I believe never can be, as efficient a system for inputting and outputting Chinese characters." For word-processing purposes, such characters are "vastly inferior" to pinyin ("linked sounds," name of the system officially promulgated in the People's Republic of China in 1958 for phonetically transcribing the characters into the Latin alphabet). Nevertheless, "most Chinese" continue to regard pinyin as an "alien script" and "prefer to work exclusively with characters."

8. Literary Devices

So far our discussion has passed from the significance of Chinese writing to the principles and structure of Literary Chinese, and from these to such physical or mechanical devices as punctuation, continuous pagination, and techniques for classification. Now the discussion departs still further from language proper as we consider some of the major stylistic devices that traditionally appear in Chinese prose composition, both literary and scholarly. At this point the possibility has to be recognized that there may be extralinguistic as well as linguistically influenced factors at work. And yet I believe that the subtle influence of the Chinese written symbols is still pervasively present.

A basic feature of Chinese traditional prose is the extensive use of quotations and allusions: more of them in literary compositions, fewer

76. An account of Chinese language reform during the first quarter-century of the People's Republic of China, from which it would appear that Mao Tse-tung himself was not anxious to phoneticize the script, is given in Milsky (1973).

in ordinary exposition, but always a possibility for which the reader should remain eternally on guard. Achilles Fang puts the matter very forcefully when he writes (probably with literary prose primarily in mind): "Practically every important piece of writing dating before 1916 (and even some subsequent to that date) abounds in allusions, clichés, parallelism, stock-in-trade emotions, and ancient tradition." He later adds that it is "not so easy to be able to recognize a quotation or allusion as such; a second sight or a sixth sense is perhaps needed for this."[77]

Quotations first occur in the *Lun yü* or *Analects* of the fifth century B.C. (some thirteen citations from the *Shih ching* and eight from the *Shu ching*), remain at about the same level in the *Mo-tzu* (approximately ten citations from the *Shih* and twenty-four from the *Shu*), increase sharply in the *Meng-tzu* or *Mencius* (approximately twenty-seven citations from the *Shih,* twenty from the *Shu,* and fourteen from the *Lun yü*), and thereafter become a flood. Though perhaps most common in Confucian writings, they are frequent enough in non-Confucian writings also, as well as in those that are totally nonphilosophical. The fact that once a quotation is in use it rarely becomes obsolete means that the store of quotations has expanded with the accumulation of literature to reach quite overwhelming proportions.[78]

Quotations may be either named or unnamed. Named quotations are naturally much easier to identify and are the kind usually found in the *Lun yü* and the other early texts mentioned above. Mencius, for example, often closes his argument by quoting, say, a stanza from the *Songs Classic* introduced by the words, "The *Song* says," and followed by a standard formula: "Such is what is meant [by what I have just been talking about.]"

Unnamed quotations are by definition much more difficult. They consist of a sentence, clause, or phrase quoted without identification from some earlier source and commonly inserted into the new text with no indication that a quotation is involved. The absence of anything like quotation marks in Literary Chinese prior to their introduction from

---

77. Fang (1953:266 and 280). Fang mentions 1916 as a turning point because it (and the year following) marked the start of the movement to replace the literary language with written *pai hua*.

78. The *P'ei-wen yün-fu* or *Rhyming Treasury of the Studio for Honoring Literature* (1711) is a gigantic repertory (but not a dictionary, for it gives no definitions) of hundreds of thousands of mostly two-word terms, culled from all stages of earlier Chinese literature. As indicated by the title, the terms are arranged according to rhyme, but fortunately the six-volume edition published by the Commercial Press, Shanghai, in 1937 (and subsequently reprinted in Taiwan) contains an index by radicals and number of strokes that makes the entries readily available.

the West early in this century made such concealment naturally easy. It is at this point that the reader, unless equipped with the encyclopedic literary knowledge traditionally taken for granted, has to develop that "second sight or a sixth sense" mentioned by Fang. Hardest of all, however, are allusions, that is to say, paraphrases or condensations of earlier writings (sometimes, as we shall see, reduced to a single word) that are inserted into the text to impress or, just possibly, to mystify the reader.

Quotations and allusions, all unnamed, are well exemplified in the passage in the *Ch'ien Han shu* (History of the Former Han Dynasty) where the historian Pan Ku (A.D. 32–92), after having devoted three chapters to the career and reign of Wang Mang, expresses his own final judgment. This "eulogy," consisting of 307 characters, contains three unnamed quotations from the *Lun yü,* two such from the *Yi ching* (Changes Classic) and one from the *Shu ching.* In addition, there is an allusion to the *Lun yü* consisting of a single character.[79]

Two of the quotations occur near the beginning of the "eulogy," where Pan Ku, to express his favorable opinion of Wang Mang during Wang's early years, says of him that he "toiled diligently for the state and 'pursued a straightforward course.' . . . Was he not one of whom it could be said that 'in his clan he is certain to be heard of, in his state he is certain to be heard of'?" The passages within single quotation marks come respectively from *Lun yü* 15.24 and 12.20.

The allusion appears in the final sentence of the "eulogy," where Pan Ku expresses his ultimate condemnation of Wang Mang by linking him with Ch'in Shih-huang-ti, the First Emperor of the Ch'in, and then saying of the two: "They were the color of purple, the sound of croaking." "Sound of croaking" involves textual problems that need not be discussed here; but *tzu,* "purple," is an allusion to *Lun yü* 17.18, where Confucius is recorded as saying: "I hate purple's usurping of redness." This is interpreted as meaning that red is a "pure" or "primary" color, purple one that is mixed or secondary, so that Confucius, by objecting to the substitution of purple for red, is really objecting to the substitution of the spurious for the genuine. Hence when Pan Ku describes the Ch'in First Emperor and Wang Mang as "purple," he is really saying that they were tyrants masquerading as sages. No doubt the parallel "sound of croaking" has a similar implication, but it escapes us today and was already obscure to the early commentators.

The fitting of quotations or allusions into passages of Literary Chinese is made particularly easy by the lapidary quality of the Chinese characters: the fact that each character normally represents a single

---

79. *Ch'ien Han shu* 99C; tr. Dubs (1938–1955:3.470–474).

word and, being uninflected, remains unchanged regardless of how it is used grammatically. For example, a quotation that in its original setting may have referred to an action performed in the past by a single person can, without any morphological change, serve equally well in its new setting to refer, if need be, to a present or future action performed by several persons. This flexibility makes it harder for the reader to identify the quotation or even recognize that one is being used. For the writer as well, however, frequent use of quotation places a heavy premium on memory, because he of course has to draw the quotation from his literary storehouse in the first place.

Quotation users the world over are all too apt to change the original meanings of quotations for their own purpose. Fang (1953:280) warns against this and a companion danger by saying:

> It often happens that the writer of the text may not have interpreted the passage in question in the same way as the translator thinks, on the best authority, it should be interpreted. Furthermore, it is quite possible that the writer was not a conscientious scholar; he may have quoted the passage indirectly from a secondhand source.

The second situation, while probably less common than the first, can sometimes have serious consequences. A striking example has to do with the *T'ai-p'ing yü-lan* encyclopedia (comp. 983). This famous work, in its section on falconry (926/2a), quotes a remark allegedly made by Li Ssu, prime minister of the Ch'in First Emperor, when Li Ssu, together with his son, was being led to execution in 208 B.C.: "Alas, however much we might want to go hunting with our yellow dogs and our falcons on our wrists, we could not do so now." Comparison with the alleged source of the remark, the biography of Li Ssu in the *Shih chi*,[80] shows that the latter says nothing whatever about falcons, though it does quote Li Ssu as speaking about hunting rabbits with yellow dogs. That the usually careful compilers of this encyclopedia should have cited the alleged statement to trace the history of falconry is startling. "They probably took it from some contemporary quotation, not bothering to look up the passage in the original text."[81]

---

80. *Shih chi* 87/5b–6a; tr. Bodde (1967:52).

81. Needham (1954:43 n. a). However, Schafer (1958:295) expresses a different opinon. After quoting the same alleged *Shih chi* quotation from the *T'ai-p'ing yü-lan*, he surmises—on the basis of very slender additional evidence—that a version of the *Shih chi* containing the reference to falcons may in fact have existed during T'ang and Sung times. From this he proceeds to what seems to me a dangerously bold conclusion, namely that falconry was practiced in China at least as early as the third century B.C.

So prevalent are quotations and allusions in Literary Chinese that many writers were probably hardly aware of the danger that through overuse they could easily become outworn clichés and faded metaphors. Here again Fang (1953:270) has forcefully stated the problem:

> Where nothing is obsolete or even obsolescent and all writers of reputation are conscious of and groan under the dead weight of the past, it is no easy matter to disentangle true sentiment from false rhetoric, and distinguish between tradition and individuality, to discriminate a sentiment of the heart from mere lip service to respectable rhetorical device—in short, to place every word in the proper perspective of space and time.

Many literary clichés, though perhaps stylistically wooden and weak, at least do no harm to the meaning of what is being said. The *Shih ching,* for example, contains a poem that has been traditionally interpreted as describing the anxiety of a ruler not to be late for the morning levée held by him for his subordinates (an institution that, throughout Chinese history, traditionally took place before dawn). The relevant section of the poem reads: "The night is not yet at an end. In the light of the courtyard torches, the lords arrive."[82] On the basis of this poem, the term "courtyard torches" *(t'ing liao)* has become a set phrase to symbolize the bustle, excitement, and bright light connected with the predawn assembly of large numbers of people—primarily of officials and others at court audiences, but also, by extension, people in general on the joyous occasion of the morning of the lunar New Year.

An example comes from the poet and scientist Chang Heng (78–139), when in his long poem describing the Later Han capital he comes to the predawn imperial audience held at the palace on New Year's morning. The account begins: "On this day of the triple beginning of the Hsia calendar, the courtyard torches *(t'ing liao)* brightly burn." A few decades later, Ts'ai Chih (d. prob. A.D. 178) writes in similar vein about the New Year's audience: "Everyone enters the imperial presence amidst the courtyard torches *(t'ing liao).*"[83] Again, some four hundred years later, in the book of festivals known as the *Ching Ch'u sui-shih-chi* (Record of the Annual Seasons in Central China), we read that on New Year's morning people assemble in their courtyards at cockcrow and there explode bamboos in a fire to drive away the evil spirits. Then the text goes on to say: "Ordinary people maintain that when bursting bamboos

---

82. *Shih ching* no. 182; tr. Legge (1960:4.294) and Karlgren (1950a:126).

83. The Chang and Ts'ai quotations are respectively cited in Bodde (1975:142 and 151–152). The former comes from Chang Heng, *Tung-ching fu* (Rhapsody on the Eastern Metropolis); tr. Knechtges (1982:265, l. 270–271) (in the larger translation by Knechtges of the *Wen hsüan*).

arise from amidst the courtyard torches *(t'ing liao),* neither in the family nor in the country will there be response to any insurrectionary movements against the sovereign."[84] In this last account, "courtyard torches" has expanded its original meaning to become an evocation of the beneficent light that on the first day of the year drives away the evils of darkness and of the past and ushers in the joy and communality associated with the New Year.

Other more extended quotations, however, when used as conventional expressions of sentiment deemed appropriate to a given historical situation, can sometimes throw the veracity of the entire story into serious doubt. In his analysis of the historiography of the *Hou Han shu* (History of the Later Han Dynasty), Bielenstein (1953–1979:1.53, 59) calls attention to two striking examples. The first concerns a general, Wu Han, whose biography (*Hou Han shu* 48) reports him as addressing his troops in A.D. 27 with the words: "Although the mass of the bandits is large, [still] they all are a group of plundering robbers. When they conquer, they do not yield to each other, and when they are defeated, they do not save each other." As Bielenstein points out, the second sentence ("When they conquer, . . .") is taken verbatim from a statement recorded in the *Tso chuan* history under the year 714 B.C.[85] Yet the biography has previously stated that Wu lacked "cultural refinement," so that it is quite unlikely that he himself would have been acquainted with this *Tso chuan* passage. "The conclusion therefore seems unavoidable," comments Bielenstein, "that the historian either completely invented these speeches or that he added embellishments to a 'kernel of truth.' "

In A.D. 23, similarly, Chang Ang, known to have been a completely uneducated man, is recorded in his biography (*Hou Han shu* 44) as drawing forth his sword, striking the ground with it, and then saying: "[He who] hesitates in undertakings has no merit. Today's conference must not have two [opinions]." Here it is the first sentence [about hesitation] that is a quotation; it comes verbatim from the forty-third chapter of the *Shih chi,* where it is attributed to Fei Yi, statesman of the state of Chao, who addressed it to the king of Chao in 307 B.C.; the *Shih chi* has in turn derived the speech (and the episode in which it occurs) from a passage in the *Chan-kuo ts'e* (Stratagems of the Warring States).[86]

When used to describe activities rather than reproduce speech, cliché

84. *Ching Ch'u sui-shih-chi* 1a; tr. Turban (1971:51–52). The basic text of this work, by Tsung Lin (ca. 498/503–ca. 561/566), is supplemented by an even longer commentary by Tu Kung-shan (fl. ca. 600). It is the latter that is here quoted.

85. Yin 9; tr. Legge (1960:5.28), Couvreur (1914:1.50).

86. *Shih chi* 43; tr. Chavannes (1895–1905:5.73). *Chan-kuo ts'e,* Chao no. 239; tr. Crump (1970:297).

quotations may be equally damaging to veracity. The following example comes once more from the *Hou Han shu,* but what it exemplifies is just as applicable to other histories and to expository writings in general. In A.D. 59, Emperor Ming of the Later Han issued a lengthy proclamation describing the great court banquet at which he, assisted by vassal kings and high officials, had entertained certain venerable men as a token of the country's respect for old age. Descriptions of the ceremony as it allegedly existed in Chou times occur in the ritualistic texts. Yet the first historically confirmed instance of its performance seems to have been that by Wang Mang in A.D. 6, and its institutionalization seems to have come only with Emperor Ming in A.D. 59.

In the course of the A.D. 59 proclamation, the emperor states: "The vassal kings set forth condiments [for the aged guests], and the [Three] Excellencies and [Nine] Ministers offered them delicacies to eat. We in person, bared, cut up [the animal victims] and [then, at the conclusion of the banquet,] served cups for them to rinse their mouths." With only minor rearrangements, this statement repeats the wording of the ceremony as described in the seventeenth chapter of the *Li chi* (Record of Ceremonial): "The Son of Heaven, bared, would cut up the animal victims. He would serve condiments [for his aged guests] to eat, and [then, at the conclusion of the banquet,] would serve cups for them to rinse their mouths." Again, two sentences further on in the proclamation, the emperor states: "Above [the steps leading up to the hall, the singers] sang 'The Call of the Deer,' and [in the court] below, the flutes played 'The New Palace.' " This sentence is a word-for-word repetition of a sentence in the *Yi li* (Observances and Ceremonies).[87]

The use, in a document purporting to give an eyewitness account of a living ceremony, of statements taken verbatim from ritualistic writings raises a problem that, with variations, sooner or later confronts anyone dealing with Chinese historical documents. This is, to what extent can statements that are interlaced with quotations or allusions be taken at face value? To what extent does such language merely convey a symbolic picture of what is being described?[88]

Clichés do not invariably have to originate as quotations. The writer can also formulate his own hyperbolical or metaphorical statements which, though not faded through earlier use, are nevertheless clichés in

87. See (a) *Hou Han Shu* 2/2b; (b) *Li chi* 17; tr. Legge (1885:28.124–125) and Couvreur (1913:2.102); (c) *Yi li* 6; tr. Couvreur (1916:209) and Steele (1917:1.147). All the passages are translated and discussed in Bodde (1975:363–364, 367, 370).

88. The questions are taken from Bodde (1975:370). For a similar instance in which a work on Han institutions uses language taken from the *Li chi* to describe a Han ceremony, see Bodde (1975:265, 268).

the sense that they use conventional literary terminology to epitomize, rather than realistically describe, the situation to which they are applied. From the *Hou Han shu* once more, Bielenstein has abstracted a considerable number of such newly created clichés of which the following are only a few examples:[89]

> The killed and wounded were innumerable. The drains and channels in the cities were all filled.
>
> The cities and suburbs were all empty, and white bones concealed the ground.
>
> Huan's son, Ying, wept blood.
>
> Banners, standards, and military equipment were not interrupted for a thousand *li*.
>
> A terrifying shouting shook heaven and earth.
>
> The tigers and leopards all trembled in their haunches.

While these exaggerated statements would be entirely appropriate in a literary account, they are out of place in what purports to be (and usually is) sober history. A moment's consideration should be enough to show that they cannot be taken literally. Yet traditional Chinese historical criticism has usually ignored the problem, thus leaving the uncritical reader to acquire unconsciously what might be called a cliché-view of history.[90]

Besides the various literary devices we have been discussing, one persistently used in Chinese argument is that of analogy: *A* is asserted to possess certain qualities because it allegedly resembles *B*, which does possess those qualities. This form of argument goes back early. One remembers, for example, the debate in the *Meng-tzu* (6a.2) on human nature, in which Kao Tzu argues that man's nature is indifferent to good and evil just as water is indifferent to the direction it may take depending on whether a channel is opened for it to the east or the west —to which Mencius replies that human nature on the contrary has a universal tendency toward goodness just as water has a universal tendency to flow downward. The uncertainty, logically speaking, of this kind of analogical argument is sufficiently evident from this example.[91]

---

89. Bielenstein (1953–1979:1.61–62) [the first three examples] and 78 [the latter three, drawn from the historian's account of the decisive battle of A.D. 23 between Wang Mang and the subsequent founder of the Later Han dynasty]).

90. For a further discussion of clichés in Chinese history, as expressed by both traditional and modern scholars (Western as well as Chinese), see Wright (1963) and Bodde (1963).

91. For a penetrating discussion of the water analogies in Mencius, see Richards (1932:45–47). Also discussed below at VII.1e, end.

Here it need merely be further remarked that the classification of things into analogical pairs, as demanded by the use of analogy, would seem, like the use of quotation and similar devices already considered, to encourage a tendency to look at phenomena in terms of set symbols rather than realities.

Another conspicuous feature in almost all traditional Chinese writing is the appeal to historical precedent. Like the analogy, this device compares one person or situation with another, but by definition does so in historical terms only. Like the quotation or allusion, it makes its appeal to a traditional authority strengthened by the fact that it comes from the past. Like the cliché, it often symbolizes rather than realistically portrays the person or situation to which it is applied. And like all these devices, its use tends to atomize rather than synthesize the subject matter at hand.

In its primitive form, the use of historical precedent involves making brief references to events or more often personages who may be either semilegendary or historical. Ideally, the bare mention of such a personage should be enough to conjure up in the mind of the reader a particular quality or attitude or a complex of qualities or attitudes. Apart from such famed sages as Yao, Shun, and Yü, or such tyrants as Chieh and Chou, a host of lesser figures serve as appropriate symbols, among them Po-yi and Liu-hsia Hui, paragons of purity; Tseng Tzu (disciple of Confucius), paragon of filial piety; Hsi Shih, ideal of feminine beauty; Robber Chih, acme of violence; and many more. Stock figures such as these begin to appear in the *Lun yü* and greatly proliferate thereafter.[92]

Toward the end of the next section, mention will again be made of the use of historical precedent when we see how it developed in postclassical times into the more complex and sophisticated device known as historical analogism.

9. Compilation versus Synthesis

We pass now from quotation, allusion, cliché, analogy, and historical precedent as building stones in the construction of small units of expository prose to the further question of how these units are in turn ordered and combined to form the larger blocks constituting essays and books. The answer, very commonly, is by "scissors-and-paste"; the larger composition is created by fitting together, usually in chronological

92. The roles of analogy and historical precedent in early Chinese argument are discussed in Bodde (1967:223–228).

sequence, a variety of relevant sources that may or may not be identified by name, but are usually quoted verbatim or, less often, in shortened paraphrase. How this is done for history, the field in which the process has been best studied, is vividly stated by Gardner in two passages of his *Chinese Traditional Historiography:*

> It follows from the general conception of history as an impersonal record of events, that any historical account which may be written assumes a kind of independent existence, a corporate personality, divorced from any kind of proprietorship by its author. And accordingly, verbatim reproduction of the records of earlier historians, no matter how extensive, is to be regarded, not as plagiarism, but rather as the natural and reasonable process by which new histories of previously recorded events should be constructed. History writing ordinarily involves, not original composition of any considerable length, but compilation of choice selections from earlier works. (Gardner 1938:70)
>
> The method of compilation employed is essentially primitive synthesis of the simplest kind, accomplished by dissection of pre-existing works and arrangement of their component fragments in chronological sequence. Respect for the integrity of texts is so strong in China that often no attempt is made to harmonize the documents which are thus brought into juxtaposition. . . . Chronological order is the accepted basis for nearly all compilation. (Gardner 1938:71)

As we read these somewhat somber judgments, it is well to counter them with the milder assessment by Edwin Pulleyblank. "We are commonly told," he writes (1961:135),

> that the method was simply one of mechanical scissors-and-paste compilation with only a primitive exercise in the choice or rejection of material. Yet, while there is a measure of truth in this judgement as it applies to such run-of-the-mill official and other history writing in China, it hardly does justice to the quality of scholarly acumen and historical thought displayed by the great Chinese historians.

This is unquestionably true. Yet it does not deny the fundamental fact that what may be called "composition through compilation" is a basic feature in the great bulk of traditional Chinese historiography, whether mechanically applied or lifted to a higher level by the truly great historians. As we shall further see, it also occurs in a great deal of other scholarly writing.

To take the biographies in the standard or dynastic histories as an example: These quite often consist in good part of writings by or to the individuals being recorded. Such writings may be quoted *in extenso* or, if overly long, in abbreviated form only; in either case, the thread of connecting narrative supplied by the historian himself may be meager. In the case of an important official, the quoted writings will probably consist of his memorials to the throne, edicts inspired by the memorials, and other official papers. If he is a creative writer, excerpts from his essays or poems may be quoted at some length. An admittedly extreme example is the *Shih chi*'s biography of the poet Ssu-ma Hsiang-ju (ca. 179–ca. 117), four-fifths of which consists of three of his *fu* poems, one proclamation, one letter, one memorial to Emperor Wu, and one essay, all quoted at length, while only one-fifth narrates the quite colorful events in his life.[93]

The most famous example of "composition through compilation" is the *Tzu-chih t'ung-chien* (Comprehensive Mirror for Aid in Government), a chronicle covering all of Chinese history from 403 B.C. to A.D. 959. It was completed by Ssu-ma Kuang in 1084 after nineteen years of effort. "In compiling the *T'ung chien* Ssu-ma Kuang took pains to use as few words of his own as possible, producing a mosaic of patches culled from all available sources; he drew upon 322 items other than the official histories published up to his time. Yet the *T'ung chien* is not a hodgepodge chronicle. . . . He had his reasons for every passage he chose to include."[94]

"Composition through compilation" is a principle probably as widely followed in nonhistorical as in historical writings. The primary difference is that in the former the quoted sources are commonly arranged chronologically in successive discrete units identified by name, whereas in histories such as the *Tzu-chih t'ung-chien* they are woven together into a continuous narrative and hence are rarely explicitly named, at least in the text itself (but see n. 94 above). A notable example of a nonhistorical work is Li Shih-chen's *Pen-ts'ao kang-mu* or *Grand Pharmacopoeia* (1596). This massive creation by "probably the greatest naturalist in Chinese history" (Needham 1986:308) discusses plants, animals, and minerals under 1,895 main entries and 374 supplementary ones. In large part its presentation consists of extensive quotations from earlier writers to which, however, Li has added his own considerable comments.[95] In the great Chinese encyclopedias the principle of quotation has been extended still further to the point where the compilers add no writing of their own; everything consists of longer or shorter quotations culled

---

93. See *Shih chi* 117; tr. Watson (1961:2.297–342) and Hervouet (1972).

94. Fang (1952:1.xviii [introduction to his translation of chs. 69–78]). In his extensive notes, Fang identifies Ssu-ma Kuang's sources, which Ssu-ma himself does not name in his text, though he discusses them in a separate 30-*chüan* section called *K'ao yi* (Examination of Differences). The *Tzu-chih t'ung-chien* will be further discussed in III.3.

95. This major work and its author are discussed at length in Needham (1986: 308–321).

from thousands of earlier works, which are chronologically arranged under topics and subtopics.[96]

With this complete triumph of "quotationism," the time has come to raise a basic question: has the reliance on quotation and the other literary devices discussed in the preceding section, as well as on compilation by scissors-and-paste, had any serious effect on scholarly writing in China?

In my opinion, the effect has been both considerable and unfavorable. First of all, we have noted the burden on memory imposed by quotations and allusions. Second, we have also seen how clichés can become facile substitutes for genuine description, and throw doubt on the reality of what is being told. Third, the ease with which quotations can be lifted *en bloc* from their original text and placed into new settings has encouraged scholars to accept them unquestioningly instead of doing the probing, analysis, and evaluation they would have to do were they to handle the same topic in their own words. The practice of filling pages with successive blocks of quotations interrupted only briefly by the writer's own discussion still lingers among a fair number of modern Chinese (and Japanese) scholars. Often it is only when such quotations happen to be translated into another language that their inherent problems become fully apparent.

Even in those learned studies in which quotations and the other devices are not conspicuously present, Chinese scholarship has traditionally tended to concern itself more with details and specifics than with wholes and generalizations. The prevalence of this tendency is demonstrated by the very large number of scholars whose reputation rests primarily on their commentaries or other bit-by-bit studies on earlier texts or topics. The tendency reveals itself very early, as, for example, in the expository techniques of the classical philosophers, only two of whom—Mo Tzu and Hsün Tzu—really succeeded in getting away from the usual miscellany of casually juxtaposed utterances to produce chapter-length essays that are systematically developed, topically centered, and add up to a manifestly coherent system of thinking.[97] Nor

---

96. Chinese encyclopedias are extensively discussed in Teng and Biggerstaff (1971: 83–129).

97. This fact in my opinion is insufficiently appreciated even by some modern Chinese scholars, who still seem influenced by the traditional Confucian bias against Mohism and (ever since Neo-Confucianism) Hsün Tzu. Here, for example, is the final judgment on Mohism pronounced by Wing-tsit Chan, a very well-known interpreter of Chinese philosophy to the West: "One thing is certain, and that is, philosophically Moism [Mohism] is shallow and unimportant. It does not have the profound metaphysical presuppositions of either Taoism or Confucianism" (Chan

does the situation greatly improve in postclassical philosophy, so that even in the case of such a giant as Chu Hsi (1130–1200), we have to derive his system from a bewildering assortment of recorded sayings, commentaries on the classics, letters to friends, and other scattered documents. There is no single *Summa* written by the master himself. The nearest approach is not an original work but an anthology—an anthology, moreover, containing nothing by Chu Hsi himself but only by his predecessors.[98]

In what today would be called scientific writings, the approach again tends to be piecemeal, even in works not consisting primarily of earlier quotations. This is true of Shen Kua (1030–1094), for example, probably the greatest of the scientifically minded Sung writers. His best-known work, the *Meng-ch'i pi-t'an* (Dream Torrent Jottings), consists of 507 separate comments or essays, almost none longer than a few hundred words, grouped very loosely under sixteen general topics such as "Ancient Matters," "Human Affairs," "Government," "Implements," "The Supernatural," and so on. Other than this, they seem to have no overall interrelation, being simply the diverse observations of an exceptionally intelligent and wide-ranging mind.[99]

So too with respect to the equally famous Ku Yen-wu (1613–1682), leader of the new school of textual criticism at the beginning of the Ch'ing dynasty. His *Jih-chih lu* (Daily Additions to Knowledge) consists of somewhat over 1,000 carefully formulated scholarly notes, usually fairly short and grouped under five humanistic topics such as the classics, government and economics, ethics and social relations, and so on. Less than a century separated their publication in definitive form (1695) from the appearance of William Gilbert's book, *On the Magnet* (1600).

1963:212). Nor is Chan's opinion of the later Mohists more favorable, despite their uniqueness as almost the only Chinese thinkers seriously interested in logic: "His followers also developed some epistemological interest and evolved a crude system of definition and argumentation" (ibid.). Concerning the uniqueness of the Mohist school and its scientific interests, more will be said at IV.2 below.

98. I refer to the *Chin ssu lu* (A Record for Reflection), compiled by Chu Hsi in 1175 with the collaboration of Lü Tsu-ch'ien. (The English title used here is that argued for by Ivanhoe [1988].) Its 622 topically arranged selections are taken from four earlier Neo-Confucian thinkers. See translation by Chan (1967), who describes the work (p. ix) as "the most important book in China for the last 750 years."

99. In his study of Shen Kua, Holzman (1958:290) writes: "He has nowhere organized his observations into anything like a general theory." Sivin (1975:12.374), in his biography and analysis of Shen Kua, writes similarly: "Shen's interests were multifarious, the record unsystematic, and its form too confining for anything but fragmentary insight."

Neither before nor after 1600 did any Chinese write an entire book, or even an extended essay, on that subject, despite the fact that magnetism (and later the compass) had been known and alluded to by a long line of Chinese savants from the third century B.C. onward, including notably Shen Kua himself (Needham 1962:229–334).

Near the end of the last section a few paragraphs were devoted to the role of analogy and historical precedent in Chinese argument. In post-classical times, and especially from the late T'ang (ninth century) onward, the use of historical precedent was gradually elaborated from mere brief citations of names or incidents into a much more sophisticated instrument. This could involve the presentation of entire complexes of historical phenomena, sometimes including institutional changes of an impersonal nature as well as the usual spectrum of human activities. Such presentations might be inserted into written statements on governmental policy as cited models to be either imitated or avoided. They were also used for purposes of education, including the education of the emperor or future emperor. Hartwell (1971), in his study of how this expanded use of historical precedent functioned during the eleventh and twelfth centuries (the Northern Sung), has coined the convenient term *historical analogism* to describe it. It was used both in the imperial seminar, established in 1033, and in the Sung civil service examinations. Notable encyclopedias and repertories of institutions were prepared as convenient compendia from which to draw historical analogies. Likewise, a didactic historical purpose was apparent in the preparation of the *Tzu-chih t'ung-chien* history, referred to earlier, which was closely connected with the imperial seminar. It was after the first chapters of this history were read in the seminar in 1067 that the emperor gave the work the title by which it came to be known, *The Comprehensive Mirror for Aid in Government* (Hartwell 1971: 694–712).

As developed in this way during the Sung dynasty, historical analogism permitted a greater degree of synthesis and abstraction than any of the other literary devices we have been discussing. In fact, as Hartwell demonstrates, it was occasionally possible for it to yield hypothetical propositions that are truly analytical. For the most part, nonetheless, it shared the tendency of the other devices to look more toward particulars than toward analysis, synthesis, and generalization. "Inference by analogy," writes Hartwell, ". . . is useful in *suggesting* hypotheses, . . . but the *formulation* of such hypotheses and their *arrangement* in an explanatory system is an independent intellectual process. The Chinese never developed the habit of framing generalizations in a hypothetical-deductive form" (Hartwell 1971:722; emphasis added).

Hartwell's further remarks, though particularly directed to the study of economics in pre-modern China, are broadly applicable to pre-modern Chinese scientific studies generally:

> The Chinese normally did not distinguish differences between the relative worth of alternative modes of logical presentation. The modern student of Chinese economic thought is constantly amazed to find in the same document equal treatment given to statements of widely varying orders of abstraction and analytical significance. . . . The failure to distinguish different orders of conceptualization severely limited the possibilities for integrating the separate ideas of economic doctrine into an explanatory system and precluded the broadening of abstraction essential to the progress of a science. This was partly owing to the habitual use of the historical-analogical method, which does not naturally lend itself to suggesting subtle differences in levels of generalization—for example, once the analogy was discovered, the problem was often deemed solved. The failure was primarily the result of neglecting to search consciously for general hypotheses. (Hartwell 1971:725–726)

Undoubtedly there are many factors, environmental and socioinstitutional as well as intellectual, that may help explain why China, despite so many scientific and technological "firsts," did not go on to develop an evolving science, one based on experimentation combined with observation rather than on uncritically accepted traditional theory. But on the intellectual side, the failure to synthesize, generalize, and hypothesize certainly lies near the heart of the problem. How much all this should properly be attributed to linguistic as against nonlinguistic factors is a subject on which scholars differ. In the next section, we shall examine some of their opinions.

10. Conclusion

The growing domination of form over content in the bureaucratic literature of late imperial China is symbolized by the literary genre popularly known as the "eight-legged essay" *(pa-ku wen).* This essay form became institutionalized in 1487, when it was made a standard part of the civil service examinations. Thereafter, with gradual modifications, it continued to be compulsory until its abolition in 1901. The themes set for it were always drawn from the "Four Books" *(Lun yü* or *Analects, Mencius, Ta hsüeh* or *Great Learning,* and *Chung yung* or *Doctrine of the Mean).* Its stylistic rules were strict, notably those requiring it always to consist of eight major divisions—hence its name. In the course of time its allowed length was gradually increased from 400 or 500 words (Ming) to 600 words or more (Ch'ing).

The theme for the eight-legged essay of 1487 was a sentence from Mencius (*Meng-tzu* 1b.3), which Legge (1960:2.31) translates as: "He who delights in Heaven, will affect with his love and protection the whole empire." Literally, the six characters mean: "Delight-in Heaven person, protect Heaven-below." Fairbank, Reischauer, and Craig (1965: 122) have described the consequences:

> In his eight-legged essay the candidate would be expected to proceed as follows—make a preliminary statement (three sentences), treat the first half ("Love Heaven person"[100]) in four "legs" or sections, make a transition (four sentences), treat the second half ("protect Heaven-below") in four "legs," make a recapitulation (four sentences), and reach a grand conclusion. Within each four-legged section, his expressions should be in antithetic pairs, such as pro and con, false and true, shallow and profound, each half of each antithesis balancing the other in length, diction, imagery and rhythm.
>
> This comparatively straightforward and systematic Ming type of treatment was further refined under the Ch'ing, with many variations of form. To avoid repetition, classical quotations might be chopped up and parts juxtaposed out of context (as if we should set as a topic from *Hamlet:* "To be or not to be, perchance to dream"). The eight-legged style, in effect, called for playing a game with the classical phrases. It stressed memory and detail without substance. Indeed, in stressing antithesis, it did violence to substance.[101]

It is no wonder that many scholars inveighed against the eight-legged essay, among them Ku Yen-wu, who wrote that its harm was as great as that inflicted by the Ch'in Burning of the Books in 213 B.C.[102] One can hardly doubt that the great wave of Ch'ing scholarly criticism and rationalism, starting with Ku Yen-wu and others, was a reaction, at least in part, precisely to the growing rigidity in official thinking typified by the eight-legged essay. That this scholarly movement, despite its originality and triumphs, ultimately failed to change official thinking, is indicated by the persistence of the eight-legged essay until the present century. No doubt the failure reflected in good part the growing crystallization of social and political institutions during the latter centuries of

---

100. "Love" is a slip here. The Chinese word *lo* means "to delight in, take pleasure in."

101. The winning eight-legged essay at the Peking examinations of September 1879 (taken by approximately 13,000 candidates, of whom approximately 330 passed) has been translated (with original Chinese text appended) by Bourne (1879–1880:352–356). Bourne's translation, minus the Chinese text, has been reproduced by Purcell (1936:31–35).

102. Ku Yen-wu, *Jih chih lu* (Daily Additions to Knowledge), 16/48 item 11; quoted in Nivison (1960:197). For a modern concurring Chinese opinion, see Ch'en Shou-yi (1961:509): "There is no question that the Eight-Legged Essay holds no place whatsoever in China's intellectual history except as a glaring example of demerit."

the Chinese empire. I believe, however, that intellectual factors long predating these last few centuries were also partially responsible—among them, I suspect, several of the various linguistic features that have been discussed here. In any event, quite aside from psychological considerations, there is no doubt that the eight-legged essay would have been technically impossible in any language other than Chinese.[103]

However, no unanimity exists among scholars as to the extent to which language has influenced Chinese patterns of thinking, or even whether such influence should be postulated at all. Before we turn to a few representative opinions, it is well to make my own views abundantly clear by recapitulating what has been said in the preceding sections. Here are the essential points:

1. Literary Chinese differs profoundly from the colloquial language in grammar, vocabulary, and style. The dichotomy is one that in style, at least, if not in other respects, goes back to the beginning of written records.

2. The Chinese characters achieved a prestige and mystique in the old China perhaps unrivaled in any other civilization. The great emphasis placed from the beginning on written documents as against oral statements probably encouraged an early and continuing tendency toward the creation of a bureaucratic form of government. Chinese writing and the literary language have been immeasurably powerful agents for conserving and diffusing Chinese cultural values, both historically and geographically. Ideologically, however, their effect has been strongly conservative, in that they made the entry of foreign ideas into China difficult, slowed down change, and discouraged cultural variation, especially within the small but dominant literate minority.

3. Linguistic features that are peculiarly Chinese include absence of inflection; a resulting freedom for individual words to assume a variety of grammatical functions according to syntax, without undergoing morphological change; the "rule of economy," that is, the option often open to the writer to eliminate, if he or she wishes, grammatically significant indicators, even though such elimination may make the text less clear; and the wide ranges of meaning—some of them seemingly unrelated—

103. Only long after these paragraphs were written did I read the detailed account of the eight-legged essay by Ching-yi Tu (1974–1975). This article is excellent when it analyzes the structure of this literary genre or translates a particular example from the late fifteenth century. It falters, however, when it attempts to defend the eight-legged essay on literary and other grounds, as in the following statement in its final paragraph: "Any literary form, no matter how rigid it may be, still allows room for the operation of imagination and originality."

that individual written words can commonly assume, depending on how they are used.

4. Stylistically, the dominant feature of Literary Chinese is its emphasis on syntactic and often semantic balance between successive cadence-units, each commonly consisting of the same number of monosyllables. The major manifestations of stylistic balance are parallelism and antithesis, whose great development in Literary Chinese is the direct result of the linguistic features enumerated in the preceding paragraph. The principle of balance, whether expressed in parallelism or antithesis, no doubt had its origin in stylistic considerations, but it became particularly significant as a grammatical device for reducing the possibility of syntactic or semantic ambiguity. At times, however, its use has resulted in verbal redundancy or the subordination of meaning to style. Ideologically, the parallelism found in Literary Chinese may have been a stimulus to Chinese analogical thinking and perhaps even to the Chinese search for cosmic order and harmony, especially as exemplified in the concept of the "harmony of man and nature." Similarly, stylistic antithesis may have encouraged Chinese dialectical thinking, which sees the universe as a never-ending flux of phenomenal change. In these respects, Chinese thinking differs sharply from that of speakers of Indo-European languages, whose tendency to look for unchanging categories in the universe may, just conceivably, have been unconsciously affected by their inflected languages, in which parts of speech are clearly demarcated.

5. The possibility of ambiguity in Literary Chinese could have been reduced and its reading considerably facilitated if punctuation had been systematically employed. Theoretically, there seems no reason why this could not have been done, since a primitive proto-punctuation occurs very occasionally already in inscriptions of the third century B.C. and a more developed, even though still primitive, system is found from time to time in printed books from the Sung period onward. The fact that punctuation nevertheless has had to wait until the twentieth century to become universal (as the result of Western influence) suggests that there existed a deeply rooted proprietary attitude toward the written language on the part of China's literate minority, who used it more as a recondite instrument for communication among themselves than for reaching large numbers of people. Combined with other factors, the net result was to emphasize the *forms* of language (syntax and style) at the expense of *content*.

6. Similar consequences resulted from the absence or only partial exploitation of certain practical devices that, if systematically used, would have considerably aided the reading of scholarly texts. They

include continuous pagination, paragraph indentations, subtitles, and catchwords on page margins. Furthermore, Chinese scholarship was considerably handicapped in its attempts to classify written materials by the unavailability of alphabetization. Substitutes for alphabetization were often awkward and placed a further burden on memory.

7. Literary devices commonly found in Literary Chinese include quotations (both named and unnamed), allusions, clichés, analogies, and appeal to historical precedent. All of them once more place a heavy premium on memory and sometimes, by substituting outworn symbols for realistic description, throw doubt on the veracity of the entire passage in which they appear. Their tendency is to fragment rather than synthesize their text.

8. The primary technique traditionally applied to the writing of histories or other significant works of scholarship has been that of scissors-and-paste, or what might be called "composition through compilation." Heavy reliance on lengthy quotations lifted *en bloc* from their original texts and inserted with only minimum discussion and analysis into their new settings exemplifies a conspicuous weakness found in a good deal of traditional Chinese scholarship: its frequent failure to synthesize, generalize, and hypothesize. Even the development of techniques of historical analogism, which could occasionally lead to the formulation of truly analytical hypothetical propositions, failed in the long run to turn the old scholarship significantly away from its overriding concern with particulars rather than systems.

These criticisms, I believe, are sufficiently detailed and comprehensive to obviate the need for quoting analogous criticisms from other scholars. Instead, four representative opinions from the opposite side will be cited. Their general tenor is that the inhibitory influence of Chinese (especially Literary Chinese) on philosophical and scholarly thinking has been slight or even nonexistent. The statement by Forrest is perhaps the most unequivocal:

> Even the least developed language is probably capable of expressing any ideas which its speakers may wish to express; and if it does not yet possess the means of expression for certain ideas, it is because the ideas themselves do not exist in the minds of its speakers. . . . A language is always sufficient to the needs of its speakers at any stage in their cultural history. (Forrest 1948:246)

Whereas Forrest writes with the spoken language primarily in mind, Dubs, in two passages, makes essentially the same point for Literary Chinese:

> In spite of the many meanings of Chinese characters, if we have a sufficiently long passage, and if the *thought* of the author was clear, there is no necessary

> ambiguity. When we come to the field of philosophy, the ambiguity of European words is notorious. . . . Yet the multitude of meanings of words does not necessarily produce ambiguity, because the context indicates which meaning is intended. . . . It is then false that the Chinese language is an inadequate vehicle for philosophical discourse because it does not possess an adequate supply of words or because these words are ambiguous. (Dubs 1928–1929:99–100)
>
> Hence we have no reason to seek in the Chinese language the cause of the failure of the Chinese to develop such philosophical systems as those of Plato or Spinoza. The Chinese language is capable of expressing whatever ideas are desired to be expressed. Such expression may be more difficult than in a European language, but great Chinese thinkers, such as Hsüntzu, have made the Chinese language express their ideas precisely. (Dubs 1928–1929:100)[104]

Graham has more than once expressed himself along similar lines (esp. in Graham 1959 and 1964). He begins an important study of the problem of "being" in Western and Chinese philosophy with the question: "To what extent are differences between Chinese and Western thought affected by grammatical differences between Chinese and Indo-European languages?" And at the end he gives his considered answer: "The present study does not encourage one to take it for granted that Chinese is either better or worse than English as an instrument of thought; each language has its own sources of confusion, some of which are exposed by translation into the other" (Graham 1959:79 and 112).[105]

Needham has also expressed an essentially similar view:

> There is a commonly received idea that the ideographic language was a powerful inhibitory factor to the development of modern science in China. We believe, however, that this factor is generally grossly overrated. It has proved possible in the course of our work to draw up large glossaries of definable technical terms used in ancient and mediaeval times for all kinds of things and ideas in science and its applications. . . . We are strongly inclined to

104. Dubs (1928–1929:100). By 1960 this distinguished translator from Literary Chinese had somewhat changed his opinion about the language. See Dubs (1960:140): "Classical Chinese, unlike European languages, is not a perspicacious language. One must think himself into the mind of the author and consider all possible nuances of thought and allusions in the Chinese text, before one can translate rightly. If a translator's attention, wearied by a lengthy work, flags for a moment, errors arise. I know of no impeccable translations from the Chinese."

105. Graham also (1989:389–428) has much of interest to say about how ideas are expressed within the grammatical framework of Literary Chinese. However, he does not explicitly address the central question of whether or not it has been an effective medium for scholarly and scientific communication.

> believe that if the social and economic factors in Chinese society had permitted or facilitated the rise of modern science there as well as in Europe, then already 300 years ago the language would have been made suitable for scientific expression.
>
> At the same time also it is wise not to underestimate the capacities of the classical language. We do not recall any instance where (after adequate consideration) we have been seriously in doubt as to what was intended by a classical or mediaeval Chinese author dealing with a scientific or technical subject, provided always that the text was not too corrupt, and that the description was sufficiently full. (Needham 1969:38).[106]

To these four distinguished scholars I offer the following rejoinders:

1. None of the four denies the fundamental fact that words in Literary Chinese commonly bear multiple meanings. This phenomenon, I contend, can at worst throw the meaning of a passage into confusion, and even when this does not happen, places a continuing burden on the conscientious reader to explicate—sometimes laboriously—what he or she is reading. In reply to Dubs's argument that the many meanings of Chinese characters are paralleled by the ambiguity of European philosophical terms, I can do no better than quote from the study by I. A. Richards on the psychology of Mencius. Apropos two highly ambiguous key terms in Mencius, Richards writes:

> But these two uncertain words by no means account for all the ambiguity of the passage. In the West we are so accustomed to explicit sentence forms that we expect this kind of vagueness only at certain specific points in discourse—at words which are recognized to be indefinite in meaning. We are ready for a good deal of vagueness there. But a sentence shocks us whose whole structure has the same kind of indefiniteness. It appears to be a mystification on a much grander scale. (Richards 1932:6)

2. When Forrest and Dubs respectively say that "a language is always sufficient to the needs of its speakers" and that "the Chinese language is capable of expressing whatever ideas are desired to be expressed," what they seem to imply is that language per se exerts little or no influence on the formulation of ideas or direction of scholarly interest. Such a thesis, I believe, is just as extreme as the counter-thesis that the influence of language is all-encompassing. Surely the truth lies somewhere in between, in the thesis that the influence between thinking and language is mutual, dialectical, and ongoing. It is no accident, in my opinion, that Pānini (possibly fourth century B.C.), the world's greatest grammarian before the nineteenth century, was an Indian dealing with a highly inflected language (Sanskrit), whereas China produced no gram-

---

106. The problems of translation are further discussed in Needham (1970). Needham's opinion is supported by Graham (1973:64–65).

marians until the present century. Or again, as suggested under point four above, I believe that there is quite possibly a relationship between the emphasis on stylistic balance in written Chinese and the emphasis on harmony and balance in the Chinese worldview.

3. It is argued that Literary Chinese is adequate for philosophical or scientific expression provided the *thought* of the writer is itself clear and expressed in precise language (Dubs) or provided the scientific description is sufficiently full (Needham). But these, I believe, are not the central issues. Even granted the possibility that Literary Chinese can, with great care, be made to express ideas clearly and precisely, the realistic fact is that, too often, satisfactory achievement of this aim is hindered by stylistic considerations. Very relevant is what Purcell (1936:93) was earlier quoted as saying in a slightly different context: "This does not mean that the language has no words to fill in the ellipses, or that there are no words to convey tense, number, or mood. It merely means that the spirit of the language is against their use." I fear that the prevailing spirit of Literary Chinese is likewise too often not conducive to clarity, precision, and simplicity.

4. No doubt it is possible, as declared above by Needham, for modern students of Chinese science to draw up large glossaries of definable Chinese technical terms. What to me seems more significant, however, is the rarity with which indigenous Chinese scholars themselves have helped that possiblity by drawing up their own systematic lists of defined terms (see also point six below). Moreover, as Needham has pointed out elsewhere (1954:36), the Chinese were linguistically handicapped, as compared with West Europeans, in their search for scientific terminology:

> In the formation of scientific terms, the West Europeans were able to draw not only upon Greek and Latin but also upon Arabic roots, adding them to languages already rather rich in Teutonic complex consonantal combinations. . . . Thus many almost synonymous words were developed, which could be given slightly different shades of meaning. No such resources were available to the Chinese.

Apropos Neo-Confucianism, Needham likewise comments (1956:491) on the unfortunate readiness of Chinese scholars and officials to continue using—even though in new philosophical contexts—the traditional terms belonging to superstitious folk religion, rather than creating a fresh terminology of their own. This tendency, which Needham attributes to the social background of Chinese bureaucratic society, could equally well, I believe, derive from the strength of the classical literary tradition and the inherent conservatism of Literary Chinese, in which "nothing is obsolete or even obsolescent" (Fang 1953:270). Writes Needham once more (1956:491):

> It may well be that when, at the end of our book, we are able to look back upon the course of Chinese thinking in its social context, we shall feel that this serious failure to elaborate new terminology instead of merely rationalizing ancient words with all their religious undertones, was one of the most unfortunate aspects of the social milieu in which Chinese science struggled for birth.

5. I have seen no answer to the contention that the propensity for using such literary devices as quotations, allusions, and the like, coupled with the inadequate development of such practical aids as punctuation, continuous pagination, marginal catchwords, and the like, and again coupled with the absence of alphabetization, have all combined to impose large—and often unnecessary—burdens of memorization and effort on Chinese scholarship.

6. Finally, as previously indicated, my most trenchant criticism of Chinese literary devices and the techniques used for scholarly creation is that they have all served to turn Chinese scholarship away from substance and toward form, away from synthesis and generalization and toward compilation and commentary. Moreover—and this is a point not previously made—I strongly suspect that the nondefinition of terms, of which Richards complains in philosophy, is part of the same broader failure to analyze. So widespread, in fact, is this nondefinition in philosophy that it causes Graham to remark (1972:164): "The later Mohist ethic is an achievement quite without parallel in Chinese philosophy, a highly rationalized ethical system in which all key terms are defined."

The conclusion reached in this chapter, therefore, is that written Chinese has, in a variety of ways, hindered more than it has helped the development of scientific ways of thinking in China. This inhibitory influence, I further believe, was evidenced already at an early period, that is, during the last three centuries of the Chou dynasty (the age of classical philosophical speculation). In making this unfavorable assessment, I must stress that I am speaking of Literary Chinese *solely* as a practical instrument for scholarly and scientific expression. In no way do I wish to deny its great beauty and power for poetic and literary expression, as well as the unique fascination and challenge inherent in the Chinese characters. As has been forcefully said, once more by Needham (1954:36):

> And it is true that this old language, in spite of its ambiguity, has a concentrated, laconic, lapidary quality, making an impression of austere elegance, pith and virility, unequalled in any other invented instrument of human communication.

III. THE ORDERING OF SPACE, TIME, AND THINGS

1. Correlative Thinking

Central to this chapter are Chinese attempts to classify the objects, phenomena, and concepts of the natural and human worlds according to correlative thinking.[1] This means grouping those items regarded as interrelated into sets of items that are the same in number and thus fall under the same numerical category. Mayers, in his *Chinese Reader's Manual* (1924:318–380; 1st ed. 1874), has listed eighteen such categories, ranging numerically from 2 to 100 and containing 317 separate sets of items. The five largest categories are those of 9 (containing 31 sets), 6 (38 sets), 4 (40 sets), 5 (63 sets), and 3 (68 sets).

Although many sets are grouped together only on the basis of sameness of number, many others are additionally felt to belong to special systems of relationships. Examples of the latter are the Five Planets (set 162 in Mayers' listing) and the Five Viscera (set 172), both members of the system of Five Elements (see below). By contrast, the Five Cardinal Relations Among Mankind (set 149) and Five Curved Portions of the Body (set 150) have nothing to do with one another save that they both number five.

In addition to the general Chinese world of numerical categories there exists a separate world of categories pertaining to Chinese Buddhism—most of them already formulated before Buddhism came to China and only a minority added within China after that event. How numerous they are is quickly apparent from the pages of the Soothill and Hodous *Dictionary of Chinese Buddhist Terms* (1937), where the largest categories are those in twos (pp. 20–31), threes (pp. 57–80), fours (pp. 169–184), fives (pp. 112–139), and tens (pp 42–54).

The fact that Buddhism, too, thus indulges in correlative thinking is a healthy reminder that there is nothing uniquely Chinese about such

---

1. When this chapter was written, I had not yet read the historical account of Chinese cosmological thinking by Henderson (1984). Henderson's first two chapters are particularly relevant to sects. 1 and 2 below and often discuss the same materials, although sometimes our approaches and interpretations differ. To take account of what Henderson says, I have occasionally modified or added to my own original presentation.

thinking. It has flourished in many societies, although conceivably nowhere with such persistence and detail as in China. A good European example comes from the astronomer Johannes Kepler (1571–1630), about whom Graham (1986:8) remarks: "On the very threshold of modern science, Kepler, whose three laws of planetary motion are the first true laws of nature since the Greeks, was trying to fit them into the symmetries of a cosmos in which sun, stars and planets correlate with the person of the Trinity."

This kind of thinking has since disappeared from the natural sciences, but in the world of everyday life its influence still persists—so much so that even in Europe today, according to Graham (1986:58), "much of ordinary practical life belongs irredeemably to correlative thinking." Whether or not this assertion is correct, we shall see at the end of this chapter that numerology still remains alive within the political life of modern China.

Before turning to China, let us consider the role of correlative thinking generally. What follows is based on two longer expositions given by Graham (1986:16–24; 1989:319–325). In these he points out that correlative thinking results from the human tendency, when speaking, to draw upon sets of concepts that have been unconsciously stored within the mind. For the purpose of correlative thinking in its simplest binary form, such items are evenly divided into two parallel groups (here labeled A and B) in such a way that any item belonging to one group stands in what Graham terms a paradigmatic relationship to the corresponding item in the other. At the same time each such item, whether in group A or group B, stands in what Graham terms a syntagmatic relationship to all other items belonging to the same A or B category:

A	Paradigm	B
	S	
1. Day	y	Night
	n	
2. Light	t	Darkness
	a	
	g	
	m	

Here, as well as following, the number of items under A and B could, in theory, be indefinitely extended. The paradigmatic (i.e., horizontal) relation is one of similarity/contrast (Day *compares* with night as light with darkness); the syntagmatic (i.e., vertical) relation is one of contiguity/remoteness (Day *connects* with light as night with darkness).

When, in a given assemblage, the tendency is toward similarity/contiguity rather than contrast/remoteness, it then becomes possible, through the figures of speech called *metaphor* and *metonymy,* for one member of a pair to substitute for another:

Metaphor	
A	B
1. King	Lion
2. Men	Beasts

Here *king* compares with *lion* as *men* with *beasts.* Hence by metaphor, the lion is the king of beasts and the king is a lion among men.

Metonymy	
A	B
1. King	Chairman
2. Throne	Chair

Here *king* connects with *throne* as *chairman* with *chair.* Hence by metonymy, the monarchy is the throne and the chairmanship is the chair.

In this and other ways, correlative thinking creates ever lengthening and more complex chains of ideas. These at first cover what is already familiar (night invariably follows day and vice versa) but then increasingly move into unknown territory (on occasion even darkness at noon may occur). And as this unknown realm expands, it generates an increasing tension in the mind of the correlationist. On the one hand, he has a desire to remain securely within the comprehensible world of correlation. On the other, he has a growing uneasy feeling that such correlation is no longer certain or adequate and that it should be checked, if possible, against causal explanations that seem to be exact, invariable, and universal rather than piecemeal ("laws of nature").

Such at least is what eventually happened to medieval and postmedieval cosmological thinking in Europe. There, beginning around 1600, the old correlative kind of cosmology, believed in even by Kepler, collapsed under the rising tide of the "discovery of how to discover." Out of the debris a new worldview gradually emerged, much more based on causality than on correlation. And yet, as Graham insists, correlative thinking today still influences many people.

In China no such triumph of causality over correlation occurred for a long time. Until the present century, as we shall see, the old correlative cosmology remained dominant. This very important difference between Europe and China (as well as other civilizations) should be kept in mind as we read further.

Let us turn now to the correlative systems of pre-modern China. The most important Chinese numerical categories have been those in twos, threes, fives, and nines. Basic to everything has been the polarity of the yang and yin principles, each with its many correlative qualities and functions. For the yang, these included brightness, heat, dryness, hardness, activity, incipience, dispersion, masculinity, Heaven, sun, south, above, roundness, odd numbers, and much else. For the yin they included darkness, cold, wetness, softness, quiescence, completion, consolidation, femininity, Earth, moon, north, below, squareness, even numbers, and much else.[2]

It is important to differentiate this kind of polarity from the Zoroastrian/Manichean struggles of light against darkness and good against evil. In Chinese thinking, the yin and yang complement rather than struggle against one another. Each is essential for the functioning of the cosmos, even though, as we shall see, the yin is hierarchically subordinate to the yang. In the repetitive cycles of seasons, days, and other phenomena, the yin and yang ever wax and wane in inverse ratio to one another, without the one ever permanently suppressing the other. Here, then, is a concept very different from the ultimate conquest of light over darkness as predicted in the Persian religions.

Also of major cosmological importance are the five entities of wood, fire, soil, metal, and water. Initially they were apparently conceived merely as the substances bearing these names, but in the course of time they gradually evolved into immaterial, all-embracing cosmic forces. This change in thinking is reflected in their changing names, apparently beginning with *wu ts'ai,* "Five Substances," then shifting to *wu te,* "Five Powers," and finally standardized (especially from Han times onward) as *wu hsing.*

The translation of this last term is not easy. Its traditional rendition as "Five Elements" reflects the similarities thought to exist between the Chinese five entities and the four elements of ancient India or Greece (earth, air, fire, water). Actually, however, *hsing* does not mean "element" at all but is fundamentally a verb, "to move, act, do." Hence the real meaning of *wu hsing* is something like "the five active entities," which is too clumsy. Other possibilities proposed by various scholars have been "five forces, agents, activities or stages of change" (listed in Kunst 1977:68). All of these have been rejected on various grounds by Major (1977:70) in favor of his own slightly earlier suggestion of "Five Phases" (Major 1976), which has since gained considerable acceptance. Yet already before Major, Porkert (1974:43ff.) had come up with "Evolutive Phases" as a rendition and had abbreviated it to "E.P.S."

---

2. For a more detailed and technical listing, see Porkert (1974:22–24 [included within his broader discussion, 8–43]).

My own reaction is that "phases"—whether or not enlarged to "Evolutive Phases" and then contracted to "E.P.S."—is too bland and colorless a word to convey adequately the dynamism of the *wu hsing* in their ceaseless interactions. Here, for example, is how Wang Ch'ung (A.D. 27–ca. 96) reports the opinion of his time: "It is said that . . . the energies *(ch'i)* of the *wu hsing* successively do violence *(tsei hai)* to one another. . . . The fact that living creatures bite and eat one another is all because the *wu hsing* energies cause them to act thus."[3]

In short, without following the pros and cons further, it seems evident to me that no really adequate translation of *wu hsing* exists. That is why, in these pages, I have decided to continue speaking of "elements"—a word probably more immediately understandable to nonspecialists than any of the suggested alternatives. Since making this decision, I have been heartened by the note on the subject by Friedrich and Lackner (1983–1985). In this they argue that objections to "element" have been too narrowly based on "a conception [of the word] limited to its use in the natural sciences," which is "only the most recent portion of its historical meaning." In conclusion, therefore, they urge that "confidence in the use of 'element' be restored, as a term not identical but strongly convergent with its Chinese analogue [*hsing*]."

How did the yin-yang and Five Elements originate and develop? Hsü Fu-kuan (1963:509–587) has presented a very scholarly survey, followed in essentials by Graham (1986:70–92), who has added important general observations, some of which have been summarized above. Unfortunately, the historical conclusions of the two men cannot always be conclusive, especially concerning the earlier stages, because several of the key texts cannot be precisely dated and others that would have been important are no longer extant. Nevertheless, Hsü and Graham have each produced lucid and usually convincing interpretations of a perplexing subject. Here is a simplified three-point summary of their main findings, to which I have added two further points of my own.

1. In their beginnings, the yin-yang and Five Elements were entirely distinct from one another. During the early Chou, when the words *yang* and *yin* first appear in the *Shih ching,* they merely mean "sunshine" and "shade" and have no cosmic significance. Likewise, the *wu hsing,* when very occasionally mentioned somewhat later (usually under other names) in the *Tso chuan* history (earliest reference is 620 B.C.), seem to be only material substances; they do not constitute a cosmic system.

2. The first to unite the yin-yang and Five Elements into a single system was apparently Tsou Yen (ca. 305–ca. 240), traditionally regarded

---

3. Wang Ch'ung, *Lun heng* (Doctrines Evaluated) 14(3/24); tr. Forke (1962:1.284–285). It should be noted that Wang himself does not subscribe to all aspects of this belief.

as the "father" of Chinese naturalistic thinking. Living as he did during the Warring States period, he gave the elements political significance by correlating each of them with a particular dynasty or reign in an endlessly recurring cycle. In this way he and his followers induced several of the Warring States rulers to institute state cults of the Five Elements in the hope of gaining the support of that particular element destined by its position in the cycle to replace the ebbing element of the Chou dynasty. Before Tsou Yen and perhaps for some time after him as well, the Five Elements seem to have been almost exclusively the concern of court astronomers, physicians, music masters, diviners, and the like, and it was to this class of men that Tsou Yen himself quite possibly belonged. By contrast, the elements are barely mentioned in sociophilosophical writings before the second half of the third century B.C.

3. However, the Five Elements finally achieved philosophical respectability through inclusion within the pages of the important eclectic work *Lü-shih ch'un-ch'iu* (Mr. Lü's Springs and Autumns), compiled in 240 B.C. There the elements are systematically correlated with the months and seasons of the year, and the ideal ruler is instructed to perform ceremonies, wear clothing, eat foods, and do much else throughout the year in a manner deemed appropriate to each season. In this way he can ensure continuing harmony between man and nature and prevent such unseasonable phenomena as cold and hail in summer or the flowering of plants and trees in autumn.[4] Not long afterward, when the First Emperor of Ch'in created China's first universal empire in 221 B.C., he officially adopted water as the element that had replaced the fire element of the by then defunct Chou dynasty. From this time onward, the Five Elements and yin-yang ideas gained general currency in all schools of thought, and the Han thinker Tung Chung-shu (ca. 179–ca. 104), in his *Ch'un-ch'iu fan-lu* (Luxuriant Spring and Autumn Dew), integrated them with Confucian moral and social values into a single, all-embracing system.

4. Following the collapse of the Han dynasty in A.D. 220, the Five Elements gradually lost much of their political significance. In cosmological thinking, however, as well as in such scientific or proto-scientific fields as medicine, biology, alchemy, and geomancy, the Five Elements, together with the yin-yang system, retained their prominence until

4. See *Lü-shih ch'un-ch'iu,* first ch. in each of the first 12 *chüan* or books; tr. Wilhelm (1928:1–3, 14–16, 41–43, and so on). The same calendrical text was later inserted into the Confucian *Li chi* (Record of Ceremonial), where it became ch. 4, called the *Yüeh ling* (Monthly Ordinances); tr. Legge (1885:27.249–310) and Couvreur (1913: 1.330–410).

recent times. This conclusion remains valid despite the growing influx of Western ideas from the Jesuits onward and the simultaneous indigenous rise of the new critical school of scholarship to which attention is called below.

5. The late Ming and early Ch'ing dynasties (primarily seventeenth through eighteenth centuries) saw the gradual formulation of a new, more exact and more empirical kind of scholarship, eventually known as the "School of Evidential Research" *(k'ao-cheng hsüeh)*. Its practitioners developed sophisticated techniques of textual and historical analysis to establish the precise texts of China's classical literature and their historical sequence. In so doing they sometimes criticized, on grounds of anachronism or improbability, passages in this literature in which traditional cosmographical ideas appeared. Henderson (1984:chs. 7–8) has performed a valuable service in bringing together a large number of such criticisms—among them, for example, the contention by Wang Fu-chih (1619–1692) that there is no absolute demarcation between one element and another but only a gradual fusing of one into another (Henderson 1984:245).

However, when we analyze the coverage and focus of these criticisms, it would appear that their influence upon the subsequent development of Chinese physical science was probably considerably less than might at first sight be supposed. First, the criticisms are scattered and unsystematic. They deal with particular cosmological references occurring in particular texts, not with the general conception of the yin-yang and Five Elements as cosmological systems. Apparently the overall reality of these systems was scarcely ever questioned. Second, the criticisms are based on the study of books within the library, not on the direct study of actual natural phenomena (through observation, experiment, and so on). These are the reasons for the conclusion reached at point four that traditional cosmological thinking retained its prominence in China until recent times—in other words, until approximately the beginning of the twentieth century.

2. Symmetry and Centrality

In his *La Pensée chinoise* (1934), Marcel Granet has forcefully propounded the theory that time and space, as viewed by the ancient Chinese, consisted of "boxed" or particularized units rather than abstract continua. Today we know that he went too far and that other early Chinese approaches to space and time also existed. Nevertheless, Granet's theory is important and deserves further exploration. We shall begin with two of his statements:

All [Chinese thinkers] prefer to see in time an assemblage of eras, seasons and epochs, and in space a complex of regions, climates and directions. In each such direction, extension [i.e., space] particularizes itself by assuming the attributes peculiar to a single climate or region. In the same way, duration [i.e., time] differentiates itself into varied time periods, each bearing the characterization appropriate to a single season or era. (Granet 1934:86)

Time and space are never conceived apart from concrete actions. . . . The words *shih* ["occasion" or "timeliness"] and *fang* ["direction" or "region"] apply respectively to all portions and parts of duration and extension—each and every one of which, however, is in each instance viewed under its own distinctive aspect. The two terms are evocative neither of space nor of time *per se*. *Shih* calls to mind the idea of circumstance or occasion (which may be either propitious or unpropitious for a given action); *fang*, that of direction or location (which may be either favorable or unfavorable for a particular instance). Thus time and space form a complex of symbolic conditions, both determining and determined; they are always imagined as an assemblage of concrete and diverse groupings of locations and occasions. (Granet 1934:88–89)

What Granet primarily has in mind here, though he does not explicitly say so, is the cosmological system dominated by the Five Elements, with their numerous correlations embracing the compass points *(fang)*, seasons *(shih)*, periods of history, and a host of other things, beings, phenomena, qualities, ideas, and emotions.[5] Still further correlations involving "boxed" space and time are those connected with the twelve cyclical "stems," the ten cyclical "branches," and especially the yin and yang principles. By Han times, as we have seen, all of these had become fused into a single, all-embracing sociocosmological system.[6]

The one other spatio-temporal system of comparable significance was that which, also in Han times, matched the eight trigrams and sixty-four hexagrams of the *Yi ching* (Changes Classic) with the months, seasons, twenty-four fortnightly divisions of the year, the compass points, and many other things and phenomena. Yet on the whole the focus of the hexagrams remained predominantly sociological, that of the yin-yang and Five Elements more directly cosmological and naturalistic. Still another sociocosmological system, also of Han date, was the one created by Yang Hsiung (57 B.C.–A.D. 18) in his *T'ai hsüan ching* (Canon of Supreme Mystery). Basic to it was a sequence of eighty-one tetragrams (graphs of four lines each), each embodying multiple symbolic meanings. Yang expounded these with a complexity and sophistication

5. The major correlations, thirty in number, appear in Needham (1956:262–263, Table 12). For a shorter table of eleven correlates, see Bodde (1957a:37).

6. For a diagram indicating the major correlations, see Fung (1953:15).

that gained his work the admiration of the few but prevented it from ever equaling the *Yi ching* in popular esteem.[7]

The major importance of these and other systems of correlative thinking during the Han should not obscure the fact, mentioned earlier, that in pre-imperial China they had not yet gained wide acceptance among social and philosophical thinkers. Very different views of time and space were expressed, for example, by some (not all) Taoists. Also different were the views of the followers of Mo Tzu (fl. 479–381), who were known after their master as Mohists or later Mohists. These men, perhaps during the first half of the third century B.C., achieved the intellectual feat of detaching time and space from their particularized matrices and analyzing them abstractly. Here is how they define time (lit. "duration," *chiu*) and space *(yü)*: "*Chiu* (duration) is pervasion of different periods *(shih)*. . . . *Yü* (space) is pervasion of different places *(so)*."[8]

The Mohists likewise denied the validity of the fixed, cyclical succession of the Five Elements—apparently virtually the only explicit doubters of this theory between their own day and the present century, although Wang Fu-chih, as we have seen, comes close. "The Five Elements," they wrote,

> do not have constant ascendancies [that is, none of them necessarily "conquers" its predecessor simply because of its position in the fixed cycle of Five Elements]. . . . That fire [under particular circumstances] melts metal is because [in this particular circumstance] there is much fire [and not because fire follows metal in the cycle]; that metal [in another particular circumstance] uses up charcoal [i.e., wood] is because there is much metal [in that particular circumstance and not because metal follows charcoal/wood in the Five Elements cycle].[9]

As we read these statements, it is vitally important to remember that the Mohist school which produced them is precisely the school that completely disappeared in Han times, leaving behind almost no later influence. From then until the present century, apparently almost nobody ever directly argued against particularized concepts of time and space

---

7. On the *Yi ching*'s correlations in Han times, see Fung (1953:106ff.). On the *T'ai hsüan ching,* see the detailed description, with extensive translations, in Nylan and Sivin (1987).

8. *Mo-tzu* 40; tr. Graham (1978:293.A40–41).

9. *Mo-tzu* 41 and 43; tr. Graham (1978:411.B43). We have seen a little above that Wang Ch'ung questioned the correlation between the mutually violent elements on the one hand and the mutually devouring animals on the other. Like everyone else, however, he did not question the existence of the elements themselves or of their cycles.

or the actuality of the Five Elements. In the next chapter, more will be said about the uniqueness of the later Mohists and their scientific interests.

How difficult it must have been for the very early Chinese to break away from particularized space-time conceptions is suggested by the written graphs for *yü* 宇 and *chou* 宙, the two words that, whether compounded as *yü-chou* or used separately within the same sentence, are the commonest respective designations for abstract space and abstract time.[10] The fact that the two written forms both contain the graphic element for "roof" supplies the clue to their concrete original meanings: the eaves of a roof *(yü)* and its (riblike) beams *(chou)*.[11] These meanings, coupled with the way the words appear in philosophical contexts, suggest certain conclusions:

1. Abstract time was more difficult for the ancient Chinese to conceive of than abstract space, to judge from the fact that a word with spatial connotations, *chou,* was used to refer to abstract time and that in philosophical contexts the sequence of words was always *yü* followed by *chou* (whereas Westerners, in speaking of "time and space," clearly give priority to the former).

2. Used philosophically, *yü* and *chou* rarely appear separately (i.e., individually in separate sentences).[12] This fact suggests that even when

10. The combination *yü-chou,* "space-time," frequently becomes a term translatable as "universe" or "cosmos." In classical texts, the words occur most often in Taoist writings. See *Chuang-tzu* 2, 22, 23, 28, 32; tr. Watson (1968:47, 244, 256, 308, 356). Also *Huai-nan-tzu* 1/1b–2a; 2/1a and 12a; 3/1a; 7/15a; 8/4a; 10/1a; 11/16a; 12/23b; 16/13a; 19/11a. The term also appears in *Hsün-tzu* 21(15/5b); tr. Dubs (1928:269). The later Mohists, as we have just seen, when they talked about space and time, replaced *chou,* "time," by *chiu,* "duration," but this word, because its etymology is uncertain, will not be discussed here.

11. *Yü* in the sense of "eaves" appears already in *Shih ching* no. 154, stanza 5; tr. Karlgren (1950a:98). *Chou,* before its combination with *yü* in the sense of "space-time," was used so rarely that one has to wait until the commentary of Kao Yu (fl. 205–212) on the *Huai-nan-tzu* to get a clear-cut statement of what was surely its original meaning. *Huai-nan-tzu* 6/7b contains an anecdote (obviously modeled on the similar anecdote in *Chuang-tzu* 1) in which a swallow and a sparrow ridicule the great *feng-huang* ("phoenix") for its inability to fly as easily as they do "between the *yü* and the *chou.*" Kao Yu's comment is unequivocal: "The *yü* are the eaves of a building *(wu yen),* the *chou* are its roof-beams *(tung liang).*"

12. A single exception occurs in *Huai-nan-tzu* 2/12a, where the Taoist True Man *(chen jen)* is said to "gallop beyond [the world of] the directions *(fang)* and rest within [the realm of] space *(yü).*" In this figurative statement, a distinction seems to be made between the immediate world of *fang,* concrete and particularized space, and the more distant reaches of *yü,* space as an undifferentiated (i.e., infinite) abstraction.

Chinese thinkers conceived of time and space abstractly, they still regarded them as an indivisible continuum in which, so to speak, time supplies an added dimension to space—a conception not without its modern analogs.

3. The metaphors involved in the equating of *yü*, "eaves," with "space" and *chou*, "roof-beams," with "time" suggest a possible derivation from the cosmological theory known as the *kai t'ien* or "Heavenly Cover." In this, the most primitive of several Chinese conceptions of the physical universe, heaven was believed to be a hemispherical dome resting upon earth, which was conceived as an almost flat square with slightly sloping sides surrounded by bodies of water known as the "four seas." In various texts, the heavenly dome is likened to an umbrella, a bamboo hat, or the umbrellalike covering of a carriage (Needham 1959: 210–216). Each of these metaphors involves a system of ribs spreading downward from a peaked center. Thus if our suggestion is correct, we may regard the *chou* or "roof-beams" as representing the heavenly canopy with its ribs, beneath which the *yü* or "eaves" constitute the outer squared edges of earth's flattened dome, peeping forth, especially at the corners, beyond the confines of the covering heavenly dome. It is natural enough to equate "space" with "earth," and the equating of "time" with "heaven" is equally natural, since it is the movements of the heavenly bodies that give man an awareness of time.

Why did the Chinese organize space and time, things, qualities, and ideas into groupings based on numerical categories? Needham (1956: 338) persuasively argues that the stimulus was societal and bureaucratic:

> Perhaps the entire system of correlative organismic thinking was in one sense the mirror image of Chinese bureaucratic society. Not only the tremendous filing system of the *I Ching*, but also the symbolic correlations in the stratified matrix world might so be described. Both human society and the picture of Nature involved a system of coordinates, a tabulation-framework, a stratified matrix in which everything had its position, connected by the "proper channels" with everything else. On the one hand there were the various . . . Ministries and departments of State (forming one division), and the Nine Ranks of officials . . . (forming the other). Over against these were the five elements or the eight trigrams or sixty-four hexagrams (forming one dimension), and all the ten thousand things divided among them and individually responsive to them (forming the other).

This theory perhaps partially explains the great burgeoning of correlative thinking that accompanied the building of the Han bureaucratic state. Yet the beginnings of correlative thinking surely go back much earlier—at least to the third century B.C., when Tsou Yen was at work on the Five Elements system, and perhaps a good bit earlier. To understand these earlier developments, therefore, it would be well to look for other stimuli.

Granet (1934:149–299) has discussed Chinese correlative thinking at great length and with much orginality. Space limitations make it impossible to go into his exceedingly involved and highly theoretical findings in detail, nor would I wish to do so, because there are too many places where I would find it difficult to agree with him. Yet I believe that some of his major conclusions, notably the typological distinction between correlations based on even and odd numbers, show great insight. In what follows, I am happy to acknowledge the stimulus of his general approach.

First of all, it should be stressed that "Chinese coordinative thinking was *not* primitive thinking in the sense that it was an alogical or pre-logical chaos in which anything could be the cause of anything else. . . . It was a picture of an extremely and precisely ordered universe, in which things 'fitted,' 'so exactly that you could not insert a hair between them' " (Needham 1956:286). This insistence on the precise fitting of things sprang, I believe, from man's intense need for psychological security in dealing with the nonhuman universe. This he could achieve by discovering regularity, simplicity, and therefore meaning in the otherwise overpowering flux of cosmic phenomena—in other words, by reducing the complexity of the flux to neatly packaged and symmetrically ordered groups of correspondences.

Symmetry is the key word here, and it is a symmetry that, though it applied to space and time alike, was particularly meaningful to the Chinese in terms of space. Above all, it was (and is) manifested in the exquisite Chinese awareness of absolute orientation. Anyone who has lived in Peking (at least before it lost its wall and became a victim of urban sprawl) will remember how even the moving of a table a few feet in a room is never carried out in the relative terminology of "left" and "right," but always based on absolute direction ("move it a little north" or "half a foot west").[13]

Symmetry, as just spoken of, is a generalized term for a generalized concept. In what follows, I wish to use it in a narrower technical sense, in apposition to which I shall also be discussing another concept I believe to be genetically derived from this technical kind of symmetry. *Centrality* will be the term used to designate this second kind of symmetry.

Before plunging into detail, it is well to ponder here the enormous importance that the concept of centrality or the mean—*chung* (lit. "mid-

13. The relationship between Chinese conceptions of right and left and absolute orientation is further discussed toward the end of this section.

dle")—has generally held in ethical as well as naturalistic Chinese thinking. From the *Lun yü* onward it has dominated the behavior of the ideal Confucian gentleman, the *chün tzu.* It means, for him, that going too far is quite as undesirable as falling short. Hence he should be neither too ceremonial nor too spontaneous, neither too pedantic nor too imaginative, and so on. In the *Yi ching,* likewise, we find centrality of behavior emphasized, often in conjunction with timeliness, that is, the idea of doing the right action at the right moment.[14] In general, the Confucian emphasis on centrality may be said to have powerfully reinforced Chinese political and social conservatism, because it is evident that a person who always cautiously follows the mean is not likely to become either a revolutionary or even a very ardent reformer.

Coming now to technicalities, symmetry in the narrow sense, whether thought of graphically or symbolically, signifies here an evenly divided duality. (Here and below, "dualism," "duality," and the like, are used nontechnically to refer to such complementary pairs as light/darkness or heat/cold. They are not used philosophically or theologically to refer to such noncomplementary opposites as mind/body or good/evil.) In its simplest form, this duality consists of a one-to-one paired relationship (A/B); but it may also be built up on the basis of any greater number of units, odd or even, so long as these are subsumed under groupings that are themselves dual or multiples of duality. This formula thus includes, for example, such basically dual appositions as AB/CD, ABC/DEF, or AB/CD:EF/GH, and so on. Thus symmetry always rests on two or multiples of two.

Centrality, on the other hand, requires the insertion of a third unit or group of units midway between the paired units or groups of units that by themselves would constitute either a simple or a multiple duality. The interposed unit or group of units must always be of the same order of magnitude as each of the units or groups of units among which it is placed. Examples would be such arrangements as A/*B*/C or A/B/*C*/D/E or AB/CD/*EF*/GH/IJ, and so on. It is self-evident that centrality always involves groupings based on three or other higher odd numbers.[15]

Symmetry and centrality may be defined still more narrowly as either linear or spatial. *Linear symmetry* means the equal bifurcation of an even

---

14. Fung (1952:69, 371–372, 338, 391).

15. Granet (1934:279) argues that *1* is not, in Chinese eyes, a number like other numbers, because by itself it is a symbol of Unity, Totality, Entirety. Added to even numerical groupings, it provides the vital pivot or nexus that changes symmetry into centrality, but it cannot by itself constitute a numerical category. Granet's interpretation has been supported, with interesting parallels from Aristotle and other Western thinkers, by Solomon (1954).

number of units arranged, as illustrated above under symmetry, in a single series (A/B or AB/CD, etc.). *Spatial symmetry* means their rearrangement two-dimensionally so that they occupy an area rather than a
C
line; an example is the group A B. Only linear symmetry is possible, of
D
course, with simple, two-unit duality, whereas spatial symmetry may occur as soon as four units or other multiples of two are employed.

Linear centrality is the same as linear symmetry except that the number of units in the series must be odd (as in the examples given above under centrality, e.g., A/*B*/C or A/B/*C*/D/E, etc.). So too, spatial centrality is the same as spatial symmetry save that among the grouped spatial units, one of them must always be placed at the center of all the others. The most common example (that of the Five Elements) is the
D
figure A B C.
E

Granet (1934) believes—and I am inclined to agree with him, although I cannot accept his mythological and societal explanations—that symmetry (meaning here primarily linear symmetry) tended to come first in Chinese thinking, followed by the more complex idea of centrality and especially of spatial centrality. Examples supporting this thesis will be presented later. Here it should be noted that long before Chinese literature recorded any of the numerical categories embodying the features we have been discussing, symmetry was already a graphically conspicuous feature in Chinese artifacts.

In the painted pottery of Neolithic China, for example, the funerary urns of the Pan Shan sequence are famed for the bold spiral designs that snake horizontally around the mantle of each vessel. Long ago it was noted that in very nearly every instance the spiral nuclei are four in number. Their arrangement is determined in the first place by the handles on opposite sides of the vessel dividing it into two general zones of decoration; these in turn are each broken down into two subzones by the demands of the decor, of which one spiral nucleus occupies each of the resulting quadrants.[16]

Equally conspicuous is the symmetry of the ogre or animal masks that appear on the sides of Shang and Chou bronze vessels, split there into mirror halves by a median vertical flange.[17] How persistent such sym-

---

16. Palmgren (1934:64) found only four exceptions to this arrangement: two urns bearing five spiral nuclei and two others with six.

17. The palaces, temples, tombs, and divination inscriptions of the Shang period were also characterized by a strong feeling for symmetry. See Chang (1977a:291).

metrical thinking has remained down to modern times is demonstrated in countless ways, not least by the rigidly symmetrical layout of the chairs, tables, vases, pictures, and the like in the living rooms of old-fashioned Chinese homes. As one enters such a room from the courtyard to its south, the paired balance of the eastern and western sides brings awareness of a median north-south axis invisibly bisecting the room into its symmetrical halves.

The supreme example of such symmetry, of course, is the system of the yin and yang and their correlates. Significantly, the original associations of the two, as we have seen, appear to have been meteorological (light and shadow, heat and cold) rather than the expected sexual analogy. Even later on, their primary focus probably remained more naturalistic than sociological, being expressed both spatially (sunny mountains and shadowy valleys, warm and cold regions, and so on) and temporally (alternation of day and night, sequence of the seasons, and so on). There can be little doubt that the ubiquity of the yin-yang system has strongly helped to perpetuate Chinese conceptions of "boxed" time and space.

Dual linear symmetry, when doubled, results in fourfold symmetries that include, spatially, those of the four compass points, the foursquare earth itself,[18] and the four seas believed to surround the earth. Linearly (which in this case means temporally, since time is one-dimensional), they include the four seasons. By far the most famous sociocosmic system based on the multiplication of even-numbered symmetry is that of the *Yi ching*'s eight trigrams, derived from the pairing of divided and solid lines and multiplied in turn into the sixty-four hexagrams.

One of the most significant examples of symmetry is that in which nature is regarded as a macrocosm and humanity as a microcosm; in this view, humanity is physiologically and psychologically a replica in miniature of the total nonhuman cosmos. The implications of such thinking for science will be discussed in Chapter VII.

The simplest road from symmetry to the more complex ideas embraced in centrality would seem to be that from duality to trinity. In the first place, 2 and 3 are the two lowest numbers for which symmetry and centrality are respectively possible. Secondly, the centrality found in 3 is linear only, not spatial, and hence less of a jump from linear symmetry in 2 than would be the full progression to spatial centrality. Prob-

---

18. The circularity of heaven is the counterpart of the squareness of earth and, like the latter, is oriented according to the compass points. Therefore its spatial symmetry is comparable to that of earth even though, being circular, it does not depend on even numbers.

ably these are the main reasons why 3, despite its lack of spatiality, is one of the most popular Chinese numerical categories.[19]

A good cosmological example of the progression from 2 to 3 is found in the concept of Heaven and Earth (a part of the yin-yang system) and its enlargement into the trinitarian concept of Heaven, Earth, and Man (*san ts'ai,* "the three powers"). Here, the added "power," Man, is intrinsically considerably less important than the earlier two, yet his median position *between* Heaven above and Earth below gives him a dynamic role in the structuring of the new concept. Graphically, too, the difference between dual linear symmetry and trinitarian linear centrality may be illustrated calligraphically as, for example, by the paired uprights in the character *k'an* 凵 (the tenth radical, picturing an open mouth and hence also an open box) as against the triple uprights in *shan* 山, "mountain." The latter's central mountain rises, as it should, above the lesser peaks to its right and left, and the effect, I venture to say, is aesthetically more pleasing than that of *k'an*'s unfilled paired uprights.[20]

The odd numbers that are really important for correlations, however, are of course those that permit spatial as well as linear centrality, that is, 5 (the Five Elements and their correlates) and 9 (spatially, as we shall see, an elaboration of the space pattern of 5). The number 7, on the other hand, contrary to its importance in the Near Eastern and European world, plays only a minor role in Chinese thinking. The reason, we may surmise, is that 7, unlike 5 and 9, can have only linear, never spatial, centrality.

We have seen that the step from linear symmetry based on 2 to linear centrality based on 3 is a natural one. Equally natural is that from fourfold spatial symmetry to fivefold spatial centrality (also portrayable as a simple cross) (see Figures 1a, b, c). In Figure 1b we should of course think of four of the elements as correlated with the four compass points,

---

19. Theoretically, a spatial (and not just linear) symmetry could be based on 3 in the form of an equilateral triangle, but in reality the triangle has played a very small part in Chinese art motifs and practically none in spatio-temporal thinking. The reason, surely, is that the triangle cannot be readily correlated with the compass points, which are so fundamental in Chinese thinking.

20. *K'an* is among the few radicals that occur only in combination with other elements, never alone. Such combination results in the filling of *k'an*'s empty space, for example in *hsiung* 凶, which, despite its unpleasant meaning (baneful), is visually more attractive than is *k'an* written alone. On the question of symmetry, centrality, and the like in Chinese characters, one should consult the treatises on calligraphy, such as Driscoll and Toda (1935), Chiang Yee (1954), and the like.

while the fifth and most important, soil, occupies the center. The concrete counterpart of Figure 1a would be any Chinese square-walled city. That of Figure 1b would be Peking (when it still retained its walls), climaxed, at its center, by the imperial walled palace enclosure known as the Forbidden City—a city within a city whose location at the center of the capital thereby places it symbolically at the very center of the civilized universe as well.[21]

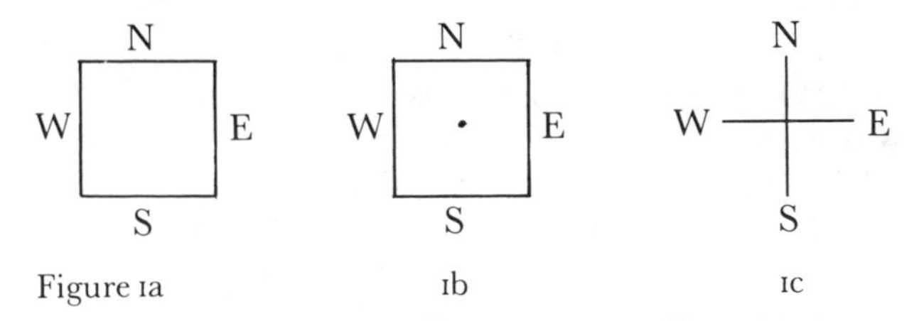

Figure 1a 1b 1c

With the number 5 we reach a vital question: why did the Chinese have five elements, the Indians and Greeks only four? Four would have fitted the compass points just as well as five fitted these points plus the center and would have saved the Chinese much trouble (see below) in trying to force the four seasons into the Procrustean bed of five elements. An easy answer—one suggested by Hsü Fu-kuan (1963:519)—is that 5 is a "natural" number because it matches the five digits of the hand and foot. However, the evidence below will show that this explanation, even if partially valid, can hardly be the whole story.

Mencius seemingly knows nothing of the Five Elements, and when he wants to group the major Confucian virtues together, he enumerates them as four instead of five. Every man, he maintains, possesses within himself four "beginnings" of moral behavior (we would call them instincts), which, if nurtured, will grow into the full-fledged virtues of *jen, yi, li,* and *chih* (conventionally translated as benevolence, righteousness, propriety, wisdom). As to *hsin,* "good faith," the fifth major virtue of later times, Mencius ignores it in his grouping, even though he refers to it separately elsewhere and it already appears often in the *Lun yü.* The reason for the omission apparently has little to do with ethical values but is caused by Mencius's eagerness to support his main argument with a physiological analogy. "Man," he declares, "has these four beginnings [of the four virtues] just as he has his four limbs *(ssu t'i).*"[22] It would

21. The idea of a cosmic or world pillar *(axis mundi)* at the very center of the world is an ancient one in Eurasia, including China. See Le Blanc (1985–1986:60) and references there given.

22. *Meng-tzu* 2a.6; tr. Legge (1960:2.203).

seem that if physiology had been the only or major factor determining the naturalistic thinkers in their choice of the number of elements, Mencius's four-limb analogy could have served the purpose quite as well as our own hypothetical five-finger analogy.

It appears from the textual references that not until the Han (no doubt under the influence of the five-element theories) did the Confucian virtues become collectively standardized as 5 (under the rubric of *wu ch'ang,* "the five constants," or *wu hsing,* "the five aspects of human nature"; this *hsing* character is different from that for *hsing* meaning "element").[23] When this happened, it is significant that the added virtue, *hsin,* was correlated with the element soil and therefore placed at the center.[24] In other words, the expansion from four virtues to five seems to reflect a growing preference for spatial centrality connected with 5.[25]

The same phenomenon, I believe, underlies the terms *ssu yüeh,* "four peaks," and *wu yüeh,* "five peaks," used in various texts to refer to China's so-called sacred mountains. The "four peaks" are associated with the compass points; the "five peaks" consist of the same four plus a fifth at the center. If so, this would be a further instance of a shift away from 4 toward 5 and the spatial centrality that goes with 5. Unfortunately, the dating of the relevant texts is so uncertain and involves such technical complexities that it is impractical to attempt a presentation here. Thus what has been suggested can only be offered as a conjecture, not as firm proof.

On the whole, nonetheless, the conclusion seems reasonable that the Chinese choice of five instead of four elements could very well have

---

23. Neither term occurs in late Chou texts, and the first clearly datable reference to the *wu ch'ang* (with the five virtues enumerated) appears in the memorial addressed by Tung Chung-shu to Emperor Wu about 136 B.C. See Tung's biography in *Ch'ien Han shu* 56/3a. See also n. 24 for a later reference.

24. The correlation of the virtues with the elements, though not greatly stressed by the Han naturalistic thinkers, appears, for example, in the *Po-hu t'ung-yi* (ca. A.D. 80) 30; tr. Tjan (1949–1952:567–568). In addition to *hsin* and soil at the center, the other correlations are *jen,* wood, east; *yi,* metal, west; *li,* fire, south; *chih,* water, north.

25. It is barely conceivable, though we have no way of knowing, that Mencius, following a kind of macrocosmic-microcosmic analogy, thought of the four human limbs (the arms and legs) as symbols of, or correlated with, the points of the compass. If such were so (and it is admittedly not too likely), the shift from four to five virtues would be even more clearly indicative of a corresponding move from spatial symmetry toward spatial centrality.

been influenced much more basically by a strong Chinese feeling for centrality than by the analogy of the five fingers or some other equally extraneous factor.[26]

The Chinese insistence on five elements despite the difficulty of making them fit the seasons is also additional evidence for the precedence of space over time in their thinking. In answer to the problem of the seasons, three solutions, all awkward, were propounded in Han and immediately pre-Han times:

1. Soil, as a central element, was in effect removed from the seasonal cycle by being said to operate equally through all the seasons without close association with any one of them.

2. Summer was divested of its final (the sixth) month, which was converted into an artificial "fifth" season over which soil was said to preside.

3. To lessen the manifest awkwardness of this truncation, the "fifth" season was reduced from a single month to the last eighteen days of the last summer month. This happened in A.D. 59, when imperial ceremonies were instituted for "welcoming the seasons" at altars built for that purpose around the periphery of the Later Han capital, Lo-yang. The solution involved the additional difficulty that the altar for the "fifth" season had to be built SSW of the capital, between the altar of fire/summer in the south and that of metal/autumn in the west. This position was chosen because it was the point on the arc of the seasons supposedly reached by the yin-yang forces just before the end of

---

26. If, however, as of course is possible, the five-finger analogy did constitute a stimulus, though a secondary one, the fact that the middle finger happens to be the longest and thus most prominent of the five (an example of linear symmetry) could have reinforced Chinese associations of five with centrality. There is a Greek parallel for this sort of thinking in the case of Philolaos of Tarentum (ca. 430 B.C.), who wanted five instead of four elements because he thought there ought to be some connection between the elements and the five known figures of solid geometry. See Needham (1956:245).

Major (1978:11–12) suggests that the Chinese choice of five instead of four elements (called by him "phases") is connected with the five visible planets, which he asserts (without providing textual evidence) were regarded as gods in China as early as the Spring and Autumn Period or even earlier. Though ingenious, I find this hypothesis unconvincing until accompanied by greater concrete evidence than provided in this article. Still another theory, advanced by Cammann (1961:47–48), explains the Chinese emphasis on 5 as deriving from the centrality of the numeral 5 in the Chinese magic square (to be discussed shortly below). Here again the difficulty is chronological: that of determining with any assurance whether or not the Chinese magic square is really older than the Five Elements.

summer in their annual round. Thus, for the special purposes of this cult, soil had to be placed in a *linear* centrality between the other four elements so that temporally it could stand at the junction between summer and autumn, the midpoint of the year. In order to do this, it had to surrender the *spatial* centrality it otherwise would have enjoyed, and which, if insisted on, would have obliged its cult altar to be built in the center of Lo-yang.[27]

Finally, we turn to the categories in 9. Spatially speaking, their basic pattern is simply an elaboration of the cruciform pattern for 5, with an added square at each of the corners between the latter's four arms. The result is a group of eight squares around a ninth, central square: ▦

The simplest concrete example is the *ching-t'ien* or "well-field" system of land tenure, alleged to have functioned during the early Chou. According to much later idealized accounts, each of the eight outer plots of land was occupied and cultivated by a separate peasant family for its own needs, while the ninth, central plot was communally cultivated by all eight families for the use of their feudal overlord. The system supposedly derives its name from the fact that the character for "well," *ching,* graphically portrays the pattern of nine fields. However, the word may also have been chosen to indicate that some kind of communal irrigation system was involved.

The same basic pattern crops up in the Chinese magic square, consisting, in simplest form, of three tiers of numbers,

4 9 2
3 5 7
8 1 6

always totaling 15 regardless of whether they are added horizontally, vertically, or diagonally.[28] Closely connected with the lore of the magic square was the Ming T'ang or Sacred Hall of the Han ritualists: a nine-chambered

27. An impossible location, because the sites for the cults of all natural forces were always outside the city. Of these three solutions, the first was propounded by Tung Chung-shu, *Ch'un-ch'iu fan-lu* 42 (11/4b). The second comes from the *Huai-nan-tzu* 5/8a. For both passages, see Fung (1952:165–166 n. 1 and 1953:21). For details of the third (cultic) solution, see Bodde (1975:192–200).

28. See Needham (1959:52–62) and the several studies by Cammann (1960, 1961, and 1963) tracing the evolution of the magic square in Han and post-Han times and its diffusion to India, the Islamic world, and Europe. See also Rickett (1985:153–158) and Major (1984:146–150). In the magic square, as in the numbers associated with the Five Elements, it is the "central" number of 5 that holds the key position, a fact emphasized by Cammann (1961–1962:47–48).

temple dwelling wherein the ruler, according to the season and month, supposedly performed rituals designed to facilitate the orderly transition of the months and to coordinate with them the human activities appropriate for each.[29]

Whereas the arrangement of the Ming T'ang, as of the Five Elements, was thus temporal as well as spatial, a purely spatial variant of the ninefold pattern is denoted by the term "nine heavens" *(chiu t'ien)*. The idea behind the term is that heaven is divided into nine expanses *(yeh)* or realms *(kai)* consisting, since heaven was believed to be circular, of a circular central realm surrounded by eight pielike sectors corresponding to the eight compass points.[30]

In all probability there is a correlation between these nine heavenly realms and the nine provinces *(chiu chou)* which, according to several retrospective accounts of much later time, constituted the Chinese *oikoumene* at the time of the Great Yü, conqueror of the flood. We are told, for example, that the provinces, like the nine heavenly realms, were oriented so that eight of them corresponded to the eight compass points, with the ninth (numerically the fifth or middle one) occupying the center of the group.[31] Other geographical features in nines that are attributed to Yü's age include the nine rivers, nine roads, nine marshes,

---

29. The subject is difficult because it is exceedingly hard to know the extent to which the elaborate lore found in Han ritualistic literature was really put into practice in the Ming T'ang that was built in A.D. 56 outside the Later Han capital. One thing of which we may be certain, however, is that the ideas of the Han ritualists definitely have nothing to do with the structure—whatever it may have been—known as the Ming T'ang in early Chou times. For specialized studies, ranging from the highly speculative to the relatively realistic, see Granet (1934:178–182, 250–273, 288ff.); Soothill (1951); Maspero (1948–1951). For the history and physical aspects of the hall of A.D. 56, see Bielenstein (1976:65–66). Henderson (1984:75–82) surveys the history of the Ming T'ang down to its possible brief revival in the mid-fifteenth century.

30. The conception is referred to in *Huai-nan-tzu* 2/11b, 3/4a, 12/22a. The same text (3/30a) also refers to another theory, according to which heaven consists of nine levels (*chiu ch'ung*; i.e., superimposed disks resembling a multiple umbrella). But although this conception appears already in the *T'ien wen* (Questions about the Heavens) poem of ca. 300 B.C. (tr. Hawkes 1959:47), it was apparently less widespread than the other and "had almost no echoes in later Chinese astronomical thought" (Needham 1959:199).

31. See *Huai-nan-tzu* 4/1b. The other earlier and somewhat less schematized accounts of the nine continents are: (1) part 1 of "The Tribute of Yü" *(Yü kung)* in *Shu ching* 3.1; tr. Legge (1960:3.92–127); (2) *Chou li* 33/6a–10a; tr. Biot (1841:2.263–279); (3) *Lü-shih ch'un-ch'iu* 62(13/1a–5a); tr. Wilhelm (1928:157–160).

and nine mountains; there were also, of course, the famed nine cauldrons.[32]

All this numerology in 9, like that of the Five Elements, probably derives much more from Tsou Yen than from any other single person. Such a conjunction of the two numerologies in one man is significant in view of the generic relationship we have noted between spatial centrality in 5 and the same in 9. The importance 9 had in Tsou's thinking is evident from the outline of his geographical ideas preserved in the *Shih chi* (74/1a–b; tr. Fung 1952:160–161). There it is said that China, according to Tsou Yen, was merely one among nine different continents, each surrounded by water; that these nine comprised one Great Continent (presumably also surrounded by water, although the account does not say this); and, finally, that there were nine such Great Continents, surrounded on their outer edges by a vast ocean separating the terrestrial earth from heaven. Thus China was but one part of the world out of eighty-one (a powerful intensification, 9 × 9, of the potency of 9).

These and other correlations in 9 are much more complex and interrelated than the foregoing somewhat bald enumeration might suggest. In an illuminating study, Major (1984) compares the several listings of nine provinces considered above with other ninefold correlations, some already mentioned, others not. The latter include the numerical configurations constituting the Lo Writing *(Lo shu)* and River Chart *(Ho t'u),* as well as the Han divination boards and gaming boards respectively known as *shih* and *liu po.* Major's conclusion is that most of these configurations were, to varying degrees, allusions to or symbols of a cosmography in which earth and surrounding space are centered on a fixed and unmoving pivot or axis, the *axis mundi* familiar to many cultures. Awareness of this cosmography enables the individual to achieve a correct ritual orientation toward the *axis mundi* with resulting personal benefit.[33]

Most striking among Chinese representations of this cosmography are the circular Han bronze mirrors known, because of their distinctive designs (resembling T's, L's, and V's), as TLV mirrors. Although the precise significance of the TLVs and other decorative details is uncertain, it seems very probable that, generally speaking, the large central boss of such mirrors represents the *axis mundi;* the square surrounding the boss represents the earth (possibly laid out in accordance with Tsou Yen's plan of 9 or 81 continents); and the concentric circular bands sur-

32. See *Shih chi* 2; tr. Chavannes (1895–1905:1.101); also for the cauldrons, Needham (1959:503–504).

33. See Major (1984:154) and, for a general discussion of the *axis mundi* concept in various forms, Eliade (1954).

rounding the square perhaps represent Tsou Yen's all-encompassing outer ocean or the circular expanse of heaven beyond this ocean or a combination of both.[34]

Why did the Chinese come to favor centrality over symmetry and spatial centrality over linear centrality? A major factor would seem to have been the strength of the *axis mundi* idea in the Chinese mind. This explanation, however, only raises the further question as to why the idea should have acquired such strength. To this I have no ready answer save the somewhat tautological one that it may possibly have been linked in some way to the exquisite Chinese awareness of absolute orientation already noted. One thinks in this connection of the early Chinese awareness of magnetism and their invention of the compass. To illustrate the importance of absolute orientation in China, it is worth noting that the honoring of left over right during most of Chinese history had nothing to do with these directions per se and everything to do with cosmology. Left was honored because the ruler was traditionally thought of as sitting on his throne at the center of the universe, his back to the cold north, his face to the warm south. His left hand was therefore directed toward the east, direction of the rising sun and of life, and his right hand directed toward the west, direction of the declining sun and of death.[35]

Let us note, finally, that despite the emphasis on space over time in Chinese thinking, it was always a two-dimensional and not three-dimensional space that commanded attention. One wonders if there is any connection here with the backwardness of geometry in Chinese mathematics.

It remains to consider what effects all this schematic thinking might have had on scientific development. Instances of harm are easy to find in East and West alike. One remembers, for example, the Greek attachment to circular movements—the multiple spheres of Eudoxus, the excentric circles of Hipparchus, the epicycles of Ptolemy—and the resulting intellectual straitjacket imposed, because of this insistence on seeming mathematical beauty, on men like Tycho, Copernicus, Galileo, and Kepler (Needham 1959:220, 399–400).

---

34. Theories by several scholars about the TLV mirrors are conveniently summarized in Loewe (1979:71–75), whose own discussion of the subject (1979:ch. 3) is of prime importance. The topic is continued in Major (1984), to whom I am very grateful for giving me a copy of his study at a time when it was not readily accessible to me.

35. Curiously, and for not entirely clear reasons, the supremacy of left over right was reversed during the Han dynasty, but thereafter was restored and maintained until modern times except for an interlude under the Mongols (the Yuan dynasty). See the survey by Demiéville (1973).

On the Chinese side, there is a much more trivial but revealing example: that of the twenty-four "fortnightly periods" found in the Chinese calendar. These solar-based periods, which go back at least to the late Chou and possibly much earlier, are diagramed in Table 1 in parallel columns to show their symmetry.

Table 1 **Twenty-four Fortnightly Periods**

Approximate Dates	Periods	Periods	Approximate Dates
22 Dec.	1a. *Winter Solstice*	1b. *Summer Solstice*	22 June
6 Jan.	2a. Lesser Cold	2b. Lesser Heat	8 July
21 Jan.	3a. Greater Cold	3b. Greater Heat	24 July
5 Feb.	4a. *Beginning of Spring*	4b. *Beginning of Autumn*	8 Aug.
20 Feb.	5a. The Rains	5b. End of Heat	24 Aug.
7 Mar.	6a. Awakening of Creatures (from hibernation)	6b. White Dews	8 Sept.
22 Mar.	7a. *Spring Equinox*	7b. *Autumn Equinox*	24 Sept.
6 Apr.	8a. Clear and Bright	8b. Cold Dews	9 Oct.
21 Apr.	9a. Grain Rain	9b. Descent of Hoarfrost	24 Oct.
6 May	10a. *Beginning of Summer*	10b. *Beginning of Winter*	8 Nov.
22 May	11a. Lesser Fullness (of grain)	11b. Lesser Snow	23 Nov.
7 June	12a. Grain in Ear	12b. Greater Snow	7 Dec.

The symmetry reveals how the listing probably evolved. First the year was split into halves based on the winter and summer solstices. These were then again halved in accordance with the spring and autumn equinoxes. The resulting quarters were then in turn divided into eighths by inserting the seasonal beginnings midway between the solstices and equinoxes. The final step was to insert two additional titles evenly in the spaces between each of the already created eight, thus resulting in twenty-four periods of approximately a fortnight each.

The major consequence of this repeated halving was to place the seasonal beginnings midway between solstices and equinoxes instead of, as in the West, six weeks later, at the equinoxes and solstices themselves. In the West, 5 February is still the height of winter, whereas in the Chinese calendar it marks the beginning of spring; 6 May is still spring, but in China it inaugurates summer, and so on. There is no question that the Western seasons are better dated as far as climatic reality is concerned. But to the Chinese, viewing the seasons in terms of a symmetry based on the waxing and waning of the yin and yang, it is inconceivable that the summer solstice could have come anywhere except at the exact middle of the summer season. Certainly this moment of annual culmination of the yang and rebirth of the yin could not possibly mark the beginning of summer.

The naming of some of the periods themselves also seems to have been more schematic than realistic. To be sure, Lesser and Greater Cold (nos. 2a and 3a) and Lesser and Greater Heat (nos. 2b and 3b) are quite suitable counterparts, both pairs coming immediately after the solstices. Other correspondences, however, appear arbitrary, notably the balancing of Lesser Fullness (of grain) and Grain in Ear (nos. 11a and 12a) against Lesser and Greater Snow (nos. 11b and 12b). The growth of grain during later May and early June is reasonable enough, but Lesser and Greater Snow are quite unlikely in the dry North China winters as early as late November and early December. It would seem that the snow designations have been inserted simply to achieve symmetry with their grain opposites, rather than for genuine meteorological reasons. These points, while individually trivial, add up to a distorted representation of natural phenomena, made in the interest of a schematic symmetry.[36]

How pervasive was the way of thinking that compartmentalized space and time and attached them, with things and ideas of all kinds, to numerical categories? "All-pervasive" would seem to be the reply, judging from the fact that practically no Chinese thinker—unless possibly he was a Buddhist monk or, much later, a Ming-Ch'ing scholar of "evidential research"—explicitly challenged the validity of the major cosmological and numerical systems from the Han dynasty until the present century. Even so, as we have seen (end of sect. 1), "evidentialist" criticisms tended to be particular rather than general.[37]

But perhaps this is not the right question, and what should really be asked is how much this sort of thinking permeated the minds not just of philosophically inclined men, but of all those others who were engaged in what today would be called science, proto-science, and technology. It can hardly be doubted that the influence was profound in fields like biology and biological technology, alchemy and chemistry, physics,

36. The above two paragraphs are based on Bodde (1975:29).

37. Even such an iconoclast as Wang Ch'ung, who attacked many magical notions that became attached to the naturalistic cosmology of his day (first century A.D.), accepted without question the basic principles of the cosmology per se (see n. 9 above). So also, and with less excuse, did most of the critically minded Ch'ing scholars sixteen and more centuries later, although sometimes, as we have seen (III.1, end), they were ready to question particular aspects of this cosmology.

meteorology, and the earth sciences; allegedly, it shaped the natural philosophy of "alchemists and acoustic experts, geomancers and pharmacists, smiths, weavers and master-craftsmen" (Needham 1965b:7).

At first sight it seems reasonable to group so many diverse professions under the umbrella of Chinese correlative thinking in view of the tremendous persistence and universality of such thinking. Further consideration, however, suggests that inclusion of the technologists ("smiths, weavers and mastercraftsmen") is probably unwarranted, because such inclusion blurs a very important distinction between pre-modern technology and pre-modern science/proto-science, namely, that science/proto-science tend to be heavily theoretical, technology to be heavily pragmatic. In other words, technicians tend to be practical people (many of them illiterate) who rely on everyday causal thinking rather than abstract theorizing to approach their problems.

Thus I find it quite unlikely that China's traditional technicians and artisans—the builders of walls, bridges, canals, or ships; the printers and the papermakers; and the many others who spanned the gap between the intellectual and the ordinary worker—would be overwhelmingly affected in their work by considerations of the Five Elements, the sixty-four hexagrams, the correlates of these and more, or the fact that the center was traditionally favored over all other directions in Chinese thinking.

Such a tendency toward particularized causal thinking has probably long separated much of technology generally from much of science and proto-science: "It may be safely assumed that technology has from the first depended on causal thinking, on discovering that when you do X the consequence is generally Y" (Graham 1986:7). Probably, too, this separation has characterized Chinese technology as much as that of the West: "In China concern for the practically useful stimulated causal thinking in technology as strongly as in the West, and contributed as much or more to material wellbeing until it was outstripped in the last few centuries" (Graham 1989:317).

3. Cyclical and Linear Time

The obvious interrelationship in Chinese eyes between segmented and cyclical time is not surprising in view of mankind's universal dependence on the movements of nature for its first temporal awareness.[38] "The Chinese conception of time is based on the cycle of natural change

38. Relevant to this subject are Sivin (1966, 1976, 1980) and Needham (1965b). Sivin (1966) surveys Chinese views of time for the nonspecialist, stressing the cyclical viewpoint. Sivin (1976 and 1980) are technical and discuss time with special refer

through the seasons, and on the regular motions of the celestial bodies" (Sivin 1966:82). Starting from the cycles of day, month, and year, the early Chinese astronomers came to formulate ever longer cycles into which the shorter cycles could be nested. In other words, each longer cycle represented, numerically speaking, the lowest common denominator into which the beginnings and ends of the several other shorter cycles could be precisely meshed. In this way the astronomers went on to a cycle of 19 years (the Greek Metonic cycle) and another of 76 years. About 5 B.C. their calculations culminated with the Great Year Cycle, which allegedly "began and ended time with a universal conjunction of sun, moon, and planets," and had a duration of 23,639,040 ordinary years.[39]

Aside from the astronomers, probably very few Chinese ever paid much attention to such enormous cycles. Nevertheless, the idea in general that time oscillates between two poles or moves in recurring cycles was widespread among early and later Chinese alike. "One yin [then] one yang: this is called the Tao." So states the *Yi ching* or *Changes Classic* (Appen. 3; tr. Legge 1899b:355; Wilhelm 1950:1.319). Or again concerning what happens when its sixty-four hexagrams reach their end: "Things cannot be used up. Therefore it [the 63rd hexagram] is followed by 'Unfinished' [*Wei chi,* the 64th], with which ends [one cycle only to be followed by another]" (Appen. 6; tr. Legge 1899b:438–439; Wilhelm 1950:2.367).

The Taoists—at least some of them—had similar thoughts. "Reversion is the action of the Tao," says *Lao-tzu* 40. And Chuang Tzu, in his famous enumeration of the successive stages of plant and animal life that culminate with human beings, concludes by saying: "Man then again reverts to the germs *(chi).* All creatures emerge from the germs. All return to the germs."[40]

On the Confucian side, Mencius believes that a sage can be expected to appear roughly every five hundred years.[41] Hsün Tzu, on the other

---

ence to Chinese alchemy. Needham (1965b), as might be expected, is the most detailed and comprehensive coverage of Chinese temporal thinking in any language. Here it will be impossible to consider all of its ramifications, some of which extend well beyond what one would ordinarily think of as concepts of time. Relevant to the remarks on Chinese historiography found in this section is the further discussion of moralism in Chinese historiography in VI.1b.

39. Sivin (1976:222) and, for more detail, Needham (1959:406–408).

40. *Chuang-tzu* 18; tr. Watson (1968:196) and Graham (1981:184). With an interpolation, the passage reappears in *Lieh-tzu* 1; tr. Graham (1960:21–22).

41. *Meng-tzu* 7b.38; also 2b.13. Needham (1959:408) points out that conjunctions of Jupiter, Saturn, and Mars recur every 516.33 years, which, he thinks, could be the basis for Mencius's belief.

hand, is virtually alone among Chinese thinkers in believing that time is static: "Past and present are the same. Things that are the same in kind, though extended over a long period, continue to have the self-same principles" (*Hsün-tzu* 5(3/59); tr. Dubs 1928:74). But Tung Chung-shu, when he synthesizes Confucianism with naturalistic thinking a century later, propounds a whole series of cycles. The smallest is a polarity (Simplicity and Refinement), followed by a trinity (Black, White, Red); then come cycles in four, five, and nine phases. All of these pertain only to human history, whereas Tung sees the yin-yang and Five Elements cycles that he adopts from the naturalist thinkers as functioning within both the natural and human worlds.[42]

In post-Han times, Buddhism brought to China a knowledge of the Indian *kalpas (chieh-po)* or world periods. The smallest of these, lasting 16,800,000 years, is evenly divided into 8,400,000 years of growth and another 8,400,000 years of decline. During the growth period, human life increases by one year per century to reach a maximum lifespan of 84,000 years, and human height increases one tenth of a foot per century to reach a maximum height of 8,400 feet. Then the period of decline reverses the process until humans have lifespans of only ten years and are but one foot tall. Twenty such small *kalpas*, each of 16,800,000 years, constitute a *kalpa* of 336,000,000 years, beyond which lies the *mahākalpa* or great *kalpa* of 1,334,000,000 years. Such a *mahākalpa* is divided into the four periods of formation, existence, destruction, and nonexistence, out of which a new *mahākalpa,* and with it a new cosmos, will emerge.[43]

From Buddhism, the revived school of Confucianism known in the West as Neo-Confucianism (ca. A.D. 1000 onward) took over the *kalpas* as an idea but reduced their length to more imaginable proportions. Shao Yung (1011–1077), the thinker most responsible for this step, used alternating multiples of 12 and 30 to formulate four nesting cycles with which, from the *Yi ching,* he coordinated twelve hexagrams. The result, comprising a single Sun Cycle, is as follows: 4,320 Zodiacal Generations (each 30 years long) are equivalent to 360 Star Revolutions (each 360 years long), which are equivalent to 12 Moon Epochs (each 10,800 years long), which in turn are equivalent to a single Sun Cycle (129,600 years long). During the first six Moon Epochs, the cosmos grows (Heaven

42. See Fung (1953:19–30, 58–69, and esp. 61) for a brief enumeration of the five historical cycles.

43. Soothill and Hodous (1937:232b); Ting Fu-pao (1970:1221); and Zürcher (1982:38–40) (discussion of the *kalpa* in relation to Chinese Buddhist messianic thinking). Besides these Buddhist *kalpas*, other *kalpa* systems are known in India itself. See Basham (1959:320–321).

comes into existence during the first Epoch, Earth during the second, mankind during the third). Then, during Moon Epochs 7–12, the cosmos declines and finally ends, together with the Sun Cycle to which it belongs. Thereafter, however, a new Sun Cycle, and with it a new cosmos, will appear and pass through the same phases. From Shao Yung, this cosmic system—totally unrelated to any actual revolutions of sun, moon, and stars—was taken over by other Neo-Confucians, notably their major thinker, Chu Hsi.[44]

Cyclical thinking, though of a more measured sort, also characterizes a great deal of Chinese historiography. It is particularly apparent in the twenty-five standard histories which—together with their Ch'ing sequel—provide an unbroken record of Chinese history from legendary beginnings down to the end of the empire in A.D. 1912. It is significant that of these twenty-six works, only five cover more than a single dynasty.[45] All the others are each restricted to a single dynasty and thus strongly reinforce a cardinal precept of Chinese historiography. This is, that the dynastic cycle is the primal movement in all history. All others, no matter how seemingly important, are subordinate to this basic ebb and flow. History, in other words, consists of a succession of dynasties, each rising and falling and punctuated at its end and beginning by briefer periods of interdynastic disorder.[46]

Among the standard histories, the most important exception to the one history, one dynasty principle is Ssu-ma Ch'ien's *Shih chi.* One would suppose that the enormous time span covered by this work—from legendary rulers of the third millennium B.C. through the Hsia, Shang, Chou, and Ch'in dynasties and so into the Han for another century—would provide Ssu-ma Ch'ien with ample opportunity to treat history linearly rather than cyclically. However, this does not happen. Instead, he uses his sources to emphasize what was to become a key element in the dynastic cycle syndrome: the belief that each dynasty comes into being because of a founder who is a sage or near sage, and that, after flourishing, it declines again because of the lessening virtue of its rulers and ends disastrously because of a degenerate or evil final ruler. As an astronomer royal himself and the son of an astronomer royal, it

44. See Fung (1953:469–474) and Birdwhistell (1989:133–144) for Shao Yung's cycles; also Fung (1953:546–547) for Chu Hsi's acceptance of them.

45. They are the *Shih chi* (Historical Records, trad. 2698–ca. 100 B.C.), *Nan shih* (Southern Histories, covering four short dynasties, A.D. 420–589), *Pei shih* (Northern Histories, covering four northern counterparts, 386–618), and *Wu-tai shih* and *Hsin Wu-tai shih* (History of the Five Epochs and New History of same, both covering the brief interregnum of 907–960).

46. On this very well-known theory, see, inter alia, Wright (1963) and Bodde (1963).

was axiomatic for Ssu-ma Ch'ien to remark: "Things that flourish then decay: such indeed is the course of transformation."[47]

Other than Ssu-ma Ch'ien's *Shih chi,* the major history to transcend dynastic divisions is that by another Ssu-ma: the *Tzu-chih t'ung-chien* (Comprehensive Mirror for Aid in Government), completed in 1084 by Ssu-ma Kuang. "This great work in 294 *chüan,* covering in chronicle form the whole history of the period 403 B.C. to A.D. 959, must rank along with Ssu-ma Ch'ien's Records of the Historian as one of the highest achievements of traditional Chinese history writing" (Pulleyblank 1961:152). And yet, within this enormous time span, Ssu-ma Kuang sees events as isolated particularities rather than interrelated elements belonging to an ongoing process:

> From our modern point of view one of the most serious limitations of Ssu-ma Kuang's method—an often-mentioned limitation of almost all traditional Chinese historians—was the restriction of attention to one isolated event at a time, with a certain amount of backward and forward glancing, generalizing about a man's character, etc., but without the attempt to see each event interwoven into a complex mesh of inter-relationships with other events. What resulted from the enormous labors of Ssu-ma Kuang and his team was a superb chronicle, but still a chronicle and not a history in our modern sense. (Pulleyblank 1961:158)

So much for cycles. Let us turn now to the linear part of our discussion. It will be remembered (sect. 2 above) that the later Mohists visualized time abstractly and questioned the fixed cyclical succession of the Five Elements. Conversely, we have seen that the Taoists sometimes viewed time cyclically. Being Taoists, however, they at other times could produce some magnificent utterances on time as an endless continuum. Here is Chuang Tzu's famed epitome of the endless regression of time and existence:

> There is a "beginning."
> There is a "not-yet-beginning to have a beginning."
> There is a "not-yet-beginning to have a not-yet-beginning to have a beginning."
> There is a "what-is."
> There is a "what-is-not."
> There is a "not-yet-beginning to have a what-is-not."
> There is a "not-yet-beginning to have a not-yet-beginning to have a what-is-not."
> Each moment: a "what-is," a "what-is-not"!

---

47. *Shih chi* 30/8b; tr. Chavannes (1895–1905:3/2.86); Watson (1961:2.82). For Ssu-ma Ch'ien's depiction of dynastic founders and last rulers, see Watson (1958:5–6). More will be said about the "bad-last ruler" complex at VI.1b.

> And I still do not know, concerning "what-is" and "what-is-not," which is "what-is" and which is "what-is-not."
>
> Now I have already referred to something. Yet I do not yet know whether there really *is* something to be referred to in what I have referred to.[48]

Confucian literature also has two well-known expositions of linear time. One is a passage in the *Li yün* (Revolutions of Ceremonial), constituting the seventh chapter in the *Li chi* (Record of Ceremonial). The other is by Ho Hsiu (129–182) and comes at the beginning of his commentary on the *Kung-yang chuan* or *Kung-yang Tradition,* itself a commentary on the *Ch'un ch'i* or *Spring and Autumn Annals.*[49]

Two successive epochs are the subject of the *Li yün* passage: (1) A very early epoch, called the Great Unity *(ta t'ung),* during which the great Tao was functioning, the world was common to all, love was not restricted to the family, everyone was cared for, and there were no robbers or other wrongdoers and (2) a later epoch of Small Tranquility *(hsiao k'ang),* extending from the founding of the Hsia to the founding of the Chou dynasty. By this time the great Tao had become obscured and people were beginning to center their attention on their own families, to build protective walls, and to engage in war. Nevertheless, civilization continued owing to the strength of the traditional mores of civilized society known as *li.* The passage ends at this point, leaving us with a view of history that is definitely linear, although downward rather than upward. There are reminders here of the Taoist Utopias that will be discussed shortly, as well as of the "sequence of the Tao" doctrine, to be presented in the fifth chapter.

The other passage, by Ho Hsiu, rests on the belief (now generally rejected) that Confucius wrote the *Ch'un ch'iu* chronicle with the moral purpose of expressing praise or blame for the events there recorded and that he expressed these judgments by the use of esoteric phraseology. According to Ho Hsiu, the *Ch'un ch'iu*'s entire span of 722–481 B.C. divides into three epochs: that of *shuai luan* or Disorder (722–627), about which Confucius learned through transmitted records; that of *sheng p'ing* or Approaching Peace (626–542), about which he learned through oral testimony; and that of *t'ai p'ing* or Grand Peace (541–481), which he per-

---

48. *Chuang-tzu* 2; tr. Watson (1968:43) and Graham (1981:55). The Chinese terms for "what-is" and "what-is-not" are *yu* and *wu,* often rendered as "existence/being" and "nonexistence/nonbeing." This passage deserves more philosophical analysis than can be given here. For a detailed discussion in which it is compared with a passage in *Huai-nan-tzu* 2/1a–b, see Le Blanc (1987).

49. For the *Li yün* passage, see tr. Legge (1885:27.364–367) and Couvreur (1913:1.497–500). For Ho Hsiu, see tr. Fung (1953:83–84). Both texts, together with other similar material, are translated and discussed by Needham (1965b:23–31).

sonally witnessed. The linearity here is evolutionary, in contrast to that of the *Li yün*. Both texts, despite this contradiction and the absence of historical reality for either of them, were to become influential in the millenarian thinking of later politico-religious revolutionary movements. In particular, some scholars see a possible connection between Ho Hsiu's *t'ai p'ing* or Grand Peace and the *T'ai-p'ing ching* or *Canon of Grand Peace* (a text allegedly going back to the first sectarian rebellion, that of the Yellow Turbans in A.D. 184). Dating problems, however, make this hypothesis doubtful.[50] But in any event, the term reached its final apotheosis as name of the Taiping Rebellion of 1850–1864, probably the world's bloodiest rebellion of all time before the present century.

Some three or four decades after the Taipings, interest in the ascending and descending ages of Ho Hsiu and the *Li yün* was revived by the political reformer K'ang Yu-wei (1858–1927), who combined them in his writings with Buddhist, Western, and other ideas. In particular, he took the term *ta t'ung* (Great Unity) from the *Li yün* to become the title of his own treatise on Utopia, the *Ta t'ung shu* or *Book of the Great Unity.*[51]

For me, however, the most interesting examples of Chinese linear thinking are those passages (more than a score in all) in which Chou and Han thinkers describe what they imagine human life to have been like during prehistoric and early historic times. To translate and discuss all these passages adequately would require a fair-sized monograph. Here only a summary can be attempted, illustrated by a couple of translated passages. The passages as a whole fall into two groups: (1) Taoist (two texts containing seven passages, as well as the *Li yün* passage, which has a Taoist flavor and possibly a Taoist origin);[52] (2) non-

---

50. On the rebellion, see IV.2, and on the *T'ai-p'ing ching,* see Kaltenmark 1979. Problems of dating make it uncertain how much of the text, if any, actually goes back to the Yellow Turbans. Pokora (1961:450–451) believes that the *Li yün* and Ho Hsiu passages were both connected with the *T'ai-p'ing ching* and that the latter supplied Ho Hsiu with the term *t'ai p'ing* rather than the other way around. Aside from the dating problem, however, there is the further difficulty that the term *t'ai p'ing* was already in use long before the Yellow Turbans. It appears, for example, about a century earlier in Wang Ch'ung's *Lun heng* 57 (tr. Forke 1962:2.192–200), where the term *t'ai p'ing* is mentioned some seventeen times.

51. See, inter alia, Fung (1953:679–691). K'ang's Utopia will reappear briefly very near the end of this chapter.

52. *Chuang-tzu* 9, 16, 29; tr. Watson (1968:105–106, 172–173, 327); also tr. Graham (1981: 204–206, 171–172, 237). *Huai-nan-tzu* 2/13b–15b, 8/1a–4a, 9/2a–3b, 13/1a–5a; tr. Morgan (1934:46–48, 80–83 [but no tr. for 9/2a–3b], 143–147). Also *Li yün* as per n. 49.

Taoist (five texts containing thirteen passages, all Legalist or semi-Legalist save those in the first and last texts).[53]

From the non-Taoist group, let us first take as an example the earliest passage, that in *Mo-tzu* 11 (with variants in 12 and 13). Here Mo Tzu (ca. 479–ca. 381) is represented as saying:

> Anciently, at the beginning of human life, when no heads of government[54] yet existed, the saying was: "[Different] men, different ideas." Thus for one man there was one idea, for two men two ideas, for ten men ten ideas—the more men, the more ideas. In this way men approved of their own ideas and condemned those of others, engaging one another in mutual condemnation. Within households, as a result, fathers, sons and brothers became hateful and estranged, unable to achieve mutual harmony. In the world outside, people used water, fire and poison to harm one another. Any extra strength for mutual aid remained unused and any surplus goods were left to rot instead of being used for mutual sharing. The Way (Tao) of goodness was kept hidden, so that it could not be used for mutual teaching. Thus the disorders of the world were like those of birds and beasts. Yet it was evident that all this disorder came only from not having heads of government. Therefore the worthiest and most able person in the world was selected to become Son of Heaven.

"Son of Heaven" was of course the traditional title of the Chinese ruler. Who selected the Son of Heaven is here unstated, leading some scholars to speculate that he was democratically chosen by the people. According to the *Mo-tzu*'s third version (ch. 13), however, the selecting was done by the divinity Heaven. The text goes on to say that because a single man, even if Son of Heaven, could not rule everything himself, successive similar selections were then made of other worthy and able persons, three of whom were appointed to be the Three High Lords of the Son of Heaven, others to be nobles holding kingdoms demarcated from the realm of the Son of Heaven, and still others to be officials under the Son of Heaven and these nobles. In short, a Chinese *oikoumene* was created not unlike what actually existed in Mo Tzu's day. As a final step, the Son of Heaven then issued a proclamation that began with the very important injunction: "On hearing good and evil, all shall report it

---

53. *Mo-tzu* 11–13; tr. Mei (1929:55–56, 59, 71). *Shang chün shu* 7, 18, 23; tr. Duyvendak (1928:225–226, 284–285, 314–315). *Han Fei-tzu* 47, 49; tr. Liao (1959:252, 275–276). *Kuan-tzu* 31; tr. Rickett (1985:412–413). Also *Kuan-tzu* 78(23/3a–b), 79(23/7a–b), 84(24/11b). *Lü-shih ch'un-ch'iu* 117(20/1a–2b); tr. Wilhelm (1928:346–348).

54. Emending "punishments and government" to "heads of government," the reading found below as well as in the parallel ch. 12.

to their superior. What the superior approves, all must approve. What the superior condemns, all must condemn."

This account, aside from the fact that Heaven selects the first ruler, is strongly reminiscent of Hobbes' *Leviathan* (1651). Its uncompromising insistence on political absolutism was soon to be taken over and propagated in more sophisticated form by the late Chou thinkers and statesmen known as Legalists. From them it later became an unstated but implicitly accepted element of Chinese imperial state dogma. Even following the end of the empire in 1912, its influence has continued within the successive regimes of republican China. Of course, as we shall see, what was absolutism in theory was often diluted in practice into something less, owing to political and social changes, the shifting personalities of rulers and officials, the ameliorating influence of Confucian humanism, and other factors.

Mo Tzu's exposition is "optimistic" in the sense that it places mankind on a path leading from savagery—the war of all against all—to the security imposed by an authoritarian government. But it is also gloomy (some would say realistic) in that this progress rests upon the willingness of those below to give unquestioning obedience to those above. The writings of the Legalists later on are similar in general direction but less blunt in their expression. Sometimes they even recognize that life may have been simpler and more pleasant in the good old days. This, however, was only because people were then fewer in number, so that there were enough of the necessities of life for everyone. Since that time, however, because of the sharpening struggle for existence caused by population growth, it has become necessary for the government to impose strict, and even harsh, controls on the people. Some of the Legalist passages also trace the development of early civilization in some detail, beginning with culture heroes like Fu Hsi (Subduer of Animals), who taught men hunting and cooking, and Shen Nung (the Divine Husbandman), who taught men agriculture; then proceeding to dynasties and rulers that are increasingly historical; and finally reaching the late Chou period of the writers themselves.

Very different in tone and approach are the Taoist passages (those listed in n. 52 above). The ninth chapter of *Chuang-tzu* contains a good example:

> During the Age of Highest Virtue, men's gait was slow and sauntering, their gaze relaxed and scattered. At this time mountains had no paths or trackings; lakes had no boats or bridges. . . . Birds and beasts formed their flocks and herds; grasses and trees stretched and grew. Thus birds and beasts could be led by cord or bridle; the nests of crows and magpies could be peered into by pulling them toward one. In this Age of Highest Virtue, people lived in sameness with the birds and beasts. . . . Who then knew anything about

> "gentleman" or "petty man"? To be in sameness and without knowledge: such is the Virtue that departs not. To be in sameness and without desire: this is called "Unwrought Simplicity."[55] . . .
>
> But then, when came the "sages," people scrambled to perform "benevolence"; they raced to perform "duty."[56] And so, for the first time, uncertainty entered the world. . . . Now to spoil Unwrought Simplicity in order to fashion tools: such was the crime of the artisans.[57] To smash the Tao and Virtue in order to fashion "benevolence" and "duty": such was the error of the "sages." . . . In the days of Ho Hsü [identity uncertain], people, when at home, did not know what they were doing; when they travelled, did not know where they were going. . . . But then, when came the "sages," came also the gyrations and bendings of the rites and music. . . . Thus for the first time people began to race after knowledge and fight endlessly for gain, unable to stop. This too was the error of the "sages."

The Taoist paradise here described reminds one irresistibly of "The Peaceable Kingdom," as painted many times by the American Quaker primitive painter A. L. Hicks (1780–1844). It also reminds us of the lines in Isaiah 11.6–9 that inspired Hicks: "The wolf also shall dwell with the lamb, and the leopard shall lie down with the kid; . . . and a little child shall lead them." The difference, of course, is that Isaiah's paradise is to be brought about by the Lord in the future, whereas the Taoist paradise belongs to the timeless and uncreated state of Nature known as the Tao; it would have continued indefinitely had not mankind destroyed it by creating its own morals and institutions. The Taoist view of history is thus "pessimistic" in that its linearity moves downward from paradise to corruption, yet this pessimism is softened by the ecstatic joy of the paradise itself. In some of the other Taoist accounts, considerably more space is given to the successive stages of man's decline.

Chinese thinkers in general tend to agree with the Taoists that the past was better than the present. They differ from the Taoists, however, in placing the beginning of decline only *after* the civilizing work of the early sage-kings had been completed, whereas for the Taoists, this "civilizing" work was itself the cause of the decline. One of the very few men to reject the doctrine entirely is Wang Ch'ung, who wrote an entire

---

55. Virtue (*te;* also sometimes translated as Power) is the attribute of the Tao, with which it is coupled in the title of Lao Tzu's *Tao te ching* (Classic of the Way and Its Virtue/Power). Unwrought Simplicity (*su p'u,* or, more usually, the single word *p'u,* as below) is the Taoist designation for man's state of primeval innocence before the corruption that arose because of the creation of human institutions.

56. These are the Confucian virtues of *jen* and *yi,* for which other renditions are also often used. The "sages" are, of course, sages of the Confucian type.

57. The artisans and their "crime" will be discussed further in Chap. VI.

chapter to prove that his own Han dynasty was superior to its predecessors (*Lun heng* 57; tr. Forke 1962:2.192–200). More will be said about the topic of decline in Chapter V.

In the Western world, too, both cyclical and linear thinking are widespread. During classical antiquity, "the theory of world-cycles was so widely current that it may almost be described as the orthodox theory of cosmic time among the Greeks, and it passed from them to the Romans. . . . The periodic theory . . . meant an endless monotonous iteration, which was singularly unlikely to stimulate speculative interest in the future" (Bury 1955:12). Later, however, Christianity contributed a more nearly linear view of history, leading from Creation to the life of Christ and from there to final Judgment. Nevertheless, the favorable potentialities of this linearity were largely stultified by the doctrine of Original Sin, according to which any amelioration of man's miserable condition could result only from divine intervention.[58]

Not until the sixteenth century, therefore, did one or two French historians, notably Jean Bodin (1530–1596), advance the radical idea that modernity was not necessarily inferior to antiquity and may in some respects even have surpassed antiquity. This bold hypothesis brought European historiography for the first time, in the words of Bury (1955:43), to the very "threshhold" of the idea of progress. Since then, of course, the idea has become a commonplace, yet even down to today, the old cyclical view has never disappeared. One has only to think of the cycles (perhaps better the spirals) of Giovanni Vico (1668–1744), the four independently rising and falling civilizations of Oswald Spengler (1880–1936), and the twenty-one challenging and responding civilizations of Arnold Toynbee (1889–1975).

Obviously, neither in China nor the West has cyclical or linear thinking held a monopoly. The question for both civilizations is not whether one or the other kind of time existed, but how much there was of each and when. For the West this is far easier to answer, because the subject has been better studied. For China, on the other hand, the data are more mixed and harder to read. Also, the problem has been made more difficult by the fact that few Chinese, either past or present, have thought about it or even been aware of its existence. This vagueness, in fact, has seemingly induced some uncertainty in Needham's own approach. Thus he writes near the beginning of his study: "In so far as the traditional natural philosophy [of China] was committed to thinking of time in separate compartments or boxes perhaps it was more difficult for a Galileo to arise who should uniformise time into an abstract geometrical co-ordinate, a continuous dimension amenable to mathe-

58. Bury (1955:20–22) and Needham (1965b:45–46).

matical handling" (Needham 1965b:8–9). Yet at the very end of the same paper (p. 52) he writes:

> The apocalyptic, almost the messianic, often the evolutionary and (in its own way) the progressive, certainly the temporally linear, these elements were always there, spontaneously and independently developing since the time of the Shang kingdom, and in spite of all that the Chinese found out or imagined about cycles, celestial or terrestrial, these were the elements that dominated the thought of the Confucian scholars and the Taoist peasant-farmers.

Regretfully, I cannot go along with this conclusion. Naturally, the evidence pro and con cannot be quantitatively weighed. Nonetheless, on the cyclical side, the evidence appears to me quite sufficient in quantity and clarity to justify the conclusion that, until quite recently, Chinese cyclical thinking was considerably more widespread and influential than was Chinese linear thinking. Conversely, evidence for the linear side seems to me harder to find, more scattered, and less convincing.

This conclusion, even if accepted by readers, leaves us with a final question: what relevance does all this have for science? Does it really matter very much, as far as science is concerned, whether most intellectuals think cyclically or linearly? Graham does not think it does. Here are his exact words:

> I must confess to a personal inability to understand why the Hindu is supposed to be paralyzed by the knowledge that no human achievement can outlast a kalpa of 4,000,000,000 years, while the Christian, cramped inside a time scheme of a few thousand years from Creation to Judgment, works hopefully at sciences that have nothing to do with his salvation in the knowledge that the Last Day has already dawned. (Graham 1973:59)

To this I can only reply that when, as in Buddhism, the *kalpa* cycles are so astronomically long as to be almost incomprehensible, it probably does not greatly matter. When, however, as in China, the compartmentalized units of time are much shorter and therefore more concrete, and are combined with compartmentalized units of space into patterns determined by direction, symmetry, centrality, and numerical correlations—the whole system adding up to a seemingly beautifully structured view of the universe that in actuality is completely at variance with the facts of nature—when all this is so, it is hard to believe that it does not severely impede the formulation of more realistic concepts of nature.

4. Quantification

The use of quantification—that is, the systematic search for measurable elements in phenomena and the application of mathematical methods to these quantitative regularities—basically distinguishes modern from pre-modern science. Aristotle's physics, for example, was only incidentally concerned with measurement, calculation, and quantitative prediction. Its primary aim was to establish "true" knowledge about the "natures" of things, based on a nonquantitative direct description of how these things were supposed to operate. The approach, in other words, was qualitative rather than quantitative. A new attitude began to emerge only in the twelfth and thirteenth centuries, when, for example, Robert Grosseteste of Lincoln (1168–1253) discussed the need to apply mathematical formulas to the study of optics; his ideas seem to have influenced Roger Bacon (1214–1292) and other thirteenth-century scientific workers. From about 1310 onward for the next three hundred years, nonetheless, no further innovations came from scholastic philosophy. Attempts to quantify natural phenomena were left to technicians and artisans, responding to the practical needs posed by gunnery, navigation, cartography, metallurgy, and other developing technologies. Finally, in the seventeenth century, there occurred the marriage between craft practice and scholarly theory that led to the Galilean experimental-mathematical method and thus the birth of modern science.[59]

On the Chinese side, can much of significance be said about quantification? Not a great deal, and yet the subject deserves comment as a sequel to the earlier discussion about the arranging of space, time, and things under numerical categories.

Remembering that China's classical thinkers belonged to the preimperial aristocratic age of "feudalism," it is striking how often they support their social and political theories with analogies drawn from the technology of presumably lowly artisans. In particular, they are fond of referring to the preciseness of measurement made possible by such instruments as the carpenter's marking-line, the water-level, the steelyard, and especially the carpenter's square and compass. The following are a few typical statements:

> The Will of Heaven is to me like the compass to the wheelwright or the square to the carpenter. The wheelwright and the carpenter take their compass and square to measure all the world's squares and circles, declaring those that fit to be correct and those that do not fit to be incorrect. (Mo Tzu)[60]

---

59. See the discussion by Needham (1959:150ff.) on mathematics and science in China and the West, as well as Crombie (1953 and 1955).

60. *Mo-tzu* 26; tr. Mei (1929:140); repeated almost unchanged in 27 and 28 (tr. Mei 1929:149, 156)). For similar references, see also 4 (tr. Mei 1929:13) and (for the Mohist *Canons*) tr. Graham (1978:316).

Just as the carpenter's compass and square are the highest perfection of squares and circles, so the sage is the highest perfection of the human relationships. (Mencius)[61]

When the marking-line is truly laid out, one cannot be deceived as to whether a thing is crooked or straight. When the steelyard is truly suspended, one cannot be cheated in weight. When the compass and square are truly applied, one cannot be mistaken as to squareness and roundness. And when the superior man *(chün tzu)* has investigated rightness of conduct *(li),* he cannot be deceived by what is false. (Hsün Tzu)[62]

Destroy and cut to pieces the arc and line, throw away the compass and square, shackle the fingers of the Artisan Ch'ui, and for the first time the people of the world will possess real skill. (Chuang Tzu)[63]

The former kings hung up balances with standard weights and fixed the lengths of the foot and the inch. Still today these are followed as models because the divisions are clear. . . . If one turns one's back on models and measures and relies on one's own private appraisal, there will always be lack of certainty. . . . This is why the former kings . . . set up models and made distinctions clear. *(Shang chün shu)*[64]

[To prevent wrongdoing,] models and distinctions,[65] calibrations and measurements, have been established. Once calibrations and measurements have been made true, there is no chance for a Po Yi [a paragon of virtue] to slip from what is right, or for a Robber Chih [a traditional exemplar of villainy] to do what is wrong. Once models and distinctions have been made clear, it becomes impossible for the talented to snatch from the stupid, the strong to aggress upon the weak, or the many to do violence to the few. (Han Fei Tzu)[66]

Even with a skilled hand, unless one follows the compass and square, it will be impossible to make squares and circles correctly. . . . Now the Tao

---

61. *Meng-tzu* 4a.2. See also 4a.1, 6a.20, 7b.5.

62. *Hsün-tzu* 19 (13/5b); tr. Dubs (1928:224–225). See also 1(1/1a) and 8(4/3b); tr. Dubs (1928:31, 97).

63. *Chuang-tzu* 10; tr. Watson (1968:111), Graham (1981:209). See also 8 and 9 (tr. Watson [1968:100, 104] and Graham [1981:201, 204]).

64. *Shang chün shu* 14; tr. Duyvendak (1928:262). See also 4 and 24 (tr. Duyvendak 1928:205, 318).

65. I insert "distinctions" *(fen)* after "models" to maintain stylistic balance with "calibrations and measurements" in the same sentence, as well as to agree with "models and distinctions" in the second sentence below.

66. *Han Fei-tzu* 26; tr. Liao (1939:267). See also 6, 14, and 27 (tr. Liao 1939:45, 129, 270).

bequeathed by the former kings likewise constitutes a compass and square for the world. (Tung Chung-shu)[67]

The feature common to these and other such quotations is their uniform concern with human affairs and their unrelatedness to natural phenomena. The only conspicuous exceptions for the classical period seem to be two passages—both representative of five-elements thinking—found in the *Huai-nan-tzu* and the *Ch'un-ch'iu fan-lu.*[68] In the *Huai-nan-tzu,* heaven is likened to the marking-line, earth to the water-level, spring to the compass, summer to the balance, autumn to the square, and winter to the steelyard. Functioning as these measuring instruments, heaven, earth, and the seasons impose their controls on the operations of the yin and yang, thereby causing these to progress harmoniously through the cycle of the year.[69] In the *Ch'un-ch'iu fan-lu,* the metaphor is less elaborate and the correlations—except for the first and last—are different from, though analogous to, those in the *Huai-nan-tzu.* Thus the element wood in the east (= spring) is said to grasp the compass and in this way engender things, fire in the south (= summer) grasps the square and causes things to grow, earth in the center (= end of summer) grasps the marking-line and patterns (i.e., matures) things, metal in the west (= autumn) grasps the balance and attacks (i.e., destroys) things, and water in the north (= winter) grasps the steelyard and stores things away.

In contrast to these naturalistic metaphors, the Legalists, more than any other philosophical school, were interested in precise mensuration as the means for establishing an efficient machinery of state. Their technical term for this conception is *shu,* literally "numbers," but more meaningfully translated as "statistics" or "statistical methods." Already in the oldest part of the *Shang chün shu* (Book of Lord Shang), there is a tendency to write quantitatively about such matters as units of reward and punishment, sizes of land area, population groups, and so on (Duyvendak 1928:96). A very mechanistic view of human nature accompanied this tendency, according to which "human conduct and emotions could be measured as quantitatively as a picul of salt or an ell of cloth" (Needham 1956:211). Recent research indicates that Shen Puhai (d. 337 B.C.), a Legalist whose writings are today lost save for fragmentary quotations, may have been particularly instrumental in pro-

67. *Ch'un-ch'iu fan-lu* 1(1/9b).

68. *Huai-nan-tzu* 5/20a–21a; tr. and discussed in Bodde (1979:147–150) and, with modifications, in Needham (1962:15–17). Also *Ch'un-ch'iu fan-lu* 58 (13/7b–9b).

69. This passage will be discussed again when we reach the topic of "laws of nature" in Chap. VII.

moting the use of quantitative techniques in government. "There is some indication," writes Creel,[70] "that Shen Pu-hai may have conceived of a government, consisting of interacting parts, as not wholly different from a machine. Certainly he used a great many mechanical similes. He said, for example, that 'the ruler must have discriminating methods and correct and definite principles, just as one suspends a weight and balance to weigh lightness and heaviness, in order to unify and organize his ministers.' "

Such thinking culminated in the reforms instituted when Ch'in created the first universal Chinese empire in 221 B.C. Included were the unification and standardization of weights and measures, of the gauge of chariot wheels, and of laws and orthography. Thus there can be no question that the consequences of Legalist quantitative thinking, both socially and politically, were enormous and lasting. The effect on the study of natural phenomena, however, was only marginal, owing to the indifference of the Legalists to most aspects of the natural world.[71]

Curiously, after the first century or more of the Han dynasty, literary references to the measuring instruments of the artisans seem to fall off sharply—a fact possibly indicative of a declining intellectual interest in the idea of quantification.[72] Another development seemingly pointing the same way is the later history of the words *kuei* and *chü,* "compass" and "square"—the two instruments mentioned far more often than any others in the early texts. In these texts, irrespective of whether the words occur separately or in tandem as *kuei-chü,* they virtually always signify the actual compasses and squares used to make accurate circles

---

70. Creel (1974a:126). See also Creel (1974b) for a book-length study of Shen Pu-hai and esp. pp. 55–56 for remarks on measurement.

71. Not all aspects, however, because, as shown by the laws of the Ch'in state excavated in 1975 from a tomb of 217 B.C., the Legalists were considerably interested in such pragmatic matters as agricultural production and conservation of natural resources. See Chap. VII as well as Bodde (1986:51–52); also Bodde (1986:56–60) for the Ch'in standardizations of 221 B.C.

72. This assertion is based on an examination of the indexes available for Han intellectual writings. On the one hand, in addition to the references to instruments in the *Huai-nan-tzu* and Tung Chung-shu (as cited above), four occur in Huan K'uan's *Yen t'ieh lun* (ca. 80 B.C.), at 3, 10, 20, 53 (beginning) and 56 (penultimate paragraph); tr. Gale (1967:23, 60, 77); no tr. available for 53 and 56. On the other hand, such references apparently scarcely occur at all thereafter in such works as Liu Hsiang's *Shuo yüan* and *Hsin hsü,* Wang Ch'ung's *Lun heng,* Pan Ku's *Po-hu t'ung-yi,* Hsün Yüeh's *Shen chien* and Ying Shao's *Feng-su t'ung-yi.* The only exception noted consists of two linked references in Wang Fu's *Ch'ien fu lun* 1/5b.

and rectangles.[73] Later, however (save in technical contexts), they almost disappear as separate words but assume new life in a very common compound, *kuei-chü*. In this compound they shed their original concrete meanings and become components of a new abstract designation for what is normative and therefore morally and traditionally proper. Thus, as defined in Mathews' *Chinese-English Dictionary*, *kuei-chü* means "custom, usage, rule, well behaved," and so on; in some contexts it covers much the same range of meaning as the ancient word *li* (ceremony, politeness, traditional mores, etc.).

No doubt this moralistic transformation of a term that under other circumstances might have been semantically useful for scientific quantitative thinking reflects Confucian influence. Just how and when the change occurred cannot be traced in detail here. Examples of the new meaning, however, occur in two adjoining sentences written by the Sung literary figure Huang Po-ssu (1079–1118).[74] In the first of them he says: "The sages of antiquity gave free rein to their minds without exceeding the proper norms *(kuei-chü)*." And a few lines later, concerning the eighth-century Wang Hsü, who was famed both for his calligraphy and the fact that he was often drunk when he produced it: "When I observe Hsü's calligraphy, I esteem the fact that despite his eccentricity, he unwittingly stayed within the proper norms *(kuei-chü)*."

Among the classical quotations cited earlier was Chuang Tzu's exhortation to "destroy and cut to pieces the arc and line, throw away the compass and square." This exemplifies the fact that the Taoists, who among the persons quoted were the only ones really interested in nature, were also, as might be expected, the ones opposed to quantification.[75]

---

73. I have encountered one possible exception, although I do not think it really is. It is the phrase *wan tuan kuei chü* found at the beginning of *Shih chi* 23; tr. Chavannes (1895–1905:3/2.202). In this phrase, I believe *kuei* and *chü* are not the usual names, "compass" and "square," but signify the configurations performed by these instruments: "the roundings *(kuei)* and squarings *(chü)* of the myriad roots [of action]." Chavannes, however—wrongly, in my opinion—translates *kuei chü* as "les règles" and then comments in n. 2: "Proprement: le compas et l'équerre; mais l'expression *koei-kiu* en est venue à ne plus signifier que les règles en générale." I believe this change happened much later than Chavannes assumes.

74. In his "Lun Chang Ch'ang-shih shu" (On the Calligraphy of Senior Secretary Chang), in his *Tung-kuan yü-lun* (Further Discussions in the Eastern Lodge), A/54a.

75. The reference here is of course to the "philosophical" Taoists, as represented in the writings attributed to Lao Tzu, Chuang Tzu, and Lieh Tzu. There were no doubt other men—the hygienists, the alchemists, the thaumaturgists, the "religious" Taoists, and others loosely referred to in later times as "Taoists"—who did not always have the same philosophical objection to quantification.

Until now, aside from bare mention of the quantitative standardizations carried out by the Ch'in in 221 B.C. we have focused on what might be called quantification as a philosophical ideal. However, another more empirical quantitative tradition has also long existed in China—one concerned in a very practical way with the precise measuring of physical objects and natural phenomena. Quite a few instances of this quantitative tradition are cited in Joseph Needham's opus. Here only three examples must suffice: "In no other part of the world [as in China] was the decimalisation of weights and measures so early and so consistent" (Needham 1962:43). Sliding calipers were known in China at least as early as A.D. 9, as proven by a dated bronze example looking remarkably like a modern wrench (minus the worm gear); one of its edges is graduated into six Chinese inches subdivided into tenths of inches. In Europe nothing similar was known before Leonardo da Vinci (Needham 1965a:61). Rain gauges are mentioned in a Chinese mathematical work of 1247, and apparently one such gauge was maintained at that period in each provincial capital, whereas in Europe the measuring of rainfall began only in 1639 (Needham 1959:471–472). It should be noted, however, that these and other examples of what might be called physical quantification all apparently pertain either to the Chinese artisanate or to those relatively few members of the official bureaucracy who served as professional astronomers, meteorologists, or technological specialists. Upon the literati as a whole the thinking underlying this tradition of physical quantification apparently made relatively little impression.

It is of course a truism that modern science everywhere uses instruments and machines that either require very exact measurement for their own manufacture or themselves provide very exact measurement for other purposes. Until recently in the West, it was not believed that instruments were known in antiquity comparable in precision to those of Renaissance times onward. Now, however, this view has had to be radically altered as the result of the brilliant reconstituting on paper of the calendar computer mechanism recovered in amorphous form from the wreck of a Greek ship of about 80 B.C. at the island of Antikythera, northwest of Crete.[76] The instrument is today recognized as a calendrical sun and moon computing mechanism, perhaps manually powered by a hand crank, whose dials were caused to indicate the daily, monthly, and yearly phases of sun and moon during the nineteen-year Metonic cycle. The movements were mediated by more than thirty large and

76. See Price (1974), who, with others, spent some twenty years, off and on, reconstituting what, when it was first recovered from its long immersion, consisted of only a few lumps of crushed and highly corroded metal.

small gear wheels, each with teeth ranging in number from approximately 15 to approximately 225. Most spectacular among the technological features was a very sophisticated differential gear assembly used to take the difference between two rotations.

Was there anything of comparable mechanical precision and complexity in China? One thinks of Su Sung's much later (A.D. 1090) great astronomical clock, a tower thirty or forty feet high, whose three great gear wheels, each with six hundred teeth, were controlled in their movements by a water-powered escapement mechanism, permitting successive "ticks," each having a duration (at least in theory) of 24 seconds or 1/3600th of a day.[77] Generally speaking, however, it would seem that exactitude of measurement was less often an absolute requisite for the instruments and machines of China than for those of Europe. It is perhaps indicative of the differences between the two civilizations that two of the most important Hellenistic devices requiring great quantitative precision—the continuously winding screw thread, male and female (as in bolt and nut), and the cylindrical worm for engaging a gear wheel—are precisely the ones that were apparently unknown in China until the seventeenth century (Needham 1965a:119).

On the Chinese side, this lack was surely in no way due to any absence of technical skill. On the contrary, the almost incredible manual virtuosity displayed by the Chinese in the creation of art and ceremonial objects, ranging from Shang bronze casting to the eighteenth and nineteenth century carving of movable ivory balls, one within another, guarantees that any instrument or machine for which a high degree of quantitative precision was needed *could* have been built if only the motivation had been there. That it was not points to differing intellectual values and economic circumstances, not to any manual deficiency.

Besides quantification as an ideal and the quantification of physical things and natural forces, the Chinese have displayed a very real interest in what might be called sociological quantification: the use of quantitative techniques in such fields as demography, political institutions, historiography, and human behavioral evaluation. This sociopolitical interest no doubt stems quite naturally from their Confucian moral background coupled with their Legalist bureaucratic tradition. Among the several examples that follow, some will probably seem admirable, others perhaps somewhat bizarre.

In the field of population statistics, the Chinese have been pioneers. Their first national census took place in A.D. 2, which coincides closely with the Augustine census of the Roman Empire of 5 B.C. On the Chi-

---

77. Needham (1965a:446–465 and, for the escapement, esp. 461–462).

nese side, the census of A.D. 2 was followed by many others through the centuries, whereas in the West the first census of a modern type was conducted in French Canada in 1665, followed by that of Sweden in 1749 (Bielenstein 1947:125). Yet admirable though Chinese census taking was as an idea, its execution was often marred by simple arithmetical errors in the addition of lesser demographic figures to make overall totals. Thus Bielenstein, by taking the trouble to add for himself the population figures, has been able to demonstrate that, for the census of A.D. 2, the officially recorded total of 59.6 million persons for all China is an arithmetical error for 57.7 million; for the census of 140, the official total of 49.2 million is an error for 48 million; and for the census of 742, the official total of 48.9 million is an error for 51.5 million.[78] This failure of execution to match conception is a phenomenon often apparent in the executing of Chinese political and economic policy.[79]

Turning to historiography, an admirable feature is the precise dating of events that is so characteristic of virtually all Chinese historical writing. Usually it is carried out so meticulously that it specifies not only the year but often the exact month and day as well. In this respect Chinese historiography is unrivaled by that of any other civilization for such a long span. A complementary feature is the similar insistence on geographical exactitude. Among the thousands of biographies in the Chinese dynastic histories, for example, care is always taken to indicate the precise place of which the person being written about is a native.

On the other hand, conventions can sometimes mar the value of these otherwise admirable techniques. Because, for example, what goes into a man's biography is largely the facts of his public career, it often happens that details of his private life—among them those particularly significant to the modern historian—are routinely omitted, including not infrequently his date of birth (because this predates his public career). Or again, the strength of family tradition triumphs over historical exactitude in those biographies in which the alleged native place of the person described turns out, on closer inspection, to be merely the locality to which his family originally belonged, whereas he himself may have been born—sometimes, to be sure, rather fortuitously—in quite a different place.

---

78. Bielenstein (1947:128 and 161 [Plate VI]). The difficulties often encountered in Chinese texts when reading numbers and statistics, including those arising from simple arithmetical errors, are well discussed in Yang Lien-sheng (1949).

79. Repeated examples are cited in Ray Huang (1981), which is an analytical study of happenings in various circles of society during a particular year (1587) of the Ming dynasty. The book is important for its broader implications concerning the functional efficiency of the Chinese system of government.

Conventions based on schematic symmetry, in which everything must fit everything else, can also, within certain recognized limits, distort the accuracy of historical chronology. Most notable is the convention that every reign and dynasty must begin and end precisely at the beginning and end of the lunar year. This means, for example, that although an emperor dies in, say, the third month, and although his heir then immediately succeeds him, the name of the existing reign period, and with it the reign itself, continues officially until the end of that calendar year. Only then does the promulgation of a new reign period officially inaugurate the new reign. This practice can easily lead to contradictions between the conventional dates of reigns and dynasties and the dates of specific events occurring near their ends or beginnings.

Comparable chronological difficulties may also arise owing to the principle of political legitimacy connected with the Mandate of Heaven *(t'ien ming)*. According to this principle, the Mandate can never, even briefly, be either interrupted or simultaneously held by two political entities. This imposes on historians the obligation of treating their material in such a way that no chronological gap exists between the official demise of one dynasty and the assumption of legitimacy by the next. On the other hand, to avoid any chronological overlap between the two, a historian must select a particular event officially marking the end of legitimacy for the one and its beginning for the other.[80]

Many examples of the quantitative approach to human activities can be found in government institutions, notably in the penal codes of successive dynasties. These are remarkable for their extraordinary detail and precision of prescribed penalties, designed to cover every variety of crime under every possible circumstance, and involving every kind of social or family status. We have seen that the Legalists were keenly interested in "statistics" as an adjunct of government. However, they also regarded law as a universal standard, to be applied equally to all. This attitude was at variance with the Confucian concern for class and family status and special circumstance, as based on the flexible *li* (traditional mores). Hence it would seem that the great particularism characteristic of the surviving law codes (the earliest complete code is that of T'ang of A.D. 653) is a marked departure from, and in fact a "Confucianization" of, the law as originally conceived by the Legalists. So numerous and minutely graduated are the specified penalties in the codes that their comparative study, dynasty by dynasty, supplies valuable indications of the slowly shifting configurations of Confucian morality, as officially defined.

---

80. The topics of these two paragraphs have been discussed in Yang Lien-sheng (1954:1–5), Bodde (1963:80), and especially Van der Sprenkel (1960:410–420).

Nevertheless, the graduated progression is sometimes more apparent on paper than in reality. As examples, let us consider the two lowest groups of punishments, those of beating with the light and heavy bamboo (five degrees of each), as listed in the Ch'ing Code of 1740. In Table 2, the "nominal number of blows" is what the code formally lists, as a heritage from earlier codes. The "actual number of blows" is the substitute number, specified in another section, which replaced the nominal number in the seventeenth century to compensate for the larger sized bamboo staves then introduced in place of the wooden sticks used under previous dynasties. The archaism of the system is shown by the fact that despite the actuality of the new numbers, all sentences throughout the Ch'ing continued to be expressed in terms of the traditional figures.

Table 2 **Punishment by Beating with the Light and Heavy Bamboo**

I. Light Bamboo: 5 degrees			II. Heavy Bamboo: 5 degrees		
	Nominal number of blows	Actual number of blows		Nominal number of blows	Actual number of blows
1.	10	4	1.	60	20
2.	20	5	2.	70	25
3.	30	10	3.	80	30
4.	40	15	4.	90	35
5.	50	20	5.	100	40

The progression governing the actual number of blows is irregular for the first and second degrees of the light bamboo, as well as different from the progression of the nominal numbers. More serious, however, is the wide gap in both sequences between the five degrees of the light bamboo and the five of the heavy. On paper the progression of nominal blows moves smoothly from the fifth degree of the light bamboo (fifty blows) to the first degree of the heavy (sixty blows). For the person experiencing the blows, however, there is obviously a huge jump from fifty light blows to sixty heavy ones. Although the gap is lessened in the sequence of actual blows, still the difference, physically speaking, between twenty light blows and twenty heavy ones remains considerable. Similar gaps (or in some cases overlaps) mark the successive transitions between the higher groups of punishment: those of penal servitude (five degrees plus three supplemental degrees), life exile (three degrees plus one preliminary degree), military exile (five degrees plus one supplemental degree), and the death penalty (two degrees plus one supplemental degree).[81]

Another striking example of institutional quantification comes from the T'ang dynasty. During its first century (before 739), if we are to believe what is recorded about the civil service system, officials both within and without the capital were annually graded by their superiors for their previous year's work by being placed in one or another of the nine ranks (another example of linear centrality; see sect. 2 above). The grading was conducted on the basis of four specified "merits" or virtues *(shan)* and twenty-seven specified "perfections" or professional skills *(tsui);* the latter differed for different kinds of officials. The four merits are described as: (1) "His virtue and his righteousness are renowned." (2) "His integrity and his circumspection are brilliantly displayed." (3) "His equity and his impartiality are worthy of being praised." (4) "His diligence and his activity never relax." Among the perfections, the first, applicable to ministers close to the emperor, reads: "He proposes what is proper, discards what is improper, repairs omissions, and fills up lacunae." The twenty-seventh "perfection," applicable to members of the Board of Astronomy, reads: "In calculating the movements of the stars and the calendar, he has observed the rules with precision and care."

To be classed in the highest rank, an official had to have all four "merits" as well as one of the "perfections." The second rank required either all four "merits" or three of the "merits" plus one "perfection." For the third rank, either three "merits" or two of the "merits" plus one "perfection" were required. The fifth rank was made up of officials possessing either a single "merit" or a single "perfection." Sixth-rank officials possessed neither one "merit" nor one "perfection." And thus the grading descended to the lowest rank of "those who have committed a proven fraud." Officials classed in the first four ranks were given a 25 percent salary increase for every rank above the fifth; those in the fifth rank were neither promoted nor demoted; those below the fifth suffered a 25 percent salary cut for every rank below, save that the bottom-rank officials were discharged from the service.

As one reads this amazing schematization, one admires its moral purpose but also wonders whether, practically speaking, it was ever carried out or was simply the product on paper of some imaginative bureaucrat. Striking is the absence of any statement as to *how* the possession or nonpossession of "merits" and "perfections" was to be determined.

---

81. On the "Confucianization" of penal law, see Ch'ü T'ung-tsu (1961:267–279) and Bodde and Morris (1967:27–38). The latter, in ch. 3, also discusses the Ch'ing system of punishments, including bambooing.

Actually, what we have here is a good example of the T'ang government's attempt to institutionalize what had long been a tradition of real or imaginary political schematization. This tradition had its origins in the highly schematized bureaucratic Utopia described in the *Chou li* or *Institutes of Chou* (a classic purporting to describe the government of early Chou but compiled only in late Chou or early Han times, and probably largely imaginary). The tendency to organize government on the basis of abstract schema rather than concrete reality has been a persistent one in Chinese imperial history. Recently it has been strongly criticized as having constituted a major impediment to the transforming of China into a modern state.[82]

The final example of sociological measurement is a monumental attempt to make a moral quantification of history. It is entitled "Table of Ancient and Modern Men," and constitutes the twentieth chapter of the *Ch'ien Han shu* (History of the Former Han Dynasty) (ca. A.D. 100). Here the historian, once more using a nine-part scale, grades and lists in chronological sequence the names of 1,955 persons belonging to history and legend, beginning with the culture hero Fu Hsi (trad. 2852–2738) and continuing to the end of the Ch'in dynasty (207 B.C.). The gradations are based entirely on moral considerations and range from "sage" at the top to morally "stupid person" at the bottom. Again nothing is said as to how the judgments were determined. According to modern count, there are 14 "sages" and 134 "stupid men" in the table, and the statistical moral average of more than 2,500 years of legend/history is 5.35, or just a trifle above the absolute mean of 5.[83]

With this we take our leave of Chinese efforts at sociological quantification. The last several examples, despite their seeming commitment to the principle of exact numerical measurement, are in actual fact as arbitrarily schematic, and as indicative of the basic Chinese desire for ready-made order and symmetry, as are the packaged numerical categories of Section 2. These seeming efforts at quantification, then, really have very little to do with any kind of science, whether natural or social.

Unfortunately for scientific progress, and despite the undoubted achievements of Neo-Confucian organic thinking, as well as the con-

---

82. See Ray Huang (1986:89) for the T'ang institutionalization and passim for the criticism. On the just described T'ang system for grading officials, see Des Rotours (1932:50–55), basing himself on the *T'ang liu tien* (Six Groups of Statutes of the T'ang Dynasty; comp. 739), 2/45b–48a.

83. The table is analyzed in Bodde (1939), which also discusses other examples of Chinese "categorical" thinking, including that of the just described T'ang civil service grading.

trarily founded empirical achievements of the Ming-Ch'ing "School of Evidential Research," the schematic kind of thinking discussed both here and earlier has remained influential down to the early years of the present century. It appears conspicuously, for example, in the writing of the famed political reformer K'ang Yu-wei (1858–1927) when, near the beginning of his remarkable Utopia, the *Ta t'ung shu* (Book of the Great Unity), he itemizes seven sets of numbers ranging from 5 to 9. They include the seven physiological kinds of human suffering, the eight kinds of suffering caused by natural disasters, the five kinds based on human relationships, the seven kinds associated with human institutions, the six kinds derived from the emotions, the five things that cause suffering because of the esteem in which they are held, and finally, the nine divisions or distinctions (those of nation, class, race, etc.) whose elimination, he declares, is needed for the eradication of suffering as a whole.[84]

That such numerological thinking is still far from dead in China is evidenced by its repeated appearance, long after K'ang Yu-wei's death, in the policies, pronouncements, and campaigns of the Marxist-Leninist People's Republic of China. Since 1949, to cite only a few examples, Mao Tse-tung and others have launched: (1) the 1950 "Five Excesses" campaign (against too many assignments, meetings, documents, etc., for lower-level Communist cadres); (2) the 1951 "Three Anti" and "Five Anti" campaigns (respectively against bureaucratic corruption/inefficiency and the bourgeoisie); (3) the 1963 "Four Cleans" investigation (to determine how Communist cadres keep accounts, distribute supplies, etc.); (4) the 1966 "Five Reds" criteria (by which the Red Guards of the Cultural Revolution were to be selected from worker, peasant, or other properly "red" families); (5) the 1975 "Four Modernizations" (of agriculture, industry, national defense, and science/technology). Other similar examples could readily be cited.[85]

We have seen that an empirical interest in the quantitative measuring of physical objects and natural phenomena long existed in China (as exemplified by the early use of sliding calipers, rain gauges, etc.). But we have also seen that this interest was narrowly limited in its social spread, and in particular, that it exerted little influence on the thinking

---

84. Note the recurrence of the number 9 at the climax of K'ang's enumeration. For a translation of this passage, see Fung (1953:687–689). Also partly translated, partly paraphrased, in Thompson (1958:73–75).

85. To modern China specialists these are of course well known, but the ones listed here come from Meisner (1977) and Harding (1981). I am indebted to Professors Adele Rickett (University of Maryland) and W. A. Rickett (University of Pennsylvania) for leading me to these books.

of the ordinary Chinese literatus. Aside from this, literary and sociological approaches to the idea of quantification certainly existed, but it must be recognized that they were usually far removed in spirit from the modern scientific ideal of quantification. If anything, they must have hindered, rather than promoted, the cause of science in China.

IV. THE ROLE OF RELIGION

1. Institutional and Diffused Religion

Religion in China is particularly difficult for the average Westerner to visualize because so much of it differs so sharply from the biblical religions with which he or she is familiar.[1] The first difficulty is the absence in Chinese of any real linguistic equivalent for the word "religion." Etymologically, that word probably derives from Latin *religare,* "to bind"; *The Shorter Oxford Dictionary* accordingly defines it as an "obligation (as of an oath)" or as a "bond between man and the gods," and so, more extendedly, as man's "reverence for the gods." The strong suggestion is that mankind is linked to the divine powers in a contractual relationship in which it holds a subordinate position.

None of these overtones attaches to *chiao* 教, "teaching," the Chinese word most commonly adduced as the Sinic counterpart of "religion." In the right context, to be sure, *chiao* can signify a body of doctrine and practice that, because it relates to supramundane forces, would quite properly be referred to by a Westerner as a religious cult or system. But the same word is also used quite often in a purely secular sense to designate an inculcated doctrine or a system of teaching—one that is often, but not invariably, ethical in content.[2] Hu Shih has written as follows about some of the word's implications:

> The Chinese word for "religion" is *chiao* which means teaching or a system of teaching. To teach people to believe in a particular deity is a *chiao;* but to teach them how to behave toward other men is also a *chiao.* The ancients did say that "the sages founded religions *(chiao)* on the way of the gods."[3] But it is not always necessary to make use of such supernatural expedients. And the Chinese people make no distinction between the theistic religions and the purely moral teachings of their sages. Therefore, the term *chiao* is applied to Buddhism, Taoism, Mohammedanism, Christianity, as well as Confucianism. They are all systems of moral teaching. Teaching a moral life is the

---

1. The primary focus of this chapter is upon the institutional aspects of Chinese religion.

2. At II.5 we encountered Confucius's famous mention of *chiao* in his ambiguous statement: "Have teaching *(chiao),* not-have category."

3. Hu Shih fails to indicate the source of this quotation, which sounds more like a paraphrase than an actual quotation.

> essential thing; and "the ways of the gods" are merely one of the possible ways of sanctioning that teaching. That is in substance the Chinese conception of religion. (Hu Shih 1934:79)

Even less suitable as equivalents of "religion," because more specialized in meaning, are several other Chinese terms that in certain contexts bear religious connotations. They include *tao,* a "way" (of life or of belief); *men,* a "gate" (leading to enlightenment and salvation); and *tsung,* a "clan" descended from a common ancestor, and hence a "group," "class," "school," or "sect." The meaning common to all of them in a religious context is that of a religious sect.[4]

In view of these semantic uncertainties, it is not surprising that scholars often disagree over what constitutes a religion in China. Least doubtful is Chinese Buddhism, for here is a tangibly structured organization possessing a clergy, ritual, dogma, and pantheon; it is significant that this most evident of Chinese religions is the one that came to China from the outside.

There is likewise little difficulty in admitting "religious Taoism" to the religious circle so long as the term is confined to certain institutionalized aspects of Taoism: the existence of monasteries and convents housing monks and nuns, of temples enshrining divinities, of a theocratic literature, and of a ritual performed by clerics. Unfortunately, the term *religious Taoism* is often stretched to cover considerably more. The real difficulty lies in how to categorize the many scattered cults and practices—dietary and sex hygiene, physical therapy, alchemy, divination techniques, shamanistic practices, and much more—that all too often are loosely termed "religious Taoism" or simply "Taoism" despite their heterogeneous origins and, at most, tangential relationship to Taoism as a theocratic institution. When, in this chapter, reference is made to "religious Taoism" (or sometimes, for the sake of convenience, simply to "Taoism"), only Taoism in its institutional form, as described above, will be intended.[5]

Most controversial of all is whether Confucianism—the third of the three *chiao* or "teachings"—should properly be regarded as a religion. In his study of Chinese religion and society, C. K. Yang (1961) has devoted a lengthy chapter to what he terms the religious aspects of Con-

---

4. The words are discussed in C. K. Yang (1961:2), whose definitions, however, I do not follow in every instance. Nor do I adduce here the modern Chinese compound term for "religion," which is *tsung chiao* (the teaching of a sect or group), precisely because it is a modern neologism (probably borrowed from Japan).

5. The confusions resulting from loose use of the word "Taoism" are forcefully discussed in Sivin (1978).

fucianism.[6] Among them he includes a belief in heaven and fate, as expressed by Confucius, Mencius, and many later Confucians; the cult of the ancestors, which long predated Confucianism but became a prominent feature of the latter's family-centered ethic; and acceptance by Confucians of divination, geomancy, and other magical practices. I would seriously question whether several of the latter—geomancy, for example—should properly be called Confucian at all. Rather, they are elements in the religious substratum common to all sectors of Chinese society, available for Confucian-minded persons, like anyone else, to accept or reject.[7]

More relevant, therefore, to the question of whether or not Confucianism is a religion are the religious functions performed by a government that, during most of its history, regarded itself as the upholder of Confucian values. These functions included not only the making of periodic offerings to the memory of Confucius himself, but also ceremonies addressed to a variety of divinities, including Heaven (restricted to the emperor), gods of the soil (both national and local), local city gods, as well as occasional ceremonies for obtaining rain, warding off disease, and achieving much else. The civil authorities who performed these ceremonies as part of their official duties were, in my opinion, no more to be therefore considered religious personages than would be a ship captain who, in the line of duty, might have to conduct a funeral service because someone had died on board the ship. In short, I am glad to agree with the overwhelming opinion of modern Chinese scholars who, while recognizing the presence of religious features in state Confucianism, do not on that ground regard it as itself a religious system.[8]

The real problem in the analysis of Chinese religions, therefore, is less that of how to classify the three *chiao* than of what to say about the many scattered beliefs and practices that, though obviously concerned with the supernatural, do not readily fit into any of the three major categories. The best solution, I believe, has been formulated by Yang, who

---

6. Yang (1961:ch. 10). See also Shryock (1932) for a historical survey of the development of Confucianism as a state cult.

7. This fact seems to be recognized by Yang himself when he writes (1961:276): "The Confucians did not constitute a group separate from the general current of religious life of traditional Chinese society. They shared with the rest of the population a basic system of religious belief in Heaven, fate, and other supernatural concepts."

8. Yang (1961:26) acknowledges as much when he says: "This present study lays emphasis on theistic religion, with the supernatural as a prominent feature. Viewed in this light, Confucianism is not treated here as a full-fledged religion in the theistic sense, but as a sociopolitical doctrine having religious qualities."

classifies Chinese religious phenomena as falling under what he terms either *institutional religion* or *diffused religion:*[9]

> Institutional religion [he writes] . . . is . . . a system of religious life having (1) an independent theology or cosmic interpretation of the universe and human events, (2) an independent form of worship consisting of symbols (gods, spirits, and their images) and rituals, and (3) an independent organization of personnel to facilitate the interpretation of theological views and to pursue cultic worship. With separate concept, ritual, and structure, a religion assumes the nature of a separate social institution, and hence its designation as an institutional religion. On the other hand, diffused religion is conceived of as a religion having its theology, cultus, and personnel so intimately diffused into one or more secular social institutions that they become a part of the concept, rituals, and structure of the latter, thus having no significant independent existence. . . . Institutional religion functions independently as a separate system, while diffused religion functions as a part of the secular social institutions. (Yang 1961:294–295)

The distinction between institutional and diffused religion came about only with the arrival of Buddhism and the beginning of religious Taoism, both in the first century A.D. What Yang calls the earlier "classical religion" was an amorphous assortment of beliefs and practices in which, however, four basic elements may be identified: ancestor worship, the worship of Heaven and other lesser naturalistic deities, divination, and sacrifice. "There was no choice in religious beliefs, but neither did it occur to the common man to make any other choice. Religious values were embodied in the traditional moral order, and religion was an integral part of communal existence." With the advent of Buddhism and religious Taoism, however, a radically new ingredient was added to Chinese religious experience: "For the first time in Chinese history, membership in a consciously organized religion was based not upon one's inherited affiliation with a community, but upon conversion—voluntary choice by the individual believer."[10]

The institutional religions of imperial times included, besides Buddhism and religious Taoism, many smaller and often fleeting groups and sects—some of them primarily Buddhist, some primarily Taoist, but more often syncretic mixtures of both. These, because they often combined religion with political activities that might be directly antidynastic, suffered persistent proscription by the government, often forcing them to lead underground or half-hidden existences. Their tenden-

9. Yang (1961) passim, but esp. ch. 12, "Diffused and Institutional Religion in Chinese Society."

10. Yang (1961:106, 108 and—for the quotations—111).

cies toward rebellion were frequently coupled with messianic and millenarian beliefs.[11]

Besides the foregoing, there was a third and least well organized sector of Chinese institutional religion, consisting of practices perpetuated from the old classical religion by diviners, geomancers, sorcerers, and other religious agents (Yang 1961:301). It is the acceptance of these poorly organized and often peripatetic practitioners as nevertheless forming part of institutional religion that may seem to some the most questionable feature of Yang's classification. His major argument is that, despite their organizational weakness, they constituted a professional (or at least semiprofessional) class of men who were especially trained for, and made a living by, their religious activities.

Under diffused religion, Yang lists a variety of cults whose common feature is the absence of professional practitioners or of an organizational structure separate from that of the social groups within which they function. They include the cult of ancestors and house gods within the family; the cult of patron gods and deified founders of trades and occupations within the guild; the cult of gods of the soil, rivers, mountains, the city, and other localized divinities within the community; and, on the national level, that of more universal divinities, either naturalistic or deified human beings. Within each category the cult is conducted primarily or entirely by nonprofessionals: the head of the family or clan, the leading officers of the guild, the community leaders. Finally, on the national level, "the performance of cultic rituals was a part of the administrative routine of the governing officials and the local gentry. The emperor and his officialdom constituted the priesthood for ethico-political worship, and the entire population of the empire was theoretically the grand congregation" (Yang 1961:297–298).

How to explain the absence of institutional religion in pre-Buddhist China? No doubt many factors are responsible, but one of the most important, I believe, was the particular emphasis placed since earliest times on ancestor worship:

> The dominance of this [ancestral] cult probably goes far to explain one of the most striking differences between early China and many other civilizations: the absence in the former of a universal church or a significant priesthood.

---

11. For eschatological and messianic thinking in Chinese Buddhism of the fifth and sixth centuries, see Zürcher (1981). See also Naquin (1976) for the much later political millenarian movement known as the Eight Trigrams, which after some three hundred years of sporadic and localized religious activity in North China burst into serious rebellion in 1813. Although the rebellion was bloodily suppressed three months after its outbreak, its forces succeeded at an earlier moment in briefly fighting their way into the Forbidden City in Peking itself.

> For this there is a two-fold explanation. In the first place, the ancestral sacrifices of each clan were necessarily offered only to its clan ancestors, not to those of any other clan. Secondly, these sacrifices, in order to be effective, had to be performed by the clan members in person, not by priestly proxies. As a result, the ancestral cult was inevitably divisive rather than unifying in its effects. It could not readily develop into a national religion with a powerful organized priesthood. (Bodde 1954:47)

Possibly this statement is overdrawn for the earliest historical period (the Shang), when the diviners of the royal family constituted an influential priestlike group; yet even then the kings themselves seem to have spent much time taking part in religious activities. Under the following Chou dynasty the prestige of diviners, wizards, exorcists, and other specialists in the supernatural greatly declined, and during the imperial age their status was far below that of the (usually Confucian) gentry.

The fact that diffused religion was all-pervasive in China inevitably meant that the role of the institutional religions was correspondingly lessened, as compared with their institutional counterparts in many other societies. How feeble Buddhism and Taoism had in fact become by the early decades of the present century is graphically illustrated by Yang in successive pages of statistical data.[12] These include various pre-1949 estimates, according to which the number of Buddhist and Taoist clerics for entire China was somewhere between half a million and one million persons, as compared with a total population of 582 million (1953 census). In other words, the clergy constituted somewhere between 0.09 and 0.18 percent of the population as a whole.[13]

On the local level, in Ting-hsien, Hopei, in 1927, there were only 39 Buddhist priests for a county population of 408,000, or less than 1 priest for every 10,000 persons. Here and elsewhere, most rural temples and shrines had no priests in attendance. In Kalgan-hsien, Chahar, for example, only 4 or 5 temples out of 570 had attendant priests. Regional samplings suggest that approximately half of all Buddhist temples were

---

12. See Yang (1961:307–339). Also the confirmatory remarks scattered in Chan (1953: 54, 145–146).

13. Welch (1967:414) offers slightly more favorable statistics for Buddhism: according to data gathered in 1928–1935, monks numbered 513,000 or 2.3 per thousand of the total male population, and nuns numbered 225,000 or 1.2 per thousand of the total female population. Welch fails to combine the male and female figures into averaged statistics for both sexes. However, the difference between his highest figure of 2.3 per thousand and Yang's lowest figure of 0.09 per hundred (i.e., 0.9 per thousand), while mathematically significant, is not overly great in terms of absolute numbers. Moreover, it must be remembered that both figures can at best be only crude estimates.

no higher in economic level than "poor family" peasants in the same communities (defined as peasants owning less than 20 *mou* or approximately 3½ acres of land). This means that the clergy hardly ever engaged in philanthropic or educational work and that in general they were looked down upon socially. For Taoism the situation was, if anything, even worse.

This sad condition of the institutional religions—the culmination of a decline that had begun some thousand years earlier—contrasts poignantly with their epoch of glory, lasting some six centuries from roughly the fall of Han (A.D. 220) to the major T'ang religious persecution of 845. The material magnificence of Buddhism at the Northern Wei capital of Lo-yang early in the sixth century is glowingly described by Yang Hsüan-chih in his *Lo-yang ch'ieh-lan chi* or *Record of Buddhist Monasteries in Lo-yang* (A.D. 547). Calculations based on this text show that the capital, with a population of 500,000 to 600,000, had no fewer than 1,367 Buddhist monasteries and temples large and small; in other words, that there was something like one such religious institution for every 366 to 439 inhabitants.[14] During this and the following centuries, Chinese Buddhism was a major economic as well as spiritual power. The commercial and industrial activities of some of its temples and monasteries included the operating of water-powered stone rolling mills and oil presses, as well as hostels, pawnshops, and large landed estates. At the same time, Buddhist welfare activities included the maintenance of hospitals, feeding stations for the hungry, havens for the aged and decrepit, bathhouses, resthouses along the routes to famous shrines, road building, bridge construction, well digging, and tree planting.[15]

As against this impressionistic picture of power and prosperity, it is well to keep certain other facts in mind. Figures recorded at various times during the T'ang dynasty make it possible, in an exceedingly rough way, to establish the following ratios of Buddhist monks and nuns to the total population: in 624 the number was around 200,000, or a ratio of 0.35 clergy for every 100 of the total population; at the beginning of the eighth century the clerical population had increased to between 250,000 and 300,000, raising the ratio to 0.58; about 845, when Buddhism reached its peak, the number of clerics had increased to 360,000, bringing the ratio for the first time to slightly over 1 percent (Gernet 1956:11).

Comparable figures for Christianity in Europe seem extraordinarily difficult to come by, no doubt owing to the absence there until the last

14. See T'ang Yung-t'ung (1938:513) and Ch'en (1964:163).

15. These and other activities, both economic and philanthropic, are examined in detail in Gernet (1956) and summarized in Ch'en (1964:261–273, 295–296).

two centuries of anything approaching the Chinese bureaucratic tradition, with its interest in population statistics. Nevertheless, according to one estimate (Moorman 1945:53), there were in thirteenth-century England about 60,000 ordained clergy out of a total population of perhaps three million persons, giving the clerics a ratio of roughly 2 percent of the population. This means, if the figures on both sides are at least roughly accurate, that Buddhism at the height of its power in ninth-century China had proportionately little more than half as many clergy as had the Christian church in thirteenth-century England.

A still more significant difference is Chinese Buddhism's lack of organizational centrality as compared with its conspicuous presence in Western Christianity—a fact that makes it misleading to refer to Chinese Buddhism as a "church." As remarked by the late Arthur Wright: "Toynbee has observed that the Mahayana is a politically incompetent religion, and we should say that its record in China bears this out. . . . It was prevented by its basic postulate of the delusive and transitory character of earthly existence from developing a comprehensive political theory."[16]

What is worse is that whatever centralization did exist in Chinese Buddhism was imposed upon it by the government rather than self-created. The history of Chinese Buddhism and, *pari passu,* of religious Taoism and the syncretic sects, is, to an important extent, the history of the efforts of the state to control and regularize them. In these efforts the state was very largely successful.

Sometimes the efforts went beyond ordinary legislative controls and culminated in active religious persecution. Such persecution might include confiscation of temples or other religious property, defrocking of the clergy, or even executions (although the latter were exceedingly rare, as we shall see in the next section). Sometimes, too, the sectarian groups, against which state prohibitions were particularly severe, might rise in bloody rebellion; this was especially common during the later empire (the Ming and Ch'ing dynasties). For the most part, however, the institutional religions accepted the governmental controls without resistance. Nor did any of them—not even Buddhism—have sufficient power or inclination to impose controls of their own upon the lay population at large. Excommunication, trials for heresy, and similar features so familiar in Western church history were unknown in China, where persecution always emanated from the state, not from a religious body. It sometimes happened, however, that governmental persecution was induced by one religious group intriguing against another at court;

16. Wright (1959:106). See, however, the remarks at the end of this chapter about Buddhism in Tibet, Mongolia, and Japan, which seem to throw some doubt on the universality of this assertion.

Taoists, for example, were involved in some of the major persecutions of Buddhism.

The broad topic of state authoritarianism will be discussed in the next chapter. In the case of Buddhism, the imposition of governmental controls began already during its period of active growth and continued all the way down to the end of the empire. The first clear-cut step was taken under the Northen Wei (the Toba Tatars), when, probably in A.D. 396, the emperor appointed a Buddhist monk to head a new office established by the government for the purpose of exercising control over the entire Buddhist clergy. Below this central office, branch offices were similarly established in other parts of the empire, each headed by a monk appointed by the Chief of Monks in the capital. In this way Buddhism became a part of the bureaucratic state organization, and the head monk in the capital, despite his Buddhist status, was obliged, as a member of the bureaucracy, to express his reverence to the emperor as the secular head of state.[17]

Similar measures were continued under later dynasties. Under the T'ang the chief innovation was complete secularization of the controls system by replacing its Buddhist appointees with ordinary non-Buddhist civil servants. During this and later dynasties, the state also periodically acted to limit the number of clergy whenever it thought them to be too numerous. During the K'ai-yüan period (713–741), for example, 12,000 clergy were initially forced to return to lay life, allegedly followed by another 30,000 later on. In 755, to raise money to combat the An Lushan rebellion, the government began to sell monk certificates to anyone desiring to become a cleric. The practice, though later halted, was renewed on a large scale under the Sung from 1067 onward, when it became a serious abuse. Still later, under the Ming and Ch'ing dynasties, it was forbidden by law to build a new Buddhist or Taoist temple or monastery without permission from the provincial governor. Individuals were forbidden to enter the Buddhist or Taoist orders if below the age of sixteen or if they belonged to a family having fewer than three sons. Buddhist and Taoist monks, for their part, had to be forty or above before they could accept novices for religious training.[18]

---

17. Ch'en (1964:146). The controversy over paying reverence to the secular ruler was a sharp one at this time, on which, for a while, the Buddhists fared better in the South than the North. In 340 and 403, for example, the Chin government twice attempted to gain Buddhist acceptance of the idea of paying reverence, but the Buddhists, for the time being, successfully resisted these efforts. See Ch'en (1964: 76–77) and, for more extended accounts, Zürcher (1959:106ff., 231ff.).

18. For the T'ang and Sung regulations, see Ch'en (1964:243, 253–257, 391–393). For those of Ming and Ch'ing, see De Groot (1903–1904:ch. 3, esp. 97–99); also, more briefly, Bodde and Morris (1967:188, 367–368).

In view of these and other limitations, how important as a dynamic spiritual force has religion—whether institutional or diffused—really been in Chinese life? C. K. Yang, in the first chapter of his study (Yang 1961:6–20), devotes considerable space to refuting the views of scholars who have tried to minimize the role of Chinese religion. Unfortunately for Yang's thesis, the evidence he later adduces to demonstrate the disintegration of institutional religion in recent times (see above) seriously weakens his argument. As he himself indicates elsewhere in his book (Yang 1961:285, 298–299), Chinese diffused religion, by its very nature, could exert little power apart from the secular institutions in which it was diffused. Its main function was to provide supernatural support for the ethical values already present in these secular institutions, rather than to serve in itself as a source of such values. Thus its role in Chinese society was deeply conservative: that of peripherally supporting already well established social institutions. The contrast is sharp with the institutional religions, which often required from their adherents a major break with the past and avowed commitment to radically new ways of life.

This distinction makes it evident that any intellectually innovative movements in Chinese religion could come only from the institutional faiths, not from the diffused religion. The decline of the faiths from the Sung onward, therefore, means a corresponding general decline in religious intellectual innovativeness. Furthermore, the fact that the state significantly curbed the institutional religions even during their earlier centuries of growth and maturity means that already then their ability to propagate their ideas and values—especially if these were socially and politically controversial—was correspondingly diminished. This situation differs profoundly from the history of Christianity in the Western world, where the church sometimes dominated the state, and even when it did not, always remained a major influence in social life in its own right. Only quite recently in certain countries has the control of state over church reached an extent comparable to the traditional situation in China.

Thus while we can fully agree with Yang's assertion that religion was a "pervading influence" in pre-modern Chinese life,[19] it by no means follows that its spiritual and material influence—except possibly during certain medieval centuries and then only to a more limited degree—was in any way comparable to the influence wielded by Christianity in Europe century after century. The true role of Chinese religion is better evaluated by Yang himself when he writes:

---

19. Yang (161:340). One must add the proviso, however, that this pervasiveness tended to be much greater among the lower and usually less well educated strata of the population than among the scholar-official elite, for whom the Confucian ethic was usually paramount.

> The role of religion [in China] in the enforcement of moral values was limited to psychological encouragement and deterrence. The weak structural position of the priesthood and religious organizations in Chinese society precluded the development of organized religious authority to enforce moral discipline among the secular population. Since religion was not the chief source of secular ethical values, religious bodies could not be the authoritative groups to pass judgment on what was right and wrong. . . . The authority for such judgment was the jealously guarded prerogative of the government and the Confucian scholar-officials. (Yang 1961:289)

2. Religion and Science

For a long time it was fashionable to regard science and religion as irreconcilably opposed. Proponents of this view have produced impressive tomes detailing the controversies about Galileo, Darwin, miracles, and countless other points of difference.[20] Following this line of thought, therefore, it could be argued that the relative weakness of religion in China as compared with the West, and the fact that Chinese moral behavior—especially upper-class morality—has been dominated by the rationalistic and sometimes agnostic code of Confucianism, have been assets rather than liabilities as far as scientific development is concerned. A further consideration, already stressed by Needham (1956: 417), is that Buddhism, the most influential of the Chinese institutional religions, was also doctrinally least favorable to scientific development, inasmuch as its doctrine of *māyā* (the illusoriness of the visible universe) inevitably discouraged serious scientific study of such a universe.

On these two counts, therefore, one would expect that the decline of Buddhism after the ninth century and its replacement in intellectual importance by Neo-Confucianism should have given impetus to scientific development—a thesis seemingly confirmed by the flourishing of science during the Sung dynasty. Thereafter, however, not only did Buddhism continue to decline, but, unfortunately for the thesis, Chinese scientific impetus weakened as well. This simultaneous decline suggests that (1) Buddhism was never a very important factor one way or the other or (2) after it declined to the point where it could no longer influence scientific development one way or the other, its once strongly inhibitory capacity was replaced by some other agent, equally potent, or (3) as seems most probable, that Buddhism was only one within a large complex of factors, so that no clear correlation can be established between its own rise and fall and that of Chinese science.

---

20. See especially Draper (1874) and White (1896).

Commenting on the broader topic of the rise of modern industrial and commercial society in the West, C. K. Yang expresses the opinion that had other circumstances been favorable, Chinese religion per se would not have hindered a similar development in China. "Chinese theistic religion," he writes (Yang 1961:80), "traditionally placed no discernable constraint on large-scale economic enterprise, for the religious values stressed the supernatural sanctioning of reliability and justice in contractual economic relations. Such values were essentially universalistic in nature." He goes on to emphasize the importance of traditional institutions as against traditional values in determining the direction of China's socioeconomic development:

> The ability of traditional-minded Chinese to develop modern commercial and industrial enterprises under another type of social structure in Southeast Asia and Latin America during the nineteenth and twentieth centuries indicates that the answer to the question of China's failure to develop an industrial type of socioeconomic order lies not only in the characteristics of her value system but also in the structural pattern of traditional Chinese society, which favored the sociopolitical dominance of the literary class over the merchant. (Yang 1961:80)

Beyond the negative conclusion that Chinese religion did not inhibit the rise of modern science, however, there looms a further question presentable in positive terms: Is it possible that Chinese religious values, far from being inhibitory, were, considered in themselves, positively favorable to the rise of a modern society in China? Rephrased and put into a Western context, what is essentially the same question has received a great deal of scholarly attention. "Religion," in Western terms, means of course Christianity. More narrowly, within the context of the question itself, it means Protestantism as against Catholicism, and above all, Calvinism and Puritanism as against Catholicism. The debate was started by Max Weber's famous essay of 1904–1905 on Protestantism and the Protestant ethic (Weber 1930). His presentation, which he himself considered preliminary and tentative, gave rise to a lively and continuing scholarly literature.[21]

Without going into technical details, the controversy may be said in a very general way to have revolved around two points:

> 1. Whether certain attitudes and values—including individualism, self-disciplined ascetism, dedication to work, denigration of poverty as indicative of

---

21. One of the best known works in English is Tawney (1926), but there have been responses by many other scholars of various countries. A convenient anthology of excerpts from Weber himself and nine of his supporters, critics, and modifiers has been assembled in Green (1959).

moral deficiency, distrust of public welfare, and others—were in fact peculiarly characteristic of seventeenth-century Protestantism and especially Puritanism, or were also current in the Catholicism of that age.

2. Whether these attitudes and values were in fact responsible for the rapid rise of capitalism, especially in the seventeenth century, or were simply "accommodations" on the part of religion, and especially of Protestantism, to the burgeoning of capitalism at that time and the secular values it brought with it.

From capitalism some scholars have directed the discussion more narrowly to science, where again a variety of opinion has been generated.[22] Generally speaking, however, it would seem that the argument has followed much the same line of cleavage as that for capitalism. Robert K. Merton, who started much of the debate in 1938 with the several chapters on Puritanism in his study of science and society in seventeenth-century England, once more raises the question in his new Preface to the 1970 reprint of his work (Merton 1970:xxviii): "To what extent did the old Puritans . . . turn their attention to science . . . because this interest was generated by their ethos, and to what extent was it rather the other way, with those having entered upon a career in science . . . subsequently finding the values of Puritanism congenial to them?" The answer, he finds, is no clearer in 1970 than it was in 1938, because "the needed data are simply not at hand." Of any society, he continues, it may commonly be said that "the relations between institutional values, interests and affiliations are typically those of reciprocal interplay." Applying this dictum to the specific case of the interplay between Puritanism and science, one can only conjecture that "both processes were at work to unknown extent."

In a book appearing two years after these words were written, Hooykaas (1972) has broadened the discussion to include not only seventeenth-century Puritanism but the entire panorama of religious and philosophical experience in the West, starting with the ancient Greeks but concentrating mainly on the Judeo-Christian tradition in both its pre- and post-Reformation phases. His conclusion is more positive than Merton's:

> Science is more a consequence than a cause of a certain [here meaning a Judeo-Christian] point of view. The confrontation of Graeco-Roman culture with biblical religion engendered, after centuries of tension, a new science. This science preserved the indispensable parts of the ancient heritage (mathematics, logic, methods of observation and experimentation), but it was directed by different social and methodological conceptions, largely stemming from a biblical world view. Metaphorically speaking, whereas the

22. See the "Selected Bibliography: 1970" in Merton (1970:266–272).

> bodily ingredients of science may have been Greek, its vitamins and hormones were biblical. (Hooykaas 1972:161–162)

Hooykaas discusses the relationship of biblical religion to science largely by tracing the history of certain theological and philosophical doctrines (God and Nature, reason and experience, nature and art, rationalism and experimentation, etc.). Some of these will be touched on later. At the moment, however, we are more particularly concerned with what may be called the *psyche* of biblical religion, as expressed in certain basic beliefs and attitudes. Among them are a sense of individual commitment and personal community with God; a further sense of faith, piety, and (sometimes) fanaticism; also asceticism and self-denial, sin and damnation, confession and repentance, divine grace and salvation. Many of these themes are particularly evident in Protestantism, but several appear in varying degrees in the entire Judeo-Christian tradition. Perhaps, indeed, they may be regarded as characteristic of the monotheistic religions generally. Basic to all of their manifestations is a high degree of emotional intensity and a totality of religious commitment.

If then, as I would argue here, some kind of relationship—perhaps one-directional but more likely reciprocal—does exist in the West between its Judeo-Christian tradition (especially in its Protestant form) and the rise of modern capitalism and science, and if this religious tradition embodies the beliefs and attitudes just enumerated, it then becomes important to determine whether, and to what extent, similar beliefs and attitudes occur in the Chinese religions. To explore this point we shall examine, very impressionistically, the incidence in China of such concepts as asceticism, salvation, sin, confession, and repentance.

Traditionally, writers on Chinese society have stressed the significance of the idea of shame ("losing face") as *the* major sanction for morality. (A good example is the study by Hu Hsien Chin 1944.) More recently, however, the role of guilt and sin in traditional Chinese society has been analyzed in the interesting book of Eberhard (1967). Basing himself on religious texts going back to about A.D. 500, as well as large numbers of much later moralistic tracts and short stories, Eberhard finds guilt and sin rather than shame to be their dominant themes. Shame, as defined by anthropologists and psychologists whom he cites, is a feeling generated in the wrongdoer when his act becomes known to society. Thus a shame society is one based exclusively on social values, that is, on values shaped by tradition plus present-day social attitudes. A wrongdoer who lives in such a society may, in theory at least, commit an offense without feeling any shame unless that offense becomes known to the community. Feelings of guilt and sin, on the other hand, arise from the consciousness of having violated either human or, above all, supernatural prohibitions. Such feelings, unlike shame, become internalized and therefore operate within the wrongdoer even when the wrongful act is not publicly revealed.[23]

This thesis, as applied to China, seems open to certain objections. In the first place, shame in China appears to be based primarily on Confucian values, and Eberhard admits that, especially among the members of the dominant elite, it was in fact possible for these values to be internalized, that is, for a wrongdoer to feel shame within her- or himself even when the wrongdoing was not revealed.

Secondly, if shame is primarily associated with Confucianism, its relative absence from the literature analyzed by Eberhard is hardly surprising, in view of the strong coloration of this literature by Buddhist and, secondarily, Taoist values (even though, as Eberhard himself points out, Confucian values become distinctly more noticeable in the later materials). A question worth keeping in mind as one assesses Eberhard's findings, even though there is no ready answer, is this: What has been more influential in shaping Chinese society? Has it been the primarily secular Confucian values of the dominant few, shared however in varying degrees by the less articulate many? Or has it been the primarily supernatural values of the non-Confucian institutional religions, most influential among the many but also varyingly shared by the few?

In the third place, Eberhard's separation of shame from guilt and sin seems to be too absolute and arbitrary. Surely in traditional China, as in other societies, all three feelings frequently entered into cases of wrongdoing, even though their proportions might vary according to the background of the particular wrongdoer.

Despite these criticisms, Eberhard's book is valuable as perhaps the first extended analysis of its subject, and especially because it clearly demonstrates the wide incidence of guilt and sin in traditional Chinese society—a phenomenon that has often been denied.

As already indicated, Eberhard explains sin as the violation of a divine code and thus a parallel to the feeling of guilt resulting from violation of a secular law code. The concept of sin, he believes, arose in China only when Buddhism introduced the novel idea of divine judgment of sinners and their resulting punishment in one of the many Buddhist hells—an idea that theistic Taoism then adopted from Buddhism. Prior to Buddhism, on the other hand, no concept of sin existed because the gods were then thought of as punishing arbitrarily and being susceptible to assuagement.

---

23. Eberhard (1967:esp. 12–13, 120–124, as well as first and last chs.).

Here again I must disagree. I see no reason why the idea of sin must derive solely from belief in a divine law court—an obvious projection of the terrestrial law courts of bureaucratic China. On the contrary, the idea of sin could equally well spring from fear of violating supernatural sanctions, even though not thought of as codified and bureaucratically administered. If one accepts this supposition, it then becomes evident that the concept of sin was already known in pre-Buddhist China, though in a form less widespread and perhaps less acutely felt than that disseminated by Buddhism and religious Taoism.

The Chinese word commonly translated as "sin" is *tsui* 罪, a word that, depending on whether it is used secularly or theistically, can mean either crime or sin; it can also designate the punishment (secular or divine) administered for committing a crime or sin. *Tsui* is a word that goes back to China's early classical writings, where it occurs quite often in both secular and religious contexts. Confucius, for example, asserts that "he who has acquired sin *(tsui)* from Heaven [i.e., has committed a sin against Heaven] has no place where he may pray" (*Lun yü* 3.13). And already well before Confucius, the word appears religiously in the *Shu ching* (Documents Classic) and *Shih ching* (Songs Classic). Here, for example, is a lament found in one of the songs: "I alone am in misery. What, then, is my guilt *(ku)* against Heaven? What is my sin *(tsui)*?"[24] Or again: "Great Heaven is very terrible, but I truly have no sin *(tsui)*. Great Heaven is vast, but I truly have no guilt *(ku)*."[25]

The following, coming from the very beginning of the Chou dynasty, is a statement issued in the name of the youthful second Chou king (trad. r. 1115–1079) by his regent uncle, the Duke of Chou, at a time when the new dynasty was threatened by rebellion. Although it does not contain the word *tsui,* it clearly expresses the idea of guilt and submission to divine authority: "Unpitying Heaven sends down injury on our house. . . . I am not perfected or wise. . . . Yes, I am but a little child. . . . In performing Heaven's service I have been remiss, and have greatly thrown trouble on my person. Yet I, the young one, have no self-pity."[26]

This is the prototype of a long series of penitential edicts in which Chinese kings and emperors, down to the early years of the present century, have held themselves accountable for all manner of natural distur-

---

24. Song no. 197; tr. Karlgren (1905a:145).

25. Song no. 198; tr. Karlgren (1950a:147). See also nos. 194, 207, 245, 265.

26. *Shu ching* 5.7; tr. Legge (1960:3.157–159) and Karlgren (1950b:36–37). Cited in Bodde (1954:49–50).

bances (floods, droughts, and so on), as well as human disorders.[27] It seems undeniable that there is an age-old tradition of guilt and penitence here, predating that inspired by Buddhism, from which it differs in one important respect. This is that in pre-Buddhist China, there being no idea of hell, all divine punishment was thought of as being inflicted within the existing world, on the living wrongdoer (or sometimes on his descendants), whereas in Buddhism and religious Taoism (which borrowed the idea from Buddhism), the punishment takes place only in the next incarnation.

The devotional, salvational, and messianic aspects of certain forms of Chinese Buddhism are well known—notably in the Pure Land Sect, in which it was taught that a single divinity, Amitābha, had vowed to save all sentient creatures from sins and depravities, and that by simply invoking his name with undivided concentration, all sins would be eradicated and rebirth in the Western Paradise would be achieved. It is recorded of a seventh century monk that he recited the name of Amitābha a record-breaking million times during a single seven-day period (Ch'en 1964:345–347).

If anything, Messianism is even more pronounced in the oft-proscribed and politico-religious offshoots of Buddhism and Taoism than in these major religions themselves. Lao Tzu, for example, became recognized as a savior god in popular Taoism of the second century. And Li Hung, originally a non-Taoist district magistrate of the first century B.C. and first century A.D., later became a messiah in the eyes of religiously inspired political rebels who repeatedly invoked his name (the latest instance is in 1112), claiming that he would soon return to earth to bring peace to the world.[28]

The earliest, and probably most significant, of the socioreligious movements connected with the rise of institutional Taoism are those occurring almost simultaneously in east and west China during the declining years of the Later Han dynasty. The one in the east, headed

---

27. See Bodde (1954:50 n. 8) for the edict of 22 August 1862 issued in the name of the youthful T'ung-chih Emperor by the co-regent Empresses Dowager:

> During the night of the fifteenth of the seventh month, a flight of shooting stars was suddenly seen moving toward the southwest. . . . That Supreme Blue One [Heaven], when thus sending down its manifestations, does not produce such portents in vain. . . . Truly, though We be of tender years, We are filled with deepest dread and apprehension. . . . These warnings, transmitted by Heaven to man, are surely indicative of present deficiencies in Our conduct of government.

28. See, for Lao Tzu, Seidel (1969:esp. 58–74) and, for both Lao Tzu and Li Hung, Seidel (1969–1970). For a much later Buddhist parallel, see Naquin (1976).

by three brothers named Chang, came to prominence suddenly in the bloody uprising of the "Yellow Turbans" in A.D. 184. That in the west had a much longer history, gradually growing, under the aegis of three generations of another Chang family, from a small Taoist group to a large semi-independent theocratic state. Whereas the Yellow Turbans were militarily destroyed within a few months of their political emergence in 184, the more remote southwestern community (popularly known as the "five-bushel rice Tao" because this was the fixed contribution demanded from each family) succeeded in retaining a separate identity for more than three decades until 215. That this major effort to fuse religion and government into a single politico-theocratic entity failed to last longer has quite properly been regarded as an event of decisive importance (Seidel 1969:117–118). Had it succeeded in surviving, the later development of religion in China might have been radically different from the course it actually took.

The all-too-scanty available information about the two movements, derived from unfriendly official and Buddhist accounts, has been repeatedly rehearsed by modern scholars.[29] These writers have differed sharply as to the extent of the interaction—both political and ideological—between the two, as well as the possible external ideological influences upon them (especially in the case of the "five-bushel rice" community). Such influences have been seen as coming from Buddhism (Levy 1956:224 n. 106; Michaud 1958:88), Tibetan culture (Eichhorn 1955:318–319), indigenous aboriginal groups (Stein 1963:35–38), and even Persian Mazdaism (Dubs 1946:286).

For our purpose it is unnecessary to go into these controversies or to present a complete picture of the religious practices. Instead, to capture the essential *spirit* of the movements, I translate from Maspero the following vivid summary:

> The administration [of the Taoist communities] was wholly religious. Law and morality were completely blended: there were no crimes but only sins, and punishment was replaced by penitence. . . . Drunkenness, debauchery, and theft were all placed on an equal footing and redeemed by confession, repentance, and acts of benificence—as, for example, the repairing of one hundred paces of road with one's own hands or at one's own expense. Only to third-time recidivists were punishments applied. Basically, however, this was not necessary, because all sin—for example theft—received, just as in the novel *Erewhon,* its natural punishment in the form of the illness which sooner or later would afflict the culprit. Furthermore, illnesses were the

29. Among them Eichhorn (1954:326–332 and 1955:310ff.), Levy (1956:216–219), Michaud (1958:76–78), Stein (1963), Maspero (1981:373–378), Demiéville (1986: 815–820), Barrett (1986).

> object of sanctions more severe than those for the sins themselves. Thus prisons, after having been abolished for ordinary crimes, were reestablished . . . for the sick; they were called Houses of Retreat *(ching shih),* and to them the sick were sent to reflect on their sins. (Maspero 1981:377)

The examples that have been cited, scattered and incomplete though they are, suffice, it is hoped, to demonstrate the existence of Chinese religious attitudes reminiscent of some found in the biblical religions—this despite the great doctrinal differences between the two groups of religions.[30] Although Buddhism is no doubt the primary example on the Chinese side, we have also noted relevant features in early political Taoism, as well as indigenous ideas of sin and divine punishment that long predated Chinese Buddhism.[31]

Of particular interest because of its completely indigenous origin is the intellectual movement known after the name of its founder, Mo Tzu (fl. 479–381), as the school of Mohists. Though grouped among the classical schools of Chinese philosophy, Mohism differs from all of them in having a religious orientation. Indeed, it sufficiently displays the characteristics of a religious movement to justify fully, in my opinion, Mei Yi-pao's assertion that it "at one time attained to the status of a reli-

---

30. Because of its Christian coloration, the great Taiping politico-religious movement, which almost succeeded in destroying the Manchu dynasty in the mid-nineteenth century, has been deliberately excluded from consideration here. However, while recognizing the immediate influence on it of a rather fundamentalist kind of Protestantism, I believe that psychologically speaking, the movement accorded closely with its many politico-religious predecessors, starting with those of the second century A.D. that have just been described.

31. Two articles, both published after these words were written, confirm and amplify what has here been said about the concepts of sin and guilt in religious Taoism and Buddhism. One of them also touches on the question of shame, discussed earlier in our critique of Eberhard's book. In the first article, Zürcher (1980:141) writes concerning Buddhist early influence on Taoism: "We may conclude that regarding ideas about sin, guilt and retribution Buddhism deeply influenced Taoist thought . . . by introducing the concept of strictly individual responsibility and personal guilt." Again, in the second article, Wu Pei-yi (1979) traces in some detail the idea of confession of sins as found in religious Taoism, Buddhism, and then—an important topic not included in our discussion—as prominently displayed by several late Ming and early Ch'ing Neo-Confucians (sixteenth and seventeenth centuries). Two statements about the latter are particularly relevant: "Seventeenth century Neo-Confucians seem to have had much in common with their Puritan contemporaries: both groups, among other things, were given to self-examination and deeply concerned with spiritual progress" (Wu 1979:34 n. 60). Again: "After 1570, guilt, at least for a significant number of Neo-Confucians, seems to have played a more important role than shame" (Wu 1979:37).

gious institution," as well as his further remark that "the Moist [Mohist] institution seems to embody all the main features of organized religion—the historical founder, the living leader, the scripture, and the community of like-minded fellows" (Mei 1934:172–173).

On the organizational side, it is well known[32] that after the death of Mo Tzu, his followers continued in being for some two and a half centuries as a highly disciplined movement headed by a succession of leaders who bore the title of *Chü-tzu,* "Grand Master." The total dedication of the movement is illustrated by various anecdotes which, though perhaps not historically exact in every instance, are true to the psychology of the group. According to one of them, a *Chü-tzu* executed his own son for murder, and according to another, a different *Chü-tzu,* after failing to maintain the territorial integrity of the princely state he had sworn to uphold, committed suicide and was followed in death by no less than 183 of his adherents.[33] It has been suggested, on the basis of stories such as these, as well as on the general militancy of the movement (never displayed aggressively, however, but only in defense of the weak against the strong), that the Mohists may have originated among the professional warriors of Warring States times known as *hsieh* or "knights-errant." These were notable for their highly developed group ethic and concepts of feudal chivalry.[34] Be this as it may, Mohist thinking displays definite populist tendencies, in contrast to the upper-class orientation of early Confucianism.

It is probably indicative of the intensity of feeling of the Mohists toward Mo Tzu himself that they are the only school of thought in Chinese antiquity—indeed, in all Chinese history—to be designated unequivocally and permanently by the name of its founder.[35] Unfortunately for their ultimate survival, the Mohists fell prey to sectarianism—itself a significant fact because probably indicative of the intensity of doctrinal belief of the school.

On the ideological side, the Mohists were ethically egalitarian in their famous doctrine of equal love for all ("all-embracing love"), but politi-

32. See, e.g., Mei (1934:ch. 9) and Fung (1952:181–184).

33. Fung (1952:83), where by mistake the figure is given as 83.

34. See Fung (1952:xxxii–xxxiii and 83–84); also Fung (1935) and James J. Y. Liu (1967).

35. "Confucianism" is a Westernism, the Chinese equivalent for which is *ju chia,* "School of the Scholars." The names of Huang-Lao (Huang Ti, the Yellow Lord, and Lao Tzu) and then of Lao-Chuang (Lao Tzu and Chuang Tzu) were successively used during Former and Later Han times to designate particular trends in Taoism. The permanent and overall name of the movement, however, was *Tao chia* or Taoist school.

cally authoritarian in their insistence on the strict hierarchical subordination of those below to those above ("identification with the superior"; see III.3). Religiously, they based their doctrine of all-embracing love on belief in a personal Heaven *(t'ien)* which, assisted by a multitude of spirits, supervises the behavior of mankind:

> Wherever shines the sun, there gets one sin *(tsui)*. How then can one escape? . . . For regardless of whether it be in unpeopled forest valleys or dark confines, Heaven, with its vision, is sure to see. . . . What then does Heaven desire and what detest? It desires righteousness and detests unrighteousness. . . . He who conforms to the intent of Heaven by embracing mutual love and interchanging mutual benefit will gain reward. He who turns against the intent of Heaven by erecting divisive hatred and interchanging mutual violence will gain punishment. (*Mo-tzu* 26; tr. Mei 1929:135–137).

> What is the reason for this [present disorder in the world]? It all comes from doubt as to the existence or non-existence of ghosts and spirits, and failure to understand their capacity to reward goodness and punish violence. . . . Regardless of whether it be in dark confines, broad expanses, mountain forests, or deep valleys, the ghosts and spirits, with their vision, are sure to know. Regardless of wealth and honor, numerical strength, daring and martial vigor, firm armor and sharp weapons, the punishments of the ghosts and spirits are sure to triumph. (*Mo-tzu* 31; tr. Mei 1929:160, 170)

Coupled with these religious ideas are Mo Tzu's insistence on utility ("How can there be anything *good* which is not useful?"[36]), on unremitting toil ("He would have men toil through life"[37]), and on ascetic denial of art and luxury ("There was to be no singing in life, no mourning at death. . . . Though men enjoy music, he condemns music"[38]). These doctrines remind us of the Puritan ethic, even though—and this difference is important—the latter stresses the individual in his or her relationship to God, whereas Mo Tzu always thinks in terms of the mass person. Even so, the similarities are striking, especially when coupled with the remarkable fact that the later Mohists were virtually the only intellectual school in Chinese history to make systematic definitions of their key terms, to be consciously and seriously interested in logic, and to deal analytically with such topics as optics, mechanics, biology, and military technology (the latter in the twenty Mohist chapters on defensive fortification techniques).[39]

---

36. *Mo-tzu* 16; tr. Mei (1929:89).

37. A description of Mo Tzu taken from the penetrating critique contained in *Chuang-tzu* 33; tr. Watson (1968:366) and Graham (1981:276).

38. Again from *Chuang-tzu* 33; tr. Watson (1968:365–366) and Graham (1981:276).

39. For the later Mohists, see the fine translation and discussion by Graham (1978). We have already (III.2, near beginning) cited the later Mohist views of time and the Five Elements.

In the time of Mencius and later, Mohism was apparently as influential as Confucianism. That it utterly disappeared after the second century B.C. just when Confucianism was becoming the official state ethic is due to many factors. Among them were the humorlessness of the Mohists, their uncompromising spirit, and certain internal intellectual contradictions (notably the emphasis on a very mundane utilitarianism as justification for the idealistic but also psychologically unrealistic doctrine of "all-embracing love"). Probably most important, however, were the school's asceticism and economic egalitarianism, which were hardly calculated to appeal to any ruling group in the way Confucianism did, with its keen appreciation of literary cultural values and its justification of hierarchical privilege. Had history followed a different course—in other words, had Mohism succeeded in becoming the ethic of the ordinary person, thereby providing a counterbalance to Confucianism—I believe that China's entire subsequent sociopolitical development would have been profoundly different. In particular, it is conceivable that a mechanistic (as opposed to an organismic) kind of science might have developed, perhaps in the end not too unlike that which eventually arose in Western Europe.[40]

That all this did not happen, and that the crucial turning point took place at the very beginning of China's imperial history, suggests that from a very early time there were important factors in Chinese society—no doubt they were *both* ideological and socioenvironmental—that already then prefigured the eventual course of Chinese society. This remark is made with full awareness of the very important and more proximate social, institutional, and economic factors that, following the Sung dynasty, then placed a brake on a development that at one time seemed to be taking China toward a new social and economic order and a modern science.

It is essential to reach a balanced perspective on the examples of Chinese religious attitudes that have been narrated. Concerning Buddhism, Hu Shih paints a vivid picture of the intensity of feeling this religion once succeeded in evoking:

> There were long periods in Chinese history when this people also became so fanatically religious that a pious monk would burn a finger, or an arm, or the

---

40. The researches of the Mohists, in the opinion of Graham (1978:53), "deserve to be classed with similar brief episodes in Greece and Mediaeval Europe among the movements which in retrospect look to us like abortive efforts in the direction of modern science." Needham has suggested that later Mohist thinking about the physical world may have verged on atomism. If this be so, such a conception would have contrasted sharply with that of the great majority of Chinese thinkers, who saw the movements of the physical world in terms of wavelike shiftings of non-atomic forces and substances (as in the alternating inversely proportional dominance of the yin and yang). See Needham (1956:194 and 1962:60).

> whole body, willingly and devoutly, as the supreme form of devotion in his Buddhist faith. There were times when every fourth man in the population would be a Buddhist monk or a Taoist priest. There were times when the court and the people spent millions of ounces of silver yearly to build grand temples and monasteries, and millions of acres of land were donated to the monasteries as voluntary offerings to the gods. (Hu Shih 1934:80)

This statement, apart from the demonstrably gross exaggeration of the second sentence, reads convincingly enough.[41] Yet I believe its general tone to be misleading, because it attributes a universality and continuity to religious phenomena that in reality probably possessed neither. Consider, for example, the monks who during "long periods" are said to have been ready to sacrifice a part or whole of their bodies as an expression of religious devotion. Two studies have been made of self-immolation by Buddhist monks and nuns in China, both based on the successive collections of their biographies that extend in time from the beginnings of Chinese Buddhism to 988.[42] The biographies record a total of 25 self-immolations by fire (the earliest in the mid fifth century), plus 12 others by other means (jumping off a high place, drowning, etc.), thus making a total of 37 suicides during some four hundred years. They also record 14 acts of self-sacrifice less than death (burning of the fingers or an arm, etc.). Of the 25 suicides by fire, 11 occurred during the fifty-year period of 451–501 (a little more than one every five years), making this by far the peak period of religious intensity. Thereafter the figures drop sharply: 2 in the sixth century, 3 in the seventh, and 4 each in the ninth and tenth.[43]

There is some reason to believe that the immolations recorded in the collections of biographies are only those of the most famous persons, leaving a good many others unrecorded. Even taking this fact into consideration, however, and therefore arbitrarily multiplying each statistic by, say, five, it still seems evident that the number of Chinese religious suicides carried out during the peak period of fervor (451–501) is proportionately far smaller than that of the seven Buddhist monks and one nun who burned themselves to death in South Vietnam during the single

41. Statistics presented in sect. 1 above indicate that the ratio of Buddhist clergy to the total population at the peak moment of 845 was slightly over 1 percent. Hu Shih's indicated 25 percent, for which he gives no source, could have held good only in a particular area (possibly a single city) and a particular time, or perhaps he derived it from some casually enunciated nonstatistical historical statement.

42. They are Gernet (1960) and Jan Yün-hua (1965). The data that follow come from Gernet (1960:529–531).

43. Gernet's chronological statistics cover only the suicides by fire, not those by other means.

six-month period of May–October 1963, for political reasons (as a protest against the Ngo Dinh Diem regime).[44]

Nor are the Chinese figures impressive when compared with the sad index of religious intensity (better, religious fanaticism) available for Europe, namely the number of persons burned to death on charges of heresy. For example, during the sixteen years of Torquemada's administration of the Spanish Inquisition (1483–1498), some 2,000 persons are said to have died at the stake, in other words, around 125 per year—this in a realm and population for smaller than those of China.

In China, conversely, governmental religious persecutions, while they could be very sweeping in terms of the numbers of people and amount of property affected, very rarely went as far as actual physical violence. Thus in the course of the major persecutions of 446, 574, 842–845, and 955, only three instances of execution (actual or intended) are recorded. In the first of these, that of 444 (a prelude to the 446 persecution), two Buddhist monks were executed. In the second, that of 446, an imperial edict called for the general execution of monks, but these, having been forewarned, went into hiding, so that it is unknown how many, if any, actually died. Finally, in 843, during the greatest of all persecutions—one covering Zoroastrianism, Manicheism, and Nestorian Christianity as well as Buddhism—an edict allegedly ordered the execution of Manichean priests throughout the empire; according to a much later (thirteenth-century) source, seventy Manicheans were actually executed at the capital.[45]

That the governmental controls and occasional persecutions inflicted on the institutional religions were accepted by the latter so readily, resulting in so little bloodshed, testifies to the unquestioned total authority in ideological matters of the emperor and the Confucian state (a point to which we shall return in the next chapter). However, it also indicates conditions and attitudes within the religions themselves: their lack of centralized structure and inability to insist on the kind of total religious commitment taken for granted in the monotheistic religions. A happy consequence is the virtual absence of religious wars in China.[46] In short, it may be said not only of China's unfocused diffused religion,

---

44. For these figures, see Jan (1965:243).

45. For the events of 444 and 446, see Ch'en (1964:149–150). The edict of 843 is not officially recorded, but it is mentioned by the Japanese monk Ennin in the diary he kept during his years in China at this time. See Reischauer (1955a:327 and 1955b:232). For the document alleging the execution of seventy Manicheans (who, very strangely, are all referred to as women), see Chavannes and Pelliot (1911–1913:301, text xxxiv).

46. A possible exception is the group of Muslim rebellions in Northwest China of the late nineteenth century, but even these were primarily rebellions against the government rather than wars between rival religious groups.

but of its institutional faiths as well, that with a few possible exceptions in particular periods and places, they were in general far more relaxed, more tolerant, and less self-demanding than have been the major monotheistic religions.

The conclusion reached in this section is that there seems to have been a significant congruency—possibly even a definite cause-and-effect relationship—between what developed intellectually in Western Christianity, especially during the past three centuries, and the concurrent rise of a new commercial-industrial society centered upon a new experimental science. If, then, similar intellectual developments could be demonstrated as having occurred in Chinese religions, one might theorize that these should or at least could have led there to somewhat similar results.

It can in fact be shown that concepts and attitudes characteristic of the monotheistic religions—guilt and sin, repentance and salvation, ascetic self-denial, and a good deal more—have manifested themselves at various times and with varying intensity among the Chinese religions. The latter, it should be added, include not only Buddhism and religious Taoism but also various politico-religious sects and messianic movements, as well as certain ideas found in the early diffused religion of pre-Buddhist China. A striking parallel with Western experience is the attempt of Mo Tzu and his followers, in pre-Buddhist China, to fuse religion, ethics, and politics into a single tightly knit socioreligious organization. The parallel is made still more striking by the deep interest of the Mohists in logic, natural science, and technology.

Nevertheless, these and other tantalizing similarities with Western religious ideas and institutions remained relatively abortive and sporadic. They were constantly discouraged, both by the repressive measures adopted in imperial times by the Confucian state and by organizational and ideological weaknesses within the religions themselves. It might be argued that this situation was inevitable in polytheistic religions that lacked the cohesion and single-mindedness inherent in monotheism. Yet this argument ignores the total role played by Lamaist Buddhism in the theocratic states of Tibet and Mongolia or the internecine militancy of the Buddhist sects in feudal Japan. These considerations cause me to believe, with respect to the Chinese religions, that it has been primarily the institutional environment in which they developed, and only secondarily their own inherent characteristics, that has so limited their growth organizationally and politically.

Without arguing this point further, it is obvious that the role of religion in China has been radically different from, and far less total than, that of religion in the West. This fact surely carries major implications for any consideration of why a modern society and modern science did not arise in China.

V. GOVERNMENT AND SOCIETY

This chapter, unsurprisingly, deals primarily with Confucianism, not only because of its vital role throughout Chinese history but especially because it was so particularly concerned with human behavior. Yet, as we shall see from time to time, important Confucian attitudes were sometimes shared in varying degrees by other schools of thought.[1]

1. Orthodoxy, Authoritarianism, and Dissent

The T'ang writer Han Yü (768–824) is far better known as a prose master than as a philosophical thinker. Nonetheless, the essay he wrote around the years 804 or 805, entitled *Yüan Tao* or "Origin of the Tao," was to have great intellectual consequences:[2]

> What I call the Tao [he wrote] is not what has hitherto been so called by the Taoists and Buddhists. Yao transmitted it to Shun, Shun transmitted it to Yü [founder of the Hsia dynasty], Yü transmitted it to T'ang [founder of the Shang], T'ang transmitted it to Wen and Wu and the Duke of Chou [founders of the Chou], Wen and Wu and the Duke of Chou transmitted it to Confucius, and Confucius transmitted it to Mencius. After Mencius died, it was no longer transmitted. Hsün [Tzu] and Yang [Hsiung] selected from it, but without reaching its essential portion.

The idea that the ancient sage-kings had originated a Tao or Way of Life embodying the highest values and institutions of civilized society and that this had been transmitted to Confucius and so to Mencius but

---

1. The first pages of the section that follows deal with the Confucian—and especially Neo-Confucian—theory of transmitted moral orthodoxy known as the *tao t'ung* or "sequence of the Tao." Among several helpful accounts are James T. C. Liu (1973: 490–491), Wing-tsit Chan (1973:73–81, the most detailed), Julia Ching (1976b:esp. 374, 379, 386), and Wing-tsit Chan (1976a:567–568).

2. In *Han Ch'ang-li ch'üan-chi* (Complete Writings of Han Yü) 11/1a–5a (and specifically 11/4b for what follows). The essay has been translated several times, e.g., in Giles (1923:115–121), Margouliès (1926:177–183), and De Bary et al. (1960:431–434). The rendition that follows is that in Fung (1953:410). For the dating, see Pulleyblank (1960:110), who thinks the essay was composed between the winters of 803 and 805, and Hartman (1986:145), who places the date as probably the summer of 805. On the intellectual and biographical background of the essay, see Rideout (1947) and Hartman (1986:145–162).

had then been interrupted was to become an article of faith among members of the Sung Confucian revival known to Westerners as Neo-Confucianism. Significantly, one of several Chinese designations of the movement is *Tao hsüeh chia* or "School of the Study of the Tao." Han Yü's role as the formulator (or rather, as we shall see, reformulator) of this theory is the primary reason why, despite his lack of philosophical depth, he is regarded as a forerunner of Neo-Confucianism.[3]

In contrast to Han Yü, who only indirectly, if at all, claimed to have revived the long dormant Tao himself,[4] claims of this sort were advanced when the first galaxy of Neo-Confucians arose in the eleventh century. The most notable early claimant, not for himself but for his brother, was Ch'eng Yi (1033–1108), younger of the Ch'eng brothers. The tomb epitaph he composed for his older brother, Ch'eng Hao (1032–1095), includes the words: "He sifted out heterodox doctrines and expunged perverse words, thereby enabling the Way of the Sages to shine gloriously forth once more in the world. Ever since Mencius, there has been only this one man."[5]

A century later, Ch'eng Yi's own name was added to that of his brother as *the* two restorers of the Tao after Mencius. This tribute came from Li Yüan-kang, a contemporary of Chu Hsi, writing in his *Sheng-men shih-yeh t'u* or *Diagrams of the Works of the School of Sages* (1173); the same work is apparently the first to use the term *tao t'ung,* "sequence of the Tao," to designate the Confucian succession. Under the first of several diagrams, entitled "Transmission of the Tao in Its Orthodox Sequence,"[6] appears the subheading "The sages and worthies of successive ages who have transmitted the greatly central and utterly orthodox

---

3. Fung (1953:408ff.). The theory owes its clear and detailed formulation to Han Yü, even though, as Pulleyblank points out (1960:97), one of Han Yü's contemporaries, Lü Wen (ca. 774–ca. 813), also expressed the general idea (in much less explicit language) that there had been an interruption of the Tao after Confucius.

4. Pulleyblank (1960:97) states that "Han Yü's innovation lay in actually assuming the role of teacher in the line of Mencius"—a seemingly extreme conclusion, based as it is on Han Yü's brief essay, *Shih shuo* (On the Teacher), in which Han discusses the teacher-student relationship and puts himself in the former role, without mentioning any historical context or even alluding to Mencius at all.

5. The epitaph is quoted at the end of Ch'eng Hao's biography in the *Sung shih* (Sung History) 427/3a. A slightly different version appears in the Neo-Confucian anthology *Chin ssu lu* (A Record for Reflection) 14.17; tr. Chan (1967:300). "Heterodox doctrines" is a rendition of that same term, *yi tuan* (lit. "strange shoots"), of which Confucius (see II.4) had once spoken so ambiguously. See also at n. 15 below.

6. *Cheng t'ung* (Orthodox Sequence) literally means "correct (or upright) sequence."

Tao,[7] making it function for ten thousand years without impairment." The names inscribed below are identical with Han Yü's list down to Mencius, save for the insertion between him and Confucius of two of the latter's disciples, Yen Tzu and Tseng Tzu, as well as his grandson, Tzu-ssu. The real innovation, however, is the addition of Ch'eng Hao and Ch'eng Yi in solitary splendor after Mencius.[8]

Among major Neo-Confucian thinkers, the first and most famous to use the term *tao t'ung* was Chu Hsi in his commentary of 1189 on the *Chung yung* or *Doctrine of the Mean.* Upon Chu's own death in 1200, the "orthodox sequence" was made complete when Chu's son-in-law, Huang Kan (1152–1221), wrote the following eulogy about the great philosopher:

> The orthodox sequence of the Tao has depended upon men for its transmission. From the Chou onward, responsibility for transmitting the Tao has been borne by a few men only, not more than one or two of whom have succeeded in making it really shine forth. After Confucius, his subtle [teaching] was perpetuated by Tseng Tzu and Tzu-ssu, but it was only Mencius who first made it shine forth. After Mencius, what had been cut off [by his death] was perpetuated by Chou [Tun-yi], the Ch'eng [brothers] and Chang [Tsai], but it was only Chu [Hsi] who first made it shine forth.[9]

With Chu Hsi the Tao apparently reached its culmination. No one after him was added to its list of transmitters.[10]

---

7. Note the emphasis on centrality, a concept already discussed in III.2.

8. *Sheng-men shih-yeh t'u* 1000–1001. With greater magnanimity than some of the other Neo-Confucians, Li Yüan-kang flanks this main list with two subsidiary groups of names, both beneath the heading "Sages and Worthies who have operated by themselves, and whose Tao, though it might have aided a single age, could not be transmitted for ten thousand generations." Among the names thus listed are those of Hsün Tzu and Yang Hsiung, the Buddha and Lao Tzu, Yang Chu and Mo Ti.

9. Quoted at the end of Chu Hsi's biography in *Sung shih* 429/6a.

10. The definitive sequence of transmitters is provided by Chang Po-hsing (1652–1725) in his *Tao t'ung lu* or *Record of the Sequence of the Tao* (1708). This list adds three mythological culture heroes to the beginning of Li Yüan-kang's enumeration and, more important, inserts two additional Neo-Confucian thinkers before Chu Hsi. Chang's complete list reads: Fu Hsi (Subduer of Animals), Shen Nung (the Divine Husbandman), Huang Ti (the Yellow Lord), Yao, Shun, Yü, T'ang, Kings Wen and Wu, the Duke of Chou, Confucius, Yen Tzu, Tseng Tzu, Tzu-ssu, Mencius, Chou Tun-yi, the Ch'eng brothers, Chang Tsai, Chu Hsi. Notably missing from the Neo-Confucian part of the list is the numerologist Shao Yung (1011–1077), whose world cycles are described above in III.3.

The official acceptance of Chu Hsi as the culminator of the Tao was by no means as readily forthcoming as our bald statement might indicate. During the latter years

Belief in a universal Tao created and transmitted by the sages did not in fact originate with Han Yü. Its beginnings go back to Mencius and, less explicitly, to Confucius. In their days, as later, the word *tao* was used in many ways, both concretely and metaphorically. Some of the usages in the Confucian *Analects* are: a roadway *(tao)* (*Lun yü* 17.14); the way *(tao)* of a father (1.11, 4.20); the good man's *tao* (11.19); a straight *tao* and a crooked *tao* (18.2); is it according to *tao* for Confucius to speak with a certain master of music? (15.41); the ancient *tao* of archery (3.16); the presence or absence of *tao* in a state (5.20, 14.1, 14.4, 15.6); the presence or absence of *tao* in the world (3.24, 16.2); the *tao* of Wen and Wu, founders of the Chou (15.28); the *tao* of the former kings (1.12); the *tao* of Heaven (which Confucius refuses to discuss) (15.12).

All of these *tao* are metaphorical except the first (a roadway), and all are human except the last (the *tao* of Heaven). This overwhelming emphasis on man is not surprising in a work that centers on Confucius; were it a Taoist text, much more would have been said about the *tao* of Heaven in addition to that of man.

What is most striking is the diversity and particularity of the many *tao* mentioned as relevant to human experience. This fact places the persons who would heed them under a continuing obligation to be watchful as they make their moral and intellectual choices. Yet from this welter of *tao* that may be either straight or crooked, significant or trivial, we see in the last several items (*tao* of the state, of the world, and of the former kings) the emergence of an idea concerning a different kind of Tao: a Tao that subsumes all the other lesser *tao* and is itself single, universal, and grounded in the achievements of the ancient sage-kings. Confucius himself does not explicitly term it single, but Mencius tells a young prince unequivocally: "The Tao is one and only one" (*Meng-tzu* 3a.1).

There is little doubt that both Confucius and Mencius conceived of such a Tao and of their own special mission to transmit it. Confucius saw it as originating during the two earliest dynasties, but best revealed in the achievements of the founders of his own dynasty, the Chou, and of his own state of Lu (which was blood-related to the Chou). "Chou," he remarks, "had the advantage of surveying the two preceding dynasties. How replete was its culture! I follow Chou." About the Chou founder Wen Wang (the "Cultured King"), he says: "The Cultured King (King Wen) having died, was not his culture *(wen)* lodged in me?"

---

of his life and for some time afterward, his ideas were under attack, so that more than a century had to pass before, in 1313, his commentaries on the classics were officially declared orthodox. Thereafter until the abolition of the civil service examinations in 1905, all participants in these examinations had to answer questions according to the Chu Hsi interpretations. For details, see, among others, Chan (1973).

And about the Duke of Chou: "How great is my decline! For long I have not dreamed, as of old, of seeing the Duke of Chou." The Tao—undefined by him but almost surely envisioned as the sum total of early Chou civilization—is revered with the words: "Having heard the Tao in the morning, if one dies that evening it will certainly be all right!"[11]

Mencius, in a famous passage, pushes the beginnings of the Tao back to the predynastic sages Yao and Shun, and from them traces its descent step by step to Confucius and thence, by implication, to himself. From Yao and Shun to the Shang dynasty founder T'ang, he says, and then from T'ang to King Wen and from Wen to Confucius, there has in each case elapsed a period of somewhat over five hundred years. In each case, too, it has been possible to see or hear "it" (undoubtedly the transmitted Tao).[12] Then he continues: "From Confucius down to today there have been one hundred and more years. So unremote [from us] is the Sage's age, and so near is the Sage's dwelling. Is there then really no one? Indeed is there no one?"[13]

The third major classical Confucian, Hsün Tzu, exalted Confucius and vigorously defended what he called the "orthodox Tao" (*cheng tao;* see below), but he also criticized Mencius severely, while making no claim himself to be the Tao's transmitter. His modesty was followed by Tung Chung-shu, despite the latter's importance as the major formulator of Han Confucianism. Only Yang Hsiung (53 B.C.–A.D. 18) boldly ventured to compare himself to Mencius as a staunch upholder of the Tao (Fung 1953:150). This claim was subsequently either ignored or denied,[14] while Confucianism, after the second century A.D., went into a kind of philosophical hibernation until the time of Han Yü.

Let us try to evaluate the intellectual consequences of the "sequence of the Tao" theory, especially on Confucianism. Its effect on classical Confucianism was comparatively slight because the theory became fully formulated only much later. With regard to Confucius, a modern study tells us that he regarded the Tao as a "continuous progress of human civilization" and believed that "the human being has an active, creative role" to play in the Tao's broadening and extending (Hall and Ames 1987:229, 231). Certainly it is true that Confucianism experienced enormous change during its classical phase; also that when it was reborn as

---

11. See *Lun yü,* respectively 3.14, 9.5, 7.5, 4.8.

12. A typical instance of the common practice in Literary Chinese of using the ambiguous third person accusative pronoun to indicate the object of a verb instead of explicitly naming the object. See II.4 at n. 20.

13. *Meng-tzu* 7b.38. See also the similar statement in 2b.13.

14. E.g., by Li Yüan-kang as quoted above.

Neo-Confucianism, the result differed vastly from earlier Confucianism because of the long intervening centuries of interaction with Buddhism and Taoism.

On the other hand, Neo-Confucian belief in the sanctity of a Tao transmitted from antiquity reinforced a tendency often found in Chinese thought, that of looking to the past as precedent for the present. In order to express their new ideas, the Neo-Confucians often attached new meanings and emphases to old words rather than trying to create a philosophical vocabulary that would break with the past. Instead of approaching the past analytically, they felt obliged to reconcile their new thinking with what had already been said and thought in the classics. Chu Hsi's success in such matters is a major reason why he became the culminating figure of the "sequence of the Tao." Yet the net result was to limit Neo-Confucian ability and desire to move into new fields, especially after its dominant Ch'eng Yi–Chu Hsi wing achieved state orthodoxy in 1313.

Fervent belief in the Tao as a body of transmitted truth may sometimes lead to strong distrust of anything thought to lie outside it. Already in Confucius we have seen the strong objection to what he calls *yi tuan,* "strange shoots"—a term traditionally but uncertainly translated in this passage as "heterodox doctrines."[15] Throughout its history, Confucianism has deprecated the use of debate as a means of advancing knowledge. So glorious is the Tao, the Confucians would say, that its validity should be self-evident without the need for argument. If, nonetheless, it has today become obscured by a welter of controversy, this is simply sad evidence of the degenerate world in which we live.

This viewpoint, as we shall see shortly, was widespread among other classical schools as well, even though the Tao for them might be quite different from the Confucian Tao. Nostalgic laments by Confucians, Taoists, and others for the loss of an allegedly unitary ancient Tao clearly reflect their dismay at the multiple changes that were then reshaping Chinese life. Among such changes were the growing political pluralism of China's contending states, the increasing specialization and restructuring of its social classes, the rise of new economic relationships based on the private buying and selling of land, the technological consequences of passing from the age of bronze to that of iron, and the increasing importance of the professional and educated class known as *shih* (discussed in sect. 3).

Some of the intellectual consequences were the development of debating techniques by practitioners of litigation, of rhetorical skills by

---

15. See *Lun yü* 2.16 and II.4, where the passage was discussed at length. See also at n. 5 above, where Ch'eng Yi uses the same term.

itinerant diplomats who attempted to persuade the rulers of one or another state to adopt one or another policy, and the creation of academies of learning by the high dignitaries of several states in which large numbers of intellectuals *(shih)* were supported and given the opportunity to discuss and debate ideas among themselves. In short, the last two centuries of pre-imperial China saw the appearance of increasingly diverse challenges to the traditional mythico-religious world view. These several developments have been perceptively described by J. L. Kroll (1985–1987:118–145).

The fact that Confucians and others so often reacted negatively to the new ideas points to a signal difference between ancient China and ancient Greece. A formalistic indication is the virtual absence in ancient Chinese philosophy of anything resembling the Socratic dialogue (meaning a reasoned discourse between two individuals pursued in order to approach closer to clarity and truth; I doubt that the Chinese debates alluded to by Kroll—of whose details we know exceedingly little—were really conducted in this manner). Dialogues, when they occur, are rarely more than three or four interchanges long and usually consist of respectful questions (not probing objections) addressed by the disciple to the master in order to facilitate his exposition.[16] In the rare cases in which sharp debate between the master and an outsider is recorded—perhaps the most famous is that between Mencius and Kao Tzu on human nature—the outcome is predetermined in the sense that after a very few exchanges, the master always has the last triumphant word while his opponent is reduced to silence.[17]

Thus, caught in the rising tide of late Chou intellectual diversity, it is

---

16. *Meng-tzu* 2A.2 is exceptional in containing no fewer than fourteen exchanges between Mencius and a disciple on the topic of courage and the *ch'i* or "vital force." However, the exchanges serve simply to provide elucidation and elaboration by the master, not doubt or contradiction by the disciple. Two unusual exceptions of later times, both involving, however, paired groups rather than paired individuals, are (1) the politico-economic debates of 81 B.C. between bureaucratic Legalists and provincial Confucians which are the basis for Huan K'uan's *Yen t'ieh lun* or *Discourses on Salt and Iron* (tr. Gale 1967) and (2) the series of late fifth- and early sixth-century debates between Buddhists and non-Buddhists on the question of the extinction or nonextinction of the soul and other philosophical topics. (See Fung 1953:284–292.) It is significant that the former occurred when Confucianism was not yet completely established as the state orthodoxy and the latter when Buddhism was strong and Confucianism weak.

17. Richards (1932:55), commenting on the Mencius–Kao Tzu arguments, observes that "the purpose of eliciting the point of difference is absent" and "the form of the opponent's argument is noticed, in the sense of *being used* in the rebuttal, but not *examined* so that the flaw, if any, may be found."

no surprise that both Mencius and Hsün Tzu should lament the need to resort to *pien,* "dispute, argument, debate." Mencius, on being informed that outsiders regarded him as being fond of argument *(pien),* replies indignantly: "How could I be fond of arguing! It is simply that I have no alternative." Then, after pointing to the world's alternating phases of order and disorder since the time of Yao, he launches into his famous diatribe against the intellectual disorder of his own day:

> The sage-kings are no longer active, the nobility give free rein to their passions, and idle scholars hold forth to one another at cross purposes. The words of Yang Chu and Mo Ti fill the world, so that what is said in the world either agrees with Yang or with Mo. . . . Fearful of these things, I build a barricade around the Way of the Sages, I repulse Yang and Mo, and I drive away obscene words, so that speakers of evil will be unable to operate. . . . I wish to extinguish evil doctrines, to repel extreme conduct, and to expel obscene words, so as to perpetuate [the heritage of] the Three Sages [Yü the Great, the Duke of Chou, Confucius]. How could I be fond of arguing! It is simply that I have no alternative. (*Meng-tzu* 3b.9)

Hsün Tzu, too, after devising three tests for the exposure of false doctrines, writes similarly though with less passion.

> All evil doctrines and perverted statements that depart from the orthodox Tao . . . may invariably be classed under these three kinds of delusion. . . . Now the people are easily united in the Tao but they should not be allowed to share in the reasons [for its functioning].[18] Therefore the intelligent ruler oversees them with his authority, leads them with the Tao, orders them with his decrees, enlightens them with his proclamations, and restrains them with his punishments. In this way his people are converted to the Tao as if by supernatural power. What need has he for argument *(pien)?* But today the sage-kings are no more, the world is in disorder, wicked words have arisen, and the Superior Man *(chün tzu)* lacks either the authority to oversee them [the people] or the punishments to restrain them. Hence the use of argument *(pien).* (*Hsün-tzu* 22; tr. Dubs 1928:289)

To modern scholars, the proliferation of ideas and schools touched on above is the crowning glory of the late Chou dynasty. The contrary Confucian viewpoint is well summed up by the historian Pan Ku (A.D. 32–92) in two passages of his "Treatise on Literature":

> Of old, when Confucius passed away, his subtle words were cut off. When his seventy disciples died, the great meaning was dissipated. Hence the *Spring and Autumn Annals* divided into five [versions], the *Songs* into four, and the

---

18. Cf. the analogous statement in *Lun yü* 8.9, of which this is an expanded paraphrase: "The people may be caused to follow it [probably meaning the Tao], but not to understand it."

Changes had several schools of transmission. The warring principalities acted at cross purposes to one another, and truth and falsity contended with one another. The doctrines of the Many Masters [the various philosophical schools] proliferated into confusion. When the Ch'in came, it was unhappy about this state of affairs and so [in 213 B.C.] burned and destroyed written documents in order to make ignorant the black-headed people. (*Ch'ien Han shu* 30/1a)

Again:

Of the ten schools of the Many Masters, only nine are worthy of regard. They all arose when, with the dwindling away of the Kingly Way,[19] the nobles ruled by force and sovereigns of the age went in different directions in their likes and dislikes. Because of this, the doctrines of the nine schools swarmed forth and became simultaneously active. Each [school] pulled out a single tip [of the Tao], emphasizing what it itself preferred. Thus they propagated their doctrines in order to gain favor from the nobles. (*Ch'ien Han shu* 30/7a)

That the Confucians were by no means alone in their yearning for intellectual oneness becomes evident when we read the very similar remarks by the anonymous Taoist whose critique of the late Chou philosophers occupies the final chapter of the *Chuang-tzu:*

The world fell into great disorder. Worthies and sages did not show themselves. The Tao . . . was no longer unified and the world's people for the most part gained only a single glimpse of it, according to whatever they themselves most liked. They were like the ear, eye, nose and mouth, each with its own perceptive faculty, not communicating with any other. So it is with the many ramifications of the Hundred Schools, none of which, though each has its own strength and utility at a particular time, is wholly sufficient or universal. . . . The Way of the Inner Sage and Outer King has become obscured and no longer shines forth. . . . Alas! The Hundred Schools go their separate ways without returning. . . . Alack! The students of this later age fail to see the unmixed purity of Heaven and Earth or the great structure of the men of old. The Tao . . . is about to be torn to pieces by the world. (*Chuang-tzu* 33; tr. Watson 1968:364, Graham 1981:364)

The word *pien,* "argument or disputation," was used pejoratively not only by Confucians but by Taoists, as for example in the *Chuang-tzu,* where members of the School of Names (the Logicians) are more than once referred to disparagingly as *pien-che,* "arguers."[20] The criticisms

---

19. *Wang tao.* Mencius tells what this Tao consisted of in *Meng-tzu* 1b.3 and 1b.6.

20. See also *Lao-tzu* 81:

> Good men do not argue *(pien).*
> Arguers *(pien che)* are not good.

directed against "the arguers" by a variety of thinkers are all remarkably similar in tone. They accuse them, among other things, of "deceiving and confusing the ignorant masses" (*Hsün-tzu* 6), "causing subtle divisions and disorder" (*Ch'ien Han shu* 30, quoting Liu Hsin), "throwing a deceiving glamor over men's minds" (*Chuang-tzu* 33), and "specializing in the definition of names but losing sight of human feelings" (*Shih chi* 120).[21]

One would naturally expect the Legalists to insist on total conformity to a narrowly defined state orthodoxy, but Mo Tzu too maintained that "what the superior approves, all must approve. What the superior condemns, all must condemn" (*Mo-tzu* 11; tr. Mei 1929:56). At the other end of the scale, Chuang Tzu, in his famous second chapter on "The Equalizing of Things," eloquently declared all argument *(pien)* to be futile because it is based on subjective man-made standards. Among the schools of the Warring States, apart from "the arguers" themselves, it is only the later Mohists who (despite the authoritarianism of their founder) recognized the inherent value of argument or disputation *(pien)* as a means for arriving at truth. "To say there is no winner in argument *(pien)*," they wrote, "surely fails to fit the fact" (*Mo-tzu* 41; tr. Graham 1978:402.B35). This somewhat touching confidence in the validity of the dialectical process is another of the many features that make the Mohists—in this case the later Mohists—seem so familiar to Westerners and therefore so "un-Chinese."[22]

Returning to the Confucians, one sometimes finds that their insistence on orthodoxy leads them to xenophobia. To cite but two prominent examples: Mencius ridiculed the non-Chinese southern accent of his anarchistic rival, the agriculturalist Hsü Hsing, by calling him "this shrike-tongued man of the southern barbarians [who] denies the Tao of the early kings."[23] In similar fashion Han Yü, with whom our discussion began, termed the Buddha a "barbarian"[24] whose "language was not the language of China, clothes were of an alien cut, and mouth did not utter the maxims of our early kings."[25]

---

21. See the quotations in Fung (1952:193–194).

22. Other "Western" and "un-Chinese" characteristics have been pointed out in IV.2, following n. 31.

23. *Meng-tzu* 3a.4. Further statements from both Mencius and Confucius, reflecting a similar attitude toward non-Chinese peoples and cultures, are presented in Miyakawa Hisayuki (1960:23–27).

24. *Yi Ti chih jen,* "man of the Yi and Ti tribes."

25. This and much more is found in Han Yü's "Memorial on a Bone of the Buddha." See *Han Ch'ang-li ch'üan-chi* 39/5a; tr. Giles (1923:127). Cf. also tr. Margouliès

In recent years, especially in the United States, scholars have increasingly tended to use such terms as "absolutist tendencies," "increasing authoritarianism," "despotic absolutism," "institutionalized terror," or just "despotism" or "absolutism" to describe the later centuries of the Chinese empire.[26] Such scholars, while generally agreeing on the Ming and Ch'ing as the periods when absolutism reached its peak, variously trace its beginnings back to the Yuan, the Sung, or even as far as the late T'ang.[27]

On the political side, evidence often cited for the Ming includes the abolition by the Ming founder of the Grand Secretariat, thereby facilitating a much greater concentration of power in imperial hands; the decline of the Censorate as an organ for addressing remonstrances to the throne and its transformation into an organ primarily concerned with maintaining imperial surveillance over the bureaucracy; the use of increasingly servile language by officials when addressing the throne; and, on several notorious occasions, the executing, exiling, beating, demoting, or otherwise punishing of hundreds and even thousands of officials on charges of political subversion.[28]

These political developments, it is further often argued, are paralleled by the growing regularization of social relationships. Subservience and obedience, whether of children to parents, the young to the old, women to men, or commoners to officials, are said to have been inculcated with increasing severity.[29] Physically symbolic of the general tightening of the social order might be cited the growth in popularity of female footbinding (tenth century onward), standing in sharp contrast to the T'ang pottery figurines of court ladies riding horseback or donkeyback and sometimes even playing polo. Intellectually speaking,

---

(1926:201), Dubs (1946–1947:8–14), and De Bary et al. (1960:428). Discussed in Hartman (1986:84–86). Other similar attacks on Buddhism are easy to find, such as that of Fu Yi (555–639), on which see Wright (1951).

26. See respectively Gernet (1982:393), Twitchett (1973:71), Hucker (1959:187), Mote (1961:41), Nivison (1959:14), Eberhard (1960:232).

27. For the latter date, see, for example, Twitchett (1973:72): "Institutionally the Ming carried further the tendency toward ever greater concentration of power in the throne which had persisted since the late T'ang period."

28. These and similar Ming developments are described in Hucker (1959), stressed in Mote (1961), and summarized in Nivison (1959).

29. For example, Ho Ping-ti (1968:34) remarks about the Ch'eng brothers and Chu Hsi that they "make the subject's loyalty, the son's filialty, and the wife's obedience almost one-sided imperatives. This was the main reason why the Ch'eng-Chu School of Neo-Confucianism was made the official orthodoxy during the Yüan, Ming, and Ch'ing periods, which witnessed an intensification of autocracy."

a similar indication of growing rigidity is the institutionalization of the "eight-legged essay" in the civil service examinations from 1487 onward.

While in no way doubting the reality of the specific data commonly adduced, I have nevertheless questioned, in a review of Nivison and Wright (1959), whether the course of Chinese absolutism has really been as unswervingly unilinear as the thesis as a whole postulates:

> The kind of government which in the eighteenth century permitted an imperial favorite like Ho-shen . . . to acquire a fortune of eighty million taels within a score of years, is paralleled by the kind of government which in the first century B.C. permitted an imperial favorite like Tung Hsien to acquire one hundred million cash within the space of ten months. . . . As against the unjust punishments suffered by many Ming and Ch'ing officials, we should remember that of the seven highest ministers who served Emperor Wu of the Han between 121 and 88 B.C., only one enjoyed a natural death; all the others were executed or committed suicide. . . . In short, further study seems necessary before we may definitely conclude that despotism was a phenomenon increasingly characteristic of late imperial times, rather than one that fluctuated with the times according to particular individuals and circumstances. (Bodde 1959–1960:448)

A general survey like this is obviously not the place for such a further study. Here it need only be noted that long stretches of China's middle imperial period were characterized either by political weakness and disunion (from the fall of Han to Sui and again during considerable spans after the An Lu-shan rebellion of 755), or by relative intellectual diversity (Buddhist and Taoist strength and corresponding Confucian weakness before the Buddhist persecution of 845), or often by both. By contrast, the period after, say, the fall of Northern Sung (1125) would seem, like that of the early empires (Ch'in and Han), to have been characterized by a very considerable degree of political and social authoritarianism; one, moreover, that in some respects—certainly not all—very possibly became increasingly severe as the post-Sung dynasties wore on.[30]

---

30. An example of "certainly not all" is the prominence of serflike subcommoner groups in pre-Sung China as against the disappearance of most of them in later times. Or again, the ranks of the bureaucracy, which had long been monopolized by a handful of aristocratic families, became open to a much wider social spectrum from Sung times onward. How very difficult it is to make any broad generalizations is illustrated, for the later dynasties, by the penalties listed in the surviving law codes, T'ang through Ch'ing, for criminal acts involving familial and social relationships. Sometimes these seem to point to a slight tightening of hierarchical differences (especially within the family), sometimes to a slight loosening (especially between social classes), and sometimes to no change at all.

Despite the great difficulty of determining political and social trends with anything like quantitative exactness, one conclusion seems indisputable: The establishment of Neo-Confucianism as the sole state orthodoxy from the early fourteenth century onward, even if it did not, as often alleged, lead to an actual narrowing of political, social, and intellectual horizons, at the very least powerfully impeded all efforts to widen these horizons. In the words of Nivison (1959:23): "Neo-Confucianism was a movement of thought in great ferment in the Sung, exhibiting great variety of opinion and range of interest. In the Ming, it was rigidified and its character as a state ideology was defined."

The Sung dynasty has sometimes been described as a Chinese "Renaissance" because of both its many social and economic changes and its revival of Confucianism.[31] Tempting though the idea is, I believe it to be misleading unless carefully qualified.[32] In the first place, the Chinese break with antiquity had never been as profound or prolonged as that of Europe, so that the analogy between the recoveries of antiquity in the two civilizations is only partially apt.[33] Second, the Sung social and economic developments, despite their magnitude, failed in the end to generate ongoing changes of the sort that, in Europe, led inexorably to our modern world. Third, and most relevant to this discussion, the Sung saw the creation of a new holistic world view in Neo-Confucianism which retained its hold as the official orthodoxy down to the end of the empire, despite the efforts of dissenting thinkers

---

Examples: (1) *Tightening of differences:* The killing of a "disobedient" child/grandchild by a parent/grandparent is punishable in the T'ang/Sung codes by a year and a half penal servitude if performed by beating, two years if performed by a knife; in either case the "disobedience" must have been to a parental order that could "properly" have been obeyed. In the Yuan/Ming/Ch'ing codes, on the other hand, the latter provision is dropped and no penalty is provided at all unless the killing has been done "inhumanely," in which case it is a mere one hundred blows of the heavy stick. (2) *Loosening of differences:* The T'ang/Sung codes specify strangulation for a slave/son who brings a public accusation of wrongdoing against a master/parent, whereas the Ming/Ch'ing codes prescribe the same only if the accusation is false; if it is true, the penalty is reduced to one hundred blows of the heavy stick and three years penal servitude. (3) *No change:* In all codes, T'ang through Ch'ing, a husband who beats his wife to death is punished by strangulation. (4) *Again no change:* In all codes, a commoner who strikes a local official suffers three years penal servitude if no real injury is inflicted, but strangulation if the striking results in a broken tooth or limb, blinded eye, or the like. See Ch'ü (1961:23–24, 107, 183, 192).

31. See, for example, Gernet (1982:298 and 330ff.).

32. Gernet (1982:298) agrees that the term "Renaissance" is open to possible criticism and can be used only to suggest a "very general parallelism."

33. This criticism has already been made by Anderson (1974:530 n. 32).

to change it.[34] This situation contrasts dramatically with the dissolution of the old holistic world view in Renaissance Europe and the intellectual liberation that resulted.

It is difficult not to see a correlation at this point between, on the one hand, the freezing of officially sanctioned thought and the drying up of scientific interest in post-Sung China and, on the other, the expansion of intellectual horizons and flowering of a new kind of science in Renaissance and post-Renaissance Europe. Here, surely, is a profound intellectual difference—one all the more impressive in that its beginnings can already be seen in the Mencian defense of the ancient Tao of the Sages in third-century B.C. China, as against the rise of the Socratic dialogue in fifth-century B.C. Athens.

But before we accept this as our final conclusion, it is well to consider the question of how intellectual persecution and censorship have operated on the two sides of Eurasia. In Europe, as everyone knows, the record of intolerance and suppression has been incredibly bloody and prolonged; though usually centered, at least ostensibly, on religion, it has often spilled over into political and other spheres. Even in ancient Athens, just praised as the home of the Socratic dialogue, men like Aeschylus, Euripides, Phidias, Aristotle, and of course Socrates himself were forced into exile, flight, imprisonment, or execution because of ideas.

In China, as we saw in the preceding chapter, religious persecutions were far fewer, less violent, less prolonged, and less based on purely doctrinal differences. Much more frequent was the use of military force to suppress or forestall the uprisings, actual or alleged, of politico-religious sects and societies. In addition, significant governmental proscriptions of literature—the great bulk of it nonreligious—occurred from time to time. Earliest and most notorious was the Ch'in Burning

---

34. I have in mind not only the reactions to the Ch'eng-Chu school of Wang Yang-ming and other Ming thinkers, but more especially the radical criticisms made by Yen Yüan, Tai Chen, and other Ch'ing thinkers. Not only did none of these men, despite their intellectual achievements, succeed in gaining official acceptance of their ideas, but their thinking, as Fung (1953:630–631) has made clear, failed in the final analysis to break away from the Sung Neo-Confucian framework. The government, for its part, retained its rigidity down to the very end, as illustrated by the edict issued by the Kuang-hsü Emperor on 18 July 1894, ordering all provincial governors to prohibit the sale of the *Ssu-shu kai-ts'o* (Correction of Errors Concerning the Four Books), a large work in 22 *chüan* compiled by the iconoclastic scholar Mao Ch'i-ling (1623–1716) when in his eighties, in opposition to Chu Hsi. The same edict warns examination candidates that they must continue rigorously to follow the Chu Hsi interpretations. See Hoang (1902:155). On Mao's opposition to Neo-Confucianism, see also Elman (1984:53).

of the Books of 213 B.C., a perfect example of Legalist political policy. Much later, under the Sui Emperors Wen (590–604) and Yang (605–617), a suppression of the Han prognostication *(ch'an)* and apocryphal *(wei)* texts was so successfully carried out that almost all of these bizarre writings were permanently destroyed aside from scattered quotations in other works (Tjan 1949–1952:105–106). Under the Yuan dynasty, the Emperor Khubilai twice (in 1258 and 1281) ordered the burning of Taoist books and woodblocks in order to halt Buddhist-Taoist controversies. And during the Ming there were many instances of individuals whose works were destroyed while they themselves suffered execution or other punishment; the usual reason was alleged subversion or *lèse majesté.* Similar cases continued during the early Ch'ing down to the beginning of the Ch'ien-lung reign (1736). The largest of them, that of 1663, led to the execution of some seventy scholars who had been involved in the writing of a private history of the Ming dynasty.[35]

But by far the most extensive of all book destructions was the literary inquisition systematically pursued by the Ch'ien-lung Emperor between 1772 and 1788. The total number of proscribed book titles amounted to 2,665, of which 2,320 were listed for complete suppression and 345 others for suppression in part. So thorough was the inquisition that of this huge total, only about 476 titles, or less than 18 percent, have survived. Primarily, the inquisition was directed against writings deemed to be either seditious or disrespectful to the Manchus or earlier conquerors of China, such as the Mongols. Few Chinese writers living between approximately 1550 and 1750 failed to have at least one title on the index.[36]

Yet this largest of all Chinese literary proscriptions,[37] effective though it was, was probably smaller than its European religious counterpart of some two centuries earlier. A counting of the items listed in the first Papal *Index Librorum Prohibitorum* of 1559, for example, reveals that it

---

35. For Ming dynasty persecutions, see Ku Chieh-kang (1938). For the Yuan and Ch'ing cases, see Goodrich (1935:3–4, 19–29). For an extensive collection of Chinese documents relating to the banning of works of fiction and drama during the Yuan, Ming, and Ch'ing dynasties, see Wang Hsiao-ch'uan (1958).

36. Goodrich (1935:esp. 61 for the statistics).

37. No statistics, of course, are available for the Ch'in book burning of 213 B.C., but it is quite unlikely that it equalled the Ch'ien-lung inquisition in size, for the simple reason that far fewer books existed in third century B.C. China than in China of the eighteenth century. Thus two hundred years after the Ch'in inquisition, the Han imperial library of the early first century A.D. contained only 677 titles, according to the catalogue reproduced in *Ch'ien Han shu* 30; 524 of these titles, or 77 percent, have since been totally lost. See Teng and Biggerstaff (1971:7–8).

condemned the writings of 601 authors in toto, as well as 441 other separate book titles (117 of them by known authors, 324 others of uncertain authorship). In addition, it proscribed 31 editions of the Bible, 12 other editions of the New Testament, and the total printings of 61 publishers.[38] Inasmuch as many or most of the totally condemned authors probably produced well over one book title each, there can be little doubt that the total number of prohibited items easily exceeded the 2,665 titles covered by the Ch'ien-lung inquisition.

Furthermore, the 1559 Index was followed by a proliferation of later editions across the face of Europe during the seventeenth and eighteenth centuries. By contrast, so quickly did the Ch'ien-lung inquisition become a dead letter after its cessation in 1788 that a mere decade later (1798), when the Ch'ien-lung Emperor was still alive, it was possible for one of the proscribed works, a book on the Sung patriot Yüeh Fei, to be republished with no apparent realization that it had ever been prohibited (Goodrich 1935:17 n. 18 and 61).

Or again, as against Europe's many massive and systematic intellectual suppressions, those of China, with only a few notable exceptions, tended to be directed rather sporadically against individuals or fairly small groups.[39] And whereas the European suppressions were usually based, at least ostensibly, on genuine intellectual and religious issues (even though sometimes seemingly trivial or tangential today), the Chinese suppressions only very rarely involved purely intellectual issues.[40] Much more often, as already indicated, they involved political charges, resting frequently on the use by the accused of seemingly obscure or

---

38. These figures are based on my own count of the items listed in the Lisbon edition of the 1559 Index, which is reproduced in facsimile in Révah (1960:183–259). Curiously, descriptions in scholarly literature of this and later editions of the Index never seem to include statistics of the actual number of authors and titles prohibited.

39. The few major examples of Chinese literary proscription have just been discussed. The major religious persecutions have been listed in IV.2 above n. 45. Literary persecution during the Ming, described in Ku (1938), was fairly frequent but almost always directed against individuals or small groups only.

40. One of the few exceptions is Hsieh Chi-shih (1689–1756), whose commentary on the *Ta hsüeh* (Great Learning), because it allegedly opposed the views of the Ch'eng-Chu school, almost led to his execution in 1729. In 1741–1742 his writings were destroyed at the orders of the Ch'ien-lung Emperor. Even in this instance, however, the persecution may have been as much political as ideological, because already in 1726 he had been sentenced to exile for having dared to accuse the governor of Honan of corruption; his exile continued until a general amnesty in 1735. See Goodrich (1935:24–25, 83–93) and Fang Chao-ying (1943a:306–307).

trivial statements alleged to bear cryptic subversive connotations.[41] A good example is a monk who lived during the transition from Yuan to Ming, and who was executed by the Ming founder because he had written in a poem: "In the days of great heat there is no place to cool oneself." This was interpreted by the emperor as "meaning that my penalties are too severe" (Ku 1938:307).

One may surmise that the real significance of religious and intellectual persecution in the West was not so much its intensity and persistence per se, but rather the situation implied by these characteristics: the fact that the persecutors often held less than total power over the persecuted, that the latter were often ready to die if necessary rather than submit, and that the issues at stake were felt on both sides to be sufficiently vital to justify the struggle. Exemplifying the fluctuating nature of that struggle in Reformation times is the fact that the successive Indexes, despite their enormous scope and circulation, were minatory only; their promulgation by no means necessarily ensured the actual elimination of the writings they censured. On the contrary, many Protestants replied by deliberately trying to acquire the books that were prohibited (McKeon 1974:1087).

Here again the contrast on the Chinese side is illuminating. In our earlier discussion of religious persecution (IV.2), it was noted that little or no attempt was ever made by the institutional religions to resist. As there suggested, this was partially the result of organizational and ideological weaknesses within the religions themselves. Even more, however, it reflected the complete acceptance on their part of the authority of the ruler and state in ideological matters. The same also seems to have been true when one turns to the nonreligious literary proscriptions: no one ever seems to have seriously resisted the Ch'ien-lung inquisition or the other destructions of literature that have been cited.

Unquestioning acceptance of the emperor's total authority was prevalent in early as well as late stages of the empire. The historian Ssu-ma Ch'ien, for example, writing in 93 or 91 B.C. about the terrible punishment of castration he suffered because he had dared to defend at court the cause of the renegade general Li Ling, refers to Wu, the responsible emperor, as "our enlightened Ruler" *(ming chu).* He states revealingly that "of my friends and associates, not one would save me; among those

41. No doubt this is an extension into politics of the theories of the school of historiography that favored the use of subtle phraseology to express moral praise or blame for the events narrated. The school goes back to Mencius and to the early commentators on the *Ch'un ch'iu* (Spring and Autumn Annals), who believed (erroneously) that this text had been composed by Confucius to express his moral judgments. See, inter alia, Fung (1952:61–62 and 1953:71ff.), as well as beginning of VI.1b below.

near the Emperor no one said so much as a word for me," and he blames only himself for having "brought upon myself the scorn and mockery even of my native village and . . . soiled and shamed my father's name."[42] Seventeen hundred years later (1625), when the Ming censor Tso Kuang-tou was being tortured to death in prison after loyal service to three emperors, he wrote despairingly to his family: "Only thus can I make recompense to the Emperor. . . . My body belongs to my ruler-father. . . . I only regret that this blood-filled heart has not been able to make recompense to my ruler." (Hucker 1959:208).

In short, viewing the situation in Reformation Europe in the light of the Chinese experience, one may conclude that the greater intensity of its persecutions, coupled with the greater resistance these provoked, far from pointing to the effectiveness and totality of thought control, indicates a healthy intellectual diversity and vigor. Such an environment would seem to be peculiarly favorable to scientific development, provided of course the forces of suppression did not become too strong.[43]

Having said this, we should not ignore the many solid scientific and technological achievements in China that could not possibly have been accomplished without governmental control and support. In technology, especially, it is obvious that the Great Wall, the Grand Canal, and numerous other massive constructions would have been unthinkable without initiative by the government. In scholarship, the same holds true for the many huge encyclopedias, dictionaries, histories, collectanea, and other works compiled under imperial auspices by large boards of scholars.

Although government control was by no means invariably applied to the nontechnological sciences, it was all-important in the case of astronomy (including the making of the calendar) because the age-old belief in the oneness of humankind and nature kept the government under a never-ending obligation to maintain careful study of the heavenly bodies.[44] This is why Chinese observations of eclipses, sunspots,

---

42. Quoted in Watson (1958:62, 66). See also Chavannes (1895–1905:1.ccxxxii, ccxxxviii). For the dating of Ssu-ma Ch'ien's famous letter to Jen An, where these sentiments appear, see Watson (1958:Appen. B).

43. Cf. the remark of Herbert Butterfield (1969:39): "If religion produced the authoritarian system, it also produced the rebellion against the system, as though the internal aspect of the faith were at war with the external. The total result over the long medieval period may have been a deepening of personality, a training of conscience, and a heightening of the sense of individual responsibility, particularly in the matter of religion itself."

44. On the "official" character of Chinese astronomy, see Needham (1959:186–195), where, among much else, it is pointed out (p. 193) that private study of astronomical

comets, novae, and other celestial phenomena are so very much more detailed and continuous than those of any other civilizations before modern times (Needham 1959:458–459). Furthermore, government support of astronomy was responsible for the construction of massive observational instruments, such as the giant masonry gnomon made in 1276 at a spot some fifty miles southeast of Lo-yang. It consisted of a truncated brick pyramid topped by a forty-foot-high gnomon (now no longer there), fronted by a template extending more than 120 feet across the ground (Needham 1959:294–302).

Nevertheless, the official character of Chinese astronomy was probably also an important reason why this science failed to progress beyond a certain point. The earliest complete surviving Chinese law code of A.D. 653 forbids the private possession of "all instruments representing the celestial bodies, astronomical charts and writings, . . . calendars of the seven luminaries [sun, moon, and five planets]," as well as military writings and various kinds of prognostication literature and divination boards. Violators of the statute are punishable by two years of penal servitude.[45] The prohibition is repeated, with slightly varying language, in all subsequent codes down to the end of the Ch'ing dynasty.

When scientific interest was at its peak during the Sung dynasty, a few bold spirits are known to have ignored the prohibition and to have kept astronomical instruments for their own use. Among them were the astronomer Su Sung (1020–1101) and Chu Hsi, both of whom possessed armillary spheres (Needham 1959:193–194). Yet these were almost surely very rare exceptions. Continuing observance of the prohibition explains why the Western mathematical treatises of Matteo Ricci were taken from him while he was on his way from South China to Peking in 1600. (He later got them back through a fortunate mistake.) To be sure, Ricci himself wrote that "in China it is forbidden under pain of death [*sic*] to study mathematics without the king's authorisation, but this law is no longer observed." (Needham 1959:194). This remark, however, should not be interpreted as meaning that the state had by then actually relaxed its hold on astronomy. Its real implication is that by 1600 most Chinese scholars had already lost interest in mathematics and astronomy, so that for them the legal prohibition had in fact become a dead letter. Important among the several causes responsible for this situation was surely the age-old governmental restriction of the study of astron-

phenomena was ipso facto strongly suspect, because it might either be connected with the preparation of nonofficial (and hence politically subversive) calendars or with the collecting of astronomical portents for purposes of political criticism.

45. See *T'ang lü shu-yi* (T'ang Code with Commentary) vol. 2(9/82), art. 110.

omy to all but a tiny handful of officials. The disastrous results for Chinese astronomy by the time of Ricci's arrival are known only too well.[46]

In this section we have seen that civilization was conceived by many Chinese thinkers as being embodied in an orthodox Tao or Way of Life that had been created in antiquity and thereafter transmitted by a line of sages. The line was at first thought to have ended with Mencius; but when Neo-Confucianism arose more than a millennium later, members of that school claimed to have recovered the Tao from its long eclipse and, through the efforts of their greatest figure, Chu Hsi, to have brought it to final glory.

This belief, of course, was particularly Confucian. In late Chou times, however, some of the other philosophical thinkers shared the view that a unitary Tao had once existed and that the intellectual diversity of their day, far from being an admirable manifestation of the human mind, was a sad result of the loss of the earlier holistic *oikoumene.* Disputation, for them, tended to be equated with sophistry or regarded as an expedient only, to be used to combat the heretical teachings of a disordered age. Not all thinkers, of course, shared this view. Exceptions were the School of Names (which was therefore often disparaged) and the later Mohists.

Insistence on the sanctity of the Tao could on occasion lead to authoritarianism, bigotry, or xenophobia. During the centuries of disunion following the Han, Confucianism was prevented by political weakness and the strength of Buddhism and Taoism from playing a dominant role. But dogmatism and xenophobia are conspicuous in what may be called Han Yü's Confucian "fundamentalism."[47] Thereaf-

46. Difficulties in accurately predicting the sun eclipses of 1528, 1540, 1584, and 1592 induced two prominent scholars to submit separate memorials strongly urging that the calendrical system be reformed. Neither one, however, led to any concrete result. See Peterson (1986:49–55). Involvement of the Jesuits began with the sun eclipse of 15 December 1610. This, like its predecessors, was miscalculated by the Imperial Bureau of Astronomy but correctly calculated by the Jesuit Didace de Pantoja (despite the Jesuit rejection then and until the end of the eighteenth century of the heliocentric theory of Copernicus). The next fifty-nine years (1610–1669) saw, among other things, miscalculation of yet another eclipse (1629), the overthrow of the Ming dynasty by the Ch'ing (1644), and a bitter intrigue launched against the Jesuits (1659 onward) by a prominent Chinese scholar, Yang Kuang-hsien. The intrigue ended in 1669 with Yang's disgrace and death, coupled with the K'ang-hsi Emperor's appointment of a Jesuit, Ferdinand Verbiest, to serve as associate director of the Astronomical Bureau. During the next century and a half (1669–ca. 1827), the bureau was always staffed by one or more Westerners who used Western techniques to calculate the Chinese calendar. See, inter alia, Fang Chao-ying (1943b).

47. This term is applied to Han Yü in De Bary (1959:34).

ter such qualities occasionally recur in the political and social thinking of the Ch'eng-Chu school, especially after it achieved undisputed orthodoxy in 1313.

It has been argued by many scholars that the post-Sung centuries were marked by a steady growth of political absolutism, accompanied by a parallel rigidifying of social mores. One may debate whether the tendencies were really as unilinear as thus postulated, also the extent to which Neo-Confucianism should be blamed for what allegedly happened. But at the least it must be recognized that the great liberation of thinking that characterized the European Renaissance failed to be duplicated in post-Sung China.

This was despite the occurrence of significant social and economic changes during the Sung itself in some ways remarkably reminiscent of those found in Europe some centuries later. It was also despite the challenges posed first by rival movements within Neo-Confucianism and then, during the seventeenth and eighteenth centuries, by iconoclastic followers of the "School of Evidential Research."[48] Throughout, the Ch'eng-Chu school continued to enjoy the official support of the governments of successive dynasties. The result is that few aspirants for official careers via the civil service examinations could hope for success in these examinations unless their answers to the questions were in substantial accord with the Ch'eng-Chu interpretations of the classics.

At the same time, in contrast to the rise of the new science that accompanied intellectual liberation in Europe, scientific thinking tended to stagnate and decline after the Sung. This phenomenon is surely connected with factors deeply rooted in the society and government of imperial China. Yet the intellectual side of the picture should likewise not be disregarded—particularly so in view of the fact that orthodoxy shows itself at such a very early stage in Chinese thinking, already before the creation of the empire was to shape a society and government that in many respects were new.

We have seen that Chinese religious and intellectual persecution, while more severe than popularly supposed, failed to reach the intensity of persecution in the West. In most cases it was motivated more by pragmatic political considerations than by purely intellectual issues. Most important of all, we have seen that the lesser intensity of Chinese persecution was probably more the consequence of intellectual acquiescence on the part of its victims than of any willingness on the part of the persecuting government to show tolerance toward unorthodoxy. The power of the emperor to suppress whatever he deemed undesirable was axiomatic to all and therefore rarely resisted. In the West, on the other

48. See III.1 and esp. VI.1f. below.

hand, the prevalence of persecution may in a curious way be interpreted as symptomatic of the intellectual vitality of the contending forces and hence as indicative of a favorable soil from which science could grow.

Chinese governmental monopoly of certain fields of science and technology led to achievements that would have been unattainable by private effort alone. On the other hand, such monopoly could raise these achievements only to a certain level and no further. Ultimately they were followed by decline. We may speculate that if, in the West, a single state or church had continued indefinitely to exercise the same control over astronomy, for example, as was taken for granted in China, the Copernican-Galilean revolution probably would not soon and possibly would not ever have taken place.

2. The Organicist View of Mankind

Confucianism very rarely looks at the human individual simply qua individual. Almost always it sees him or her as functioning within a series of concentrically larger social units: the family, the local community, the several greater jurisdictional entities culminating in country or empire, and finally the known human world, *t'ien hsia,* "all-under-heaven." Often, especially in early times, empire and *t'ien hsia* are thought of as coterminous. Beyond all these, yet inextricably interwoven with them, lies the greater world of nature. Between each smaller and larger entity, notably the family and the state, the human and the natural worlds, there exists a paired relationship: the family is a microcosm, the state a macrocosm; the human world is a microcosm, the natural world a macrocosm.

This kind of thinking, in which each social entity is organically related to every other and all cooperate on differing levels to achieve the integrated functioning of the total organism, is an expression of the same organicist world view that Joseph Needham has so often stressed in his writings about China (see, e.g., Needham 1956:287f., 294, etc.). Its man/nature aspects will be discussed in a later section (VII.2b; see esp. n. 66). Here we are only concerned with its purely human aspects.

On the human side, organicist thinking was no doubt initially encouraged by Confucian nostalgic memories—some might call them idealizations—of the early Chou political system. Under this system, the Chinese *oikoumene,* ruled by the Chou Son of Heaven, was divided into numerous principalities, many of them small enough to be administered by hereditary nobles on a semipersonal basis with only a minimum of administrative machinery; these were in turn subdivided into smaller and therefore even more personally administered domains and estates.[49]

By the time the empire, with its bureaucratized government and vast territorial and demographic expansion, came into being (221 B.C.), the Confucian organicist view of society had already become firmly established. Thereafter it simply shifted its focus from the king, principalities, and domains of feudal China to the emperor, provinces, and districts or counties *(hsien)* of the empire.

But of course it was the family, far more than any political factors, that was basic in shaping the Confucian view of society. "Microcosm" and "macrocosm" are convenient Westernisms for describing the family/state relationship as long as we remember that they suggest only a dimensional superiority of the latter over the former, not a priority in importance or time. In actual fact, the hierarchical relationships within the patriarchal family were felt by Confucians to be so utterly "natural" and instinctive to man as to constitute inevitable prototypes for the more extended hierarchical relationships in society at large. That such thinking actually predates Confucianism is demonstrated by the occurrence twice in the *Shih ching* of the term "people's father and mother" *(min chih fu mu)* to designate the ruler. The first occurrence reads:

> Happy be the lord,
> The light of our state.
> Happy be the lord,
> The father and mother of the people.[50]

Not only did similar phraseology continue to be applied to the emperor throughout imperial times. Perhaps more important is that the magistrates of the thirteen hundred and more districts into which imperial China was divided—men who, among the emperor's surrogates, had closest contact with the ordinary people—were known in popular parlance as *fu mu kuan*, "father-and-mother officials."

Confucius (in *Lun yü* 12.11) has given the briefest, as well as one of the most forceful, enunciations of the microcosm/macrocosm relationship between family and state. It consists of only four separate words, each repeated twice. Good polity, he maintains, obtains when *chün chün, ch'en ch'en; fu fu, tzu tzu.* This literally means:

---

49. The system is studied in detail in Creel (1970:ch. 11), whose general conclusions suggest that the Confucian view of early Chou government may have been less idealized than is often supposed.

50. *Shih ching* no. 172; tr. Karlgren (1950a:116). See also the similar language in no. 251; tr. ibid. 208.

[The] ruler rules, [the] minister ministers;
[The] father fathers, [the] son sons.

Stylistically, this apothegm derives its power from three factors: its extreme brevity; the repetition of each word twice, changed each time from a noun to a verb; and above all, the absolute quantitative and qualitative parallelism maintained between the two four-word cadence units, each subdivided in turn into paired antithetical phrases. Graphically, the balanced structure can be represented by the formula: Aa:Bb ::A′a′:B′b′. Spatially, the sentence provides an example of dualistic symmetry.[51]

Although, in the equation, the state is placed before the family, no doubt because of its larger size, there is no indication elsewhere that Confucius viewed the state as the more important of the two. On the contrary, his preference for the family emerges clearly in the well-known story in which he is told about a person so upright that when that person's father stole a sheep, the person informed the authorities of the fact. "With us," Confucius comments dryly, "uprightness is different from this. The father conceals the son and the son conceals the father. Therein lies uprightness" (*Lun yü* 13.18).

In imperial times, the primacy of the family was enshrined in the law codes by a provision exempting close relatives from testifying in court against one of their members. More remarkable is the further provision, known from Han times onward, that a son who brings an accusation of parental wrongdoing before the authorities is thereby deemed to be unfilial and hence subject to heavy punishment. Under the Ch'ing Code of 1740, for example, a son's accusation, if false, was punished by strangulation; even if true, it brought to the son three years of penal servitude plus one hundred blows of the heavy stick. "Probably China is the world's only country where the true reporting of a crime to the authorities could entail legal punishment for the reporter." Only in serious cases of treason and rebellion was the right of family concealment denied; in such situations, close relatives of the offender were either executed or suffered permanent exile. "Thus we see that when the Confucian state felt its existence to be *really* threatened, it was willing to forgo its Confucian precepts."[52]

Confucian thinkers rarely say much, however, about the possibility of conflict between family and state loyalty. Apparently they prefer to think that it would not arise if people would only carry out more ear-

---

51. Discussed in III.2. Stylistic parallelism and antithesis are the subject of II.5.

52. Quotations and discussion from Bodde and Morris (1967:40–41). See also Ch'ü (1961:70–74).

nestly the social duties enjoined on them by the ancient sages. Mencius stresses that people are above all social beings, whose awareness (if not always fulfillment) of the social relationships and their obligations distinguishes them from animals. "To be without ruler or father," he says, "is to be a bird or a beast." Again: "The sage is the apogee of the human relationships *(jen lun)*." These relationships, five in number, are enumerated by Mencius for the first time in Confucian literature in what was to become their standard sequence. Together with the qualities that he says should properly characterize each, they are: father and son (affection), ruler and minister (correct conduct), husband and wife (differentiation), old and young (hierarchical distinction), and friend and friend (good faith).[53] As has often been remarked, only the last of the relationships is based on social equality. Striking also to the modern mind is the choice of the word "differentiation" *(pieh)* to typify the marital relationship. Within its patriarchal context, it of course denotes the wife's inferior status, socially as well as physically.

In classical Confucianism, Hsün Tzu more than anyone else is interested in giving Confucian social theory a rational basis. His key term is *fen*, "social distinctions." Observance of the *fen* is essential, he argues, for the functioning of any society:

> If people live together but are without social distinctions *(fen)*, there will be strife. . . . To get rid of . . . such calamity, there is nothing like clarifying the social distinctions when forming social aggregations. If the strong coerce the weak, the intelligent terrorize the stupid, and the people below rebel against their superiors; if the young insult the aged and the government is not guided by virtue—if this be the case, then the aged and weak will suffer the misfortune of losing their subsistence, and the strong will suffer the calamity of division and strife. . . . This is why wise men have created social institutions on their behalf. (*Hsün-tzu* 10; tr. Dubs 1928:152–153)

When Tung Chung-shu formulated the curious amalgam of socioethical and naturalistic thinking that constitutes Han Confucianism, he reinforced the family and social relationships by linking them hierarchically to the cosmic yin and yang principles: "In each correlation there is the yin and the yang. . . . Thus the relationships between ruler and subject, father and son, husband and wife, are all derived from the principles of the yin and yang. The ruler is yang, the subject yin; the father is yang, the son yin; the husband is yang, the wife yin" (*Ch'un-ch'iu fan-lu* 53[12/5b–6a]).

In post-Han times, these and other relationships were regularized in the penal codes of successive dynasties by provision of penalties that

53. For these three statements, see respectively *Meng-tzu* 3b.9.para. 3; 4a.2; 3a.4.para. 8.

varied according to the relative social or family status of the offender vis-à-vis his victim. Illustrative on the social side is the statute in the Ch'ing Code covering a beating that results in no wound. (A wound is defined as discoloration of the skin or other more serious injury.) This act, when committed between social equals (a commoner striking a commoner or a slave a slave), is punishable by the standard penalty for this offense of twenty blows of the light bamboo; when committed by a slave against a commoner, the penalty is increased by one degree to thirty blows; but when committed by a commoner against a slave, it is diminished by one degree to only ten blows (Bodde and Morris 1967:33).

Within the family, offenses were precisely measured in terms of the five degrees of mourning. This system, originating in the classical ritualistic literature, specified that each member of a family should mourn the death of any other member for a period of time and wearing a kind of mourning garb that differed according to the closeness of relationship. The first or highest degree (a twenty-seven-month mourning period) applied to only four major relationships: son (or unmarried daughter) who mourns his father, wife who mourns her husband, concubine who mourns her master, father who mourns the eldest son by his principal wife. The fifth or lowest degree (three months), on the other hand, extended to more than forty relationships, including such unlikely possibilities as male ego's (self's) grandfather's spinster first cousin or female ego's (self's) husband's grandnephew wife. The degrees were not necessarily reciprocal: a husband stood toward his wife in a degree 1 relationship (she would mourn him for twenty-seven months), whereas a wife, because of her inferior status, stood to her husband only in a 2a relationship (he would mourn her for only one year). Legal penalties were fixed accordingly: a wife who struck her husband would, under the Ch'ing Code, receive a minimum punishment of one hundred blows of the heavy bamboo, whereas a husband who struck his wife would receive no punishment at all unless he inflicted significant injury and she lodged a formal complaint with the authorities, in which case he would be liable to eighty blows of the heavy bamboo.[54]

It would be easy to adduce many more Confucian statements of social theory, but they would not add substantially to what has already been said. Instead, we turn for a final very expressive utterance to a Taoist—one, however, who as a member of the third and fourth centuries A.D. revival of philosophical Taoism known as Neo-Taoism represents a kind of Taoism deeply influenced by centuries of Confucianism. Kuo Hsiang (d. A.D. 312), commenting on a passage in the *Chuang-tzu,* writes:

---

54. Bodde and Morris (1967:36–37) and, for a more detailed account of the five-degree mourning system, Ch'ü (1972:313–317).

> Error arises when one has the qualities of a servant but is not satisfied to perform a servant's duties. Hence we may know that [the relative positions of] ruler and subject, superior and inferior, . . . conform to a natural principle of Heaven and are not really caused by man. . . . Let the servants simply accept their own lot and assist each other without dissatisfaction. Let them assist each other like the hand and foot, the ear and eye, the four limbs and the hundred other parts of the body, each having his own particular duty and at the same time acting on behalf of others. . . . Let those whom the age accounts worthy be the rulers, and those whose talents do not meet the requirements of the world be the subjects, just as Heaven is naturally high and Earth naturally low. . . . Although there is no [conscious] arrangement of them according to what is proper, the result is inevitably proper.[55]

Society as a cooperative organism is *the* basic premise from which Confucianism, throughout its history, has launched its attack on other schools of thought. Mencius bases himself squarely on it when he criticizes both Yang Chu for his ultra-egoistic "acting for self" *(wei wo)* and Mo Tzu for his ultra-altruistic "all-embracing love" *(chien ai).* The former, he says, destroys, by his excessive narrowness, the broader allegiance owed by every subject to his ruler; the latter, by his excessive looseness, destroys the particular affection owed by every son to his father. But Confucianism, by following a middle course, ensures proper recognition of both obligations (*Meng-tzu* 3b.9.para. 9).

Ssu-ma T'an (d. 110 B.C), father of the great historian Ssu-ma Ch'ien and himself the initiator of the latter's *Shih chi,* was a man of Taoist sympathies. Nevertheless, his pungent critique of the Legalists so perfectly sums up the Confucian point of view that it deserves quotation here:

> The Legalists make no distinction between kindred and strangers, nor do they differentiate the noble from the humble. All such are judged by them as one before the law, thereby sundering the kindliness expressed in affection toward kindred and respect toward the honorable. (*Shih chi* 130; tr. Watson 1958:46)

Chu Hsi, in a passage providing a good example of "organismic" thinking, strongly criticizes Buddhists and Taoists alike for their repudiation of the normal family and social relationships. He points out that despite this repudiation, they find it impossible to escape the compelling need of all men to live in organized groups, as shown by the fact that they create monastic communities with hierarchical relationships similar in character to those within the family.[56]

In sum, it is evident that Confucian social thinking, by its constant

55. Kuo Hsiang's commentary on *Chuang-tzu* 2; tr. Fung (1953:227).

56. *Chu-tzu yü-lei* (Classified Conversations of Chu Hsi) 126/4829; tr. Fung (1953: 569). See also Sargent (1955) for an extended presentation of Chu's criticisms of Buddhism.

emphasis on human relationships rather than separate human beings, is strongly anti-individualistic from a Western standpoint. When it talks about the individual, it is concerned with the person's social *responsibilities,* not his or her individual *rights.* Indeed, the word "rights" has no real semantic equivalent in Literary Chinese. Confucian social thinking encourages harmony and cooperation, decries conflict and competition, and at this point unexpectedly joins forces with Taoism in lauding humility and yieldingness *(jang).*[57]

Between Confucian social thinking and that of the early Greeks there are both similarities and differences. As Bertrand Russell has pointed out (1945:598): "The philosophers of Greece, down to and including Aristotle, were not individualists in the sense that I use the term. They thought of a man as essentially a member of a community. . . . [But] with the loss of political liberty from the time of Alexander onwards, individualism developed, and was represented by the Cynics and Stoics." More specifically, Russell has this to say about Aristotle's political thinking:

> In order of time, the family comes first; it is built on the two fundamental relations of man and woman, master and slave, both of which are natural. Several families combined make a village; several villages, a State. . . . The State, though later in time than the family, is prior to it, and even to the individual, by nature; for "what each thing is when fully developed, we call its nature," and human society, fully developed, is a State, and the whole is prior to the part. The conception involved here is that of *organism:* a hand, when the body is destroyed, is, we are told, no longer a hand. The implication is that a hand is defined by its purpose—that of grasping—which it can only perform when joined to a living body. In like manner an individual cannot fulfill his purpose unless he is part of a State. (Russell 1945: 185–186)

As against the similarities suggested by these words, it is easy to find equally significant dissimilarities: the relative disinterest of the Greeks in the family (so that Plato was even ready in his *Republic* to abolish it), and, instead, the Greek concentration on the city-state as the basic sociopolitical unit (so that Aristotle, the tutor of Alexander, proclaimed the best state to be the one that can be seen in its entirety from a hilltop). This view contrasts sharply with that of the Chinese, who at least as early as the beginning of the Chou dynasty thought of the entire human world, the *t'ien hsia* (all-under-Heaven), as being ruled by a sin-

---

57. On Confucian and Taoist emphasis on yieldingness, see Needham (1956:61–63). For an exceptional analysis of the role of compromise in Chinese culture, see Creel (1987). Individualism will be discussed further in VI.3 below.

gle Son of Heaven.[58] But most significant of all, the Greeks, who originated democracy, also justified slavery on grounds that certain individuals or races are by nature inferior (Aristotle), whereas the Confucians, despite their acceptance of an authoritarian and nondemocratic form of government, rejected a frozen caste system.

For this saving grace, Confucianism was indebted above all to Mencius, with his insistence that all individuals are potentially morally perfectible, and hence potentially capable of passing from one class to another. Yet it should be stressed that acceptance of the idea of individual social mobility was by no means confined to Mencius or even to Confucianism as a whole. During the Warring States period, social thinkers often advocated the doctrine that the appointment of individuals to government office should be determined by their moral and intellectual capacities, irrespective of birth. This idea paved the way for the development of the civil service examination system in imperial times.[59]

The real Western parallel to Confucian social organicism, then, does not lie in Greek thinking but in the holistic worldview formulated by the Christian Church during the Middle Ages. The following brief summary by R. H. Tawney reads almost as if it might have been written by a Confucian:

> The facts of class status and inequality were rationalized in the Middle Ages by a functional theory of society. . . . Society, like the human body, is an organism composed of different members. Each member has its own function. . . . Each must receive the means suited to its station, and must claim no more. Within classes there must be equality. . . . Between classes there

58. See Creel (1970:93–94, 418, 441). Also the famous lines from *Shih ching* no. 205, tr. Karlgren (1950a:157–158), as modified by Creel (1970:418):

> Under the vast heaven
> There is no land that is not the King's.
> To the very borders of the earth
> There are none who are not the King's servants.

59. See Hsü Cho-yun (1965:141–146) for a discussion of philosophical attitudes toward social mobility during the Warring States period. Hsü quotes Mo Tzu, for example, as saying: "Therefore in administering the government, the ancient sage-kings ranked the morally excellent high and exalted the virtuous. If capable, even a farmer or an artisan would be employed. . . . If a person is capable, promote him; if incapable, lower his rank. . . . Here, then, is the principle." See *Mo-tzu* 8; tr. Mei (1929:32–33). This statement should be compared with Mo Tzu's insistence (in III.3) on the need for the absolute intellectual conformity of inferior to superior. Such a curious amalgam of sociopolitical authoritarianism with belief in the moral and intellectual potentialities of persons of humble social origin is found not only in Mo Tzu and Confucianism, but in much Chinese thinking generally.

> must be inequality. . . . Peasants must not encroach on those above them. Lords must not despoil peasants. Craftsmen and merchants must receive what will maintain them in their calling, and no more. (Tawney 1926:22–23)

The analogy of the human body (like Aristotle's of the human hand) reminds us of Kuo Hsiang's reference to the parts of the body. The four enumerated social classes are almost the same as those of Confucianism (the sole difference being the replacement of the Confucian scholar by the European feudal lord), and very remarkably, even the order of enumeration is closely similar (the sole difference being that the Confucian sequence places the scholar ahead of the peasant).[60]

The first major breach in the medieval European synthesis of dogma, law, and custom came from Protestantism, which, by questioning the infallibility of General Councils, encouraged individuals to exercise their own judgment. "To determine the truth thus became no longer a social but an individual enterprise. . . . Meanwhile individualism had penetrated into philosophy. Descartes' fundamental certainty, 'I think, therefore I am,' made the basis of knowledge different for each person" (Russell 1945:598–599). In the case of modern science, the role of individualism, though less conspicuous than in some other fields, proved to be vital:

> The outlook of the typical scientific discoverer has perhaps the smallest dose of individualism. When he arrives at a new theory, he does so solely because it seems right to him; he does not bow to authority. . . . At the same time, his appeal is to generally received canons of truth, and he hopes to persuade other men, not by his authority, but by arguments which are convincing to them as individuals. In science, any clash between the individual and society is in essence transitory, since men of science, broadly speaking, all accept the same intellectual standards, and therefore debate and investigation usually produce agreement in the end. This, however, is a modern development; in the time of Galileo, the authority of Aristotle and the Church was still considered at least as cogent as the evidence of the senses. This shows how the element of individualism in scientific method, though not prominent, is nevertheless essential. (Russell 1945:599)

In China, as we saw in the preceding section, the breach in "dogma, law and custom" failed to come until the end of the empire. The Neo-Confucian synthesis, despite attacks made on it by some, retained its hold down to the beginning of this century, and its social thinking became, if anything, more rather than less rigid as the centuries wore on. Not only did it oppose attempts by particular individuals to seek for knowledge outside the framework of transmitted authority. In addition, its insistence on the primacy of social harmony and stability above all

---

60. These matters are the subject of the next section.

else meant an equally firm general opposition to almost any kind of change. By 1600, when science in the West was beginning its great qualitative transformation, scientific progress in China, with only a few exceptions, had ground to a halt.

3. Social Classes

Traditionally, the Chinese have thought of society as consisting of four major categories or classes. In descending order of prestige these are *shih, nung, kung, shang,* or "scholars, farmers, artisans, merchants." Collectively, the four are known as the *ssu min,* "four peoples," that is, "four categories of people." The term sometimes appears in book titles in the sense of "everyone", for example in the now fragmentary almanac, the *Ssu min yüeh ling* (Monthly Ordinances for the Four Categories of People), by Ts'ui Shih (d. ca. A.D. 170).

The four-step sequence, with scholars in first place and merchants last, has remained fixed since Han times as a Confucian scale of values. As we shall soon see, however, it probably has a Legalist origin. The classification omits at least four other groups that would be significant in the West: clergy, nobility, military, and slaves.[61] We have noted previously (see IV.1) that the absence of a well-organized priestly class in ancient China distinguishes it from many other early civilizations. Furthermore, as also noted in the preceding section, Confucian scholars usually had a low opinion of the Buddhist and Taoist clergy.

The omission of a hereditary nobility is not surprising because, though all-important in pre-imperial China, these men (aside from the emperor himself) were rarely allowed, under the empire, to hold much political power; their once dominant role was replaced by that of the scholar-gentry known as *shih,* about whom more directly. Already in Warring States times, in fact, as we saw in the last section, political thinkers were advancing the radical doctrine that moral and intellectual abilities, rather than good birth alone, should be important considerations when awarding governmental office.

As to the military, their exclusion from the traditional listing no doubt reflects the age-old Confucian antipathy to warfare and violence, as well as the Confucian conviction that a military class does not properly belong to a truly well-ordered state. A further pragmatic consideration is that Chinese military forces were inherently impermanent because

61. Ch'ü T'ung-tsu, in his study of Han social structure (1972), conforms to Chinese tradition by excluding both clergy/priests and military/soldiers from his index. On the other hand, he devotes much space to both nobility and slaves.

they were usually made up of short-term peasant conscripts rather than long-term mercenaries.

Finally, it is striking how little is said by Confucian and other social thinkers about slavery.[62] A major reason, perhaps, is that this institution, generally speaking, was not economically very significant and the proportion of slaves to the total population was usually quite small. Nevertheless, there is good evidence (aside from Tung Chung-shu's memorial just cited in n. 62) that slavery became a matter of some political concern during the Former Han dynasty. The decisive indication is Wang Mang's edict of A.D. 9, prohibiting the buying and selling of slaves (which, however, had to be rescinded three years later owing to the opposition it aroused).[63] Further, as also noted earlier in this chapter, the surviving law codes of later dynasties (especially T'ang and Sung) pay considerable attention to slaves and other subcommoner groups. I therefore suspect that the real reason for their exclusion from the traditional social ranking is the conviction that they, like the military, are anomalies in the well-ordered state and hence should not be recognized as a distinct social group.

The names of the four categories—*shih, nung, kung, shang*—are used in the texts with much latitude and a minimum of definition. For Chou times, when the word *shih* first appears, its rendition as "scholar" is anachronistic. In such a context, therefore, it will here be rendered as "gentleman"—with the understanding, however, that it is primarily a term of social status, in contrast to *chün tzu,* which, as we shall see shortly, can also be rendered as "gentleman," but with a strongly moral connotation.

It is questionable whether during the earliest centuries (the Western Chou), the *shih* could really be thought of as a recognized social class at all. Most often the word then seems to mean "an official" (who no doubt would have had to be of patrician origin). In other instances, however, it seems merely to signify "a male (as distinguished from a female) person" (Creel 1970:331–333). But by the Spring and Autumn period (722–481) a definite social stratum seems to be intended, consisting of the younger sons and peripheral branches of ministerial and official families, who thus stood at the bottom of the aristocratic ladder. Having no fixed hereditary positions of their own, these men had to live

62. Tung Chung-shu, in a memorial to the throne of about 100 B.C., recommended the elimination of slavery, but this was not one of his philosophical writings. See *Ch'ien Han shu* 24A; tr. Swann (1950:183). I do not agree with Swann that Tung merely wanted slavery to be reformed, not abolished.

63. On Han slavery there is the extensive study by Wilbur (1943), in addition to the discussion in Ch'ü (1972).

by their wits, and hence commonly possessed professional skills and a modicum of learning. Some were warriors, some stewards in charge of the fief towns and estates of the higher nobility, and some conceivably even tilled landholdings of their own in common with tenant serfs. During the later Chou (the Warring States period), when Chinese society was undergoing radical social, political, and economic changes, many of the *shih* succeeded in gaining positions at the expense of the declining high nobility. This dissolution of the old fixed social order paved the way, when the empire came, for the *shih* to change from petty Chou aristocrats into the "scholar-gentry"—rooted economically in rural landlordism—from whom China's officialdom was thereafter primarily recruited during most of imperial history.[64]

The second term, *nung*, "farmer," is loosely used to designate anyone working the land, whether as independent farmer, tenant cultivator, or hired laborer. It does not include "gentleman farmers," who do not cultivate the land themselves, for these would be members of the *shih* or "gentry." *Kung*, "artisans," likewise covers a wide range. Primarily the word refers to skilled craftsmen and technologists (what we in the latter case would call engineers, architects, and the like), but it also, to a lesser degree, includes any kind of moderately skilled or completely unskilled manual workers, aside from those in agriculture. Finally *shang*, "merchants," similarly covers a wide range of mercantile and industrial activities, all the way from large-scale traders, entrepreneurs, industrialists, and exploiters of natural resources down to petty shopkeepers and itinerant peddlars. Occasionally, *shang* is coupled with another word, *ku*, or even replaced by it, and when a distinction is insisted on, *shang* is said to refer to itinerant merchants and *ku* to those who remain in one place. In the course of time, however, this distinction faded, and *shang* became the comprehensive term for all forms of mercantile-industrial activity.

Not until the late Chou did the four-part social classification become clearly formulated. Previously the only recognized distinction was that between the small group of "superiors" (*shang*; not the same *shang* as that for "merchants") and the much larger mass of "inferiors" (*hsia*)—that is, in modern parlance, between the "haves" and the "have-nots." The two basic terms used in the early texts to express this dichotomy are *chün tzu*, "lord's son," and *hsiao jen*, "small (or petty) man."[65] Other

64. On the evolving status of the *shih* down to the creation of the Ch'in empire in 221 B.C., see Hsü (1965:7, 37, 89–90, 150–151).

65. See, for example, *Shih ching* no. 203 (tr. Karlgren 1950a:154), which says of the great highways of Chou:

> That is where the noblemen *(chün tzu)* tread
> And where the commoners *(hsiao jen)* look on.

parallel terms serving to designate the "superiors" are *pai hsing* or "the hundred surnames"[66] and *ta jen,* "great man." For the "inferiors," the most common word is *min,* "the people," either used alone or prefixed by descriptive epithets in terms like *shu min,* "the multitudinous people," *chung min,* "the numerous people," or *li min,* "the black-haired people";[67] some of the same prefixes may also be joined to *jen,* "person," as in *shu jen,* "the multitudinous persons" (Vandermeersch 1965:134). Another exceedingly common but ambiguous word, known from earliest times, is *ch'en,* "servant, subject, minister." According to context, it can equally well refer either to the total population (all of whom are "servants" or "subjects" of the ruler) or, more particularly, to the highest ministers (who par excellence are the "servants" or "subjects" of the ruler).

Beginning with Confucius, the meanings of *chün tzu* and *hsiao jen* shifted from designations of status to designations of moral worth.[68] Yet even much later, mention of the *chün tzu* (Legge's "superior man") would at the same time evoke, like its English counterpart "gentleman," strong suggestions of its aristocratic origin.[69]

The traditional two-layer division of high and low continued through Mencius, who has presented its classical formulation in two famous statements:

> Some labor with their brains and some labor with their brawn. Those who labor with their brains govern others; those who labor with their brawn are governed by others. Those governed by others, feed them. Those who govern others, are fed by them. This is a universal principle in the world. (*Meng-tzu* 3a.4.para. 6)

> If there were no men of superior quality *(chün tzu),* there would be no one to rule the country folk. If there were no country folk, there would be no one to support the men of superior quality. (*Meng-tzu* 3a.3.para. 14)[70]

---

66. In early times only the nobles possessed surnames, so that it was natural for *pai hsing* to have an upper-class connotation. But later, when surnames became common to all, the term reversed its meaning to become the everyday designation of the common people.

67. We cannot go into the tangled question here of whether *li,* "dark," really means "black-haired" in this term, as one tradition has it, or is equivalent to *chung,* "numerous." Karlgren (1944:39.Gloss no. 430) favors the latter interpretation.

68. In the *Lun yü* the two terms are often used in moral apposition, as in 2.14: "The superior man *(chün tzu)* is catholic and not partisan; the small man *(hsiao jen)* is partisan and not catholic."

69. How the meaning of *chün tzu* shifted from early to late Chou times is discussed in detail in Hsü (1965:158–173).

70. Similar statements by other thinkers of the same general period are quoted in Ch'ü (1957:236).

How and when, then, did the four-part social division originate? We can attempt an answer only after evaluating a mass of data that, because it is technical, has been relegated to the Appendix. Thus what follows in the next six paragraphs is only a summary of what is said in this Appendix. The presentation there rests on a table listing twenty-nine passages from Chou and Han texts, each containing references to the four social classes (or variants of them).

Among these twenty-nine passages, the twenty-three that are of Chou date contain so many variants and inconsistencies as to indicate clearly the uncertainty with which the new four-part conception was advanced during the latter part of the Warring States period. Especially revealing are the averages of the twenty-three passages as a whole, prepared by assigning numerical values to each of the four categories within each passage. The resulting profile indicates a wide gap between a *shih/nung* cluster at the top and a *kung/shang* cluster below. Furthermore, it also indicates that in the *shih/nung* cluster there is little preference for *shih* over *nung* (their numerical values are respectively 69 and 63) and in the *kung/shang* cluster there is none at all for *kung* over *shang* (both have values of 41). In short, the profile not only demonstrates the tentative nature of the late Chou four-part formulation but also the absence of any marked bias then in favor of the artisan as against the merchant.

All six Han texts, on the other hand, follow the "standard" sequence precisely, with an even progression that unmistakably places *shih* at the top and *shang* at the bottom. Thus we may justifiably conclude that the standard sequence of *shih, nung, kung, shang* assumed final form only in Han times (ca. 200 B.C. onward) and that only then did the prestige of the merchant drop sharply below that of the artisan.

This anti-mercantilism may in part have been connected with unfavorable governmental reaction to the considerable growth of private trade and industry that characterized the Former Han, despite efforts by the government to control it. Even more important, however, may have been the increasing Confucianization of official thinking during the Han. As we shall see, Confucian objections to the merchant were often more moral than economic.

In the Chou dynasty, however, it is unlikely that Confucian thinking played any role in stimulating the formulation of the four-part classification system, nor does that system appear in either Mohism or Taoism. A much more probable source is the thinking of the political scientists known as Legalists, who were interested in quantitative techniques (see III.4) and wanted to create a powerful state by imposing strict controls upon the people. To this end, they might well have been interested in efforts to classify the population along socioeconomic lines.

Surprisingly, however, the four-part system scarcely appears in two of the major Legalist texts: only once in the *Shang chün shu* (Book of Lord Shang) and not at all in the *Han Fei-tzu.* The most probable source, therefore, seems to be the economically oriented chapters of the eclectic text known as the *Kuan-tzu.* These chapters, in contrast to the harsh dictates of the *Shang chün shu* and *Han Fei-tzu,* display a softer and more humane kind of Legalism. Their central focus is upon techniques for centralized government control designed to increase agricultural output and generate a stable economy—in short, techniques for creating what would seem to be an authoritarian but relatively humane welfare state.

In the Appendix's table of twenty-three Chou textual references to the four classes, it is significant that six of them come from the *Kuan-tzu* and two others from a text narrating the same particular event found in one of the *Kuan-tzu* passages. As to the authorship of the relevant *Kuan-tzu* chapters, the most widely held theory today maintains that they were probably written during the first half of the third century B.C. by scholars then belonging to the Chi-hsia Academy in the state of Ch'i.

Let us see now how the *Kuan-tzu* looks at the four categories. Generally speaking, it recognizes the primacy of agriculture over the other two nonintellectual activities as supplying the basic food needs of the state. Yet it also recognizes that all four categories, and not just agriculture, have their useful contributions to make. This viewpoint appears in such passages as the following:

> Farmers *(nung)* should begin cultivating the soil as soon as the snow melts away. . . . Gentlemen *(shih)* should listen and pay heed to those who are broadly learned. . . . Merchants [reading *ku* here instead of the usual *shang*] should be aware of the rise and fall of prices and should daily go to the marketplace. . . . Artisans *(kung)* should devote themselves to the form and function of their products. (*Kuan-tzu* 5[1/21a–b]; tr. Rickett 1985:123)

> Gentlemen *(shih),* farmers *(nung),* artisans *(kung)* and merchants *(shang):* these four categories of people *(ssu min)* are the pillars (lit. "stones") of the state. (*Kuan-tzu* 20[8/6b]; tr. Rickett 1985:325)

> Let the gentlemen [i.e., officials] be incorrupt, the farmers simple-minded, and the merchants and artisans honest. (*Kuan-tzu* 30[10/13a]; tr. Rickett 1985:403)

There is no evident hostility here toward artisans and merchants despite their joint ranking below the scholar-officials and farmers. Rather, what the *Kuan-tzu* authors really fear is that an imbalance will develop between the three nonintellectual occupations, leading to the neglect of farming and overdevelopment of the other two:

> If the sovereign does not hold dear the basic occupation [agriculture], the secondary activities [those of artisans and merchants] will then not be kept in

> check. If these secondary occupations are left unchecked, the people will then become lackadaisical in their seasonal [i.e., agricultural] activities and indifferent to the profits of the land. . . . If merchants stand in the court, goods and wealth will flow upward [in the form of bribes]. (*Kuan-tzu* 3[1/10b]; tr. Rickett 1985:95)

These words no doubt reflect the growing social fluidity of late Chou times, in contrast to the old fixed social system in which most of the people had been attached as serflike peasants to the land they cultivated. Private land ownership, which Shang Yang (d. 338 B.C.), as prime minister of Ch'in, had apparently initiated in that state, became universal for all of China a century and some decades later under the Ch'in empire. The reform, however, does not seem in the long run to have improved the lot of the farmer, especially when, under the empire, it was coupled with a very low land tax. This encouraged wealthy persons to buy land and become a new landlord class, while farmers who had been free were forced into tenantry.[71]

Many who could do so, in consequence, apparently left the land and tried to seek other livelihood. From the third century B.C. downward through the Former Han dynasty, the cry is repeatedly raised that farmers have lost their former simplicity, that they are no longer willing to continue their hard way of life, and that they are therefore abandoning the primary occupation *(pen)* of agriculture in order to go into the easier and more profitable secondary occupations *(mo)* of trading and production of luxury goods. The *Kuan-tzu* itself explains:

> Today those who engage in the secondary occupations . . . can feed themselves for five days through one day's work, whereas the farmer's work through the entire year is not enough for him to feed himself. This is why the people are abandoning the fundamental occupation for the secondary ones, . . . with the result that the fields are left fallow and the state becomes impoverished. (*Kuan-tzu* 48[15/14b]; tr. T'an and Wen 1954:94)

In 178 B.C. the Han statesman Ch'ao Ts'o wrote very similarly in a memorial to the throne:

> The great traders and merchants hoard their stocks to make one hundred percent profits; the lesser ones range themselves in their stalls to sell what has been released. . . . Therefore the men [in merchant families] neither plow nor weed and their women neither tend silkworms nor weave. Yet their clothing is always embroidered and variegated and their food always includes fine grain and meat. They have none of the farmer's bitter toil, yet gain profits of ten- or a hundred-fold. . . . The laws today demean the merchant, yet the

---

71. See, inter alia, Duyvendak (1928:53).

merchant enjoys wealth and honor; they honor the farmer, yet the farmer suffers poverty and lowliness.[72]

The efforts of the Former Han government to cope with the agrarian crisis by imposing restrictions on mercantile activity are well known. On the whole they probably did little to close the economic gap between the merchant and the farmer. The latter's only consolation was his realization—if indeed he realized the matter at all—that he was regarded by his betters as a person morally superior to the merchant, and one whose work was supremely important to society.

In early Han times, antimercantile feeling seems to have been particularly strong among those traditionally minded Confucians who distrusted "bigness of government" and looked nostalgically back at what they idealistically believed to have been a simpler and more honest age. These are the men who very possibly were responsible for making antimercantile sentiment a lasting element in Confucian thinking. Their opposition, however, rested more on moral than on economic grounds. By contrast, the Legalistically minded administrators who often dominated the Former Han government (despite that government's official dedication to Confucianism, beginning around 136 B.C.) regarded the growth of trade and industry as desirable and inevitable within the context of an expanding empire, provided only it was subjected to close government control. In short, these Han statesmen seem to have been very much the ideological heirs of the economically oriented kind of Legalism advocated in the *Kuan-tzu*.

The differences between the two viewpoints emerge clearly in Huan K'uan's remarkable book, the *Yen t'ieh lun* (Discourses on Salt and Iron). This work, though its text has no doubt been "prettified" to some extent both in content and style, provides a valuable record of the debate on economic and social issues held in 81 B.C. at government initiative between high members of the central government and nonofficial Confucian scholars from the provinces.[73] The following representative utterances from the government side manifestly recognize the reciprocal role played by different social groups in the national economy:

Without the output of the artisan, the needs of agriculture will be left unsatisfied. Without the output of the merchant, the supply of valuable commodities will be cut off. When agricultural needs are left unsatisfied, grain will not

72. Quoted in *Ch'ien Han shu* 24A; tr. Swann (1950:164–166). Also tr. in Duyvendak (1928:54–55). For a similar memorial by Chia Yi, also of 178 B.C., see *Ch'ien Han shu* 24A; tr. Swann (1950:154).

73. For a convenient brief summary of the major arguments recorded in this work as having taken place between the two sides, see Loewe (1974:91–112).

> be planted. When valuable commodities are cut off, material needs will be unmet. (*Yen t'ieh lun* 1; tr. Gale 1976:6)

> Artisans, merchants, makers of wooden utensils, and carpenters all serve the needs of the state. . . . The exchanges between farmers and merchants serve to benefit both what is basic *(pen)* and what is secondary *(mo)*. (*Yen t'ieh lun* 3; tr. Gale 1967:22–23)

The replies of the literati are typically moralistic and backward looking:

> The ancients esteemed physical effort, devoted themselves to the basic occupation, and so the plantings of seeds and trees multiplied. They personally cultivated the soil and rushed to their seasonal tasks, and so food and clothing were sufficient. (*Yen t'ieh lun* 2; tr. Gale 1967:13)[74]

> Anciently, merchants circulated goods without any prearranged scheming and artisans displayed solid honesty without any trickery. . . . [But now] merchants are good at deception and artisans cover up their trickery. (*Yen t'ieh lun* 2; tr. Gale 1967:15)

In a much later chapter, the government speakers, perhaps remembering how the literati had praised the manual labors of the ancients, sharply censure them for their own present neglect of farming:

> Today the literati discard their plow handles and shares to study unverifiable statements, to which they devote the livelong day without any benefit to truth. They come and go in aimless wandering, enjoying food without tilling the soil and wearing clothes without rearing silkworms. They make a clever appearance of being decent people, so as to take from the farmers and interfere with the government. Indeed they are a tribulation to the present age! (*Yen t'ieh lun* 20; tr. Gale 1967:167–168)

To this remarkable diatribe the literati reply with surprising meekness. They cite the example of the great Yü, who was too busy coping with the flood to participate in person in agriculture. Then they go on to paraphrase Mencius' dictum (see above) that just as it is the duty of the brain worker to govern the manual worker, so it is equally that of the manual worker to feed the brain worker. "It would not be proper," they conclude, "for such a one [the brain worker] to till and weave like an ordinary man or woman. Were the superior man *(chün tzu)* to till the soil and not study, this would be the road to anarchy."

The doctrine of the separation of intellectual from manual work goes

---

74. "The ancients" can scarcely mean members of the *shih* class here, because these, at least in Han thinking, would not themselves expect, or be expected, to work with their hands. See immediately below.

back to Mencius,[75] and through him to Confucius. When a disciple once asked the Master to be allowed to study husbandry and then to study gardening, Confucius informed him in no uncertain tones: "I am not as good for that as an old farmer" or "as an old gardener." And when the disciple had walked out, the Master indignantly turned to the other disciples and termed the departed one "a small man" *(hsiao jen)*. Then, after mentioning some of the moral qualities that should be developed by the "superior man," the *chün tzu,* he explained that, following the acquisition of these qualities, "the people will come to him from all quarters, bearing their children on their backs. What use has he for husbandry?" (*Lun yü* 13.4).

The idea that the man at the top, the *chün tzu,* must be broadly trained in moral and humanistic values and avoid becoming a technical specialist is summed up by Confucius in the famous apothegm: "The *chün tzu* is not a utensil" (*Lun yü* 2.12). *Ch'i,* "utensil," epitomizes the specialization of the subordinate, and Confucius uses the word again, this time verbally, in another passage (*Lun yü* 13.25), in which he contrasts the *chün tzu* with the small man *(hsiao jen)*. The *chün tzu,* he says, "when he employs men, utensils them" (*ch'i*'s them); in other words, he assigns to each subordinate that particular task for which the subordinate is best suited. Conversely, the small man, "when he employs men, expects completeness in them"; in other words, he expects each person, no matter how lowly and specialized, to be able to do anything. The consequences of the doctrine of nonspecialization will be discussed in the next section.

From Confucius down to the twentieth century, the feeling of innate superiority with which the Confucian literati viewed members of the other social classes remained essentially unchanged. The statement that follows happens to come from the sixth century writer Yen Chih-t'ui, but it might equally well have been made in any other century. "I have often seen scholars," he writes, "who felt it disgraceful to associate with farmers and merchants, or to exert themselves like an artisan or craftsman."[76]

4. Consequences of Class Attitudes

In the three subsections that follow, we shall successively focus upon the scholar-official *(shih),* the merchant, and the artisan. Nothing more will

75. See *Meng-tzu* 3a.4, where Mencius, in rebutting the views of the agriculturalist Hsü Hsing, argues that the ruling of a state is too important to be combined with the ruler's personal cultivation of the soil. See also the two quotations at n. 70.

76. Yen Chih-t'ui (531–591), *Yen-shih chia-hsün* (Family Instructions for the Yen Clan) 8; tr. Teng Ssu-yü (1968:52).

be said about the farmer save incidentally because, in my opinion, his mode of life only indirectly influenced, and in turn was influenced by, the course of Chinese scientific and technological development. For the most part, despite his endlessly patient role as the economic provider of Chinese society, he was the recipient, rather than the generator, of whatever social forces might be at work. Occasionally, when conditions became unbearable, he might rebel (almost always in concert with disaffected intellectuals), and the rebellion might lead to bloody warfare or even, very occasionally, to a change of dynasty. Then for a while such bloodletting might slightly improve the conditions of rural life for those who survived. Yet in the longer run the farmer always seemed to end up very much where he had started.

4a. The Chün Tzu *as a Generalist*

Let us start with the motto quoted from Confucius above: "The *chün tzu* is not a utensil." These words naturally helped to perpetuate the gap between intellectual and nonintellectual. But more than this, they also encouraged the scholar to concentrate his efforts on broadly humanistic studies—the classics, the major histories, the major literary works—at the expense of other more specialized and more scientific subjects, such as mathematics, astronomy, law, medicine, biology, or civil and hydraulic engineering. Thus the overwhelming majority of educated men who entered the civil service did so on the basis of what would today be called a "liberal arts" education.[77]

The distinction is seen clearly in the workings of the examination system from the T'ang dynasty onward.[78] On the one hand, the system provided a broadly literary kind of examination, known as the *chin shih* or Doctor in Letters;[79] on the other, it included a series of more specialized examinations, such as those on particular classics (humanistic, like the *chin shih* examination, but more specialized) and on mathematics, law, and writing (primarily etymology rather than calligraphy). Still

77. In the case of astronomy, it will be further remembered (V.1 at n. 45) that the government itself prohibited its study by anyone but a relatively small number of selected government students.

78. On the T'ang examinations, the classic study by Des Rotours (1932:esp. 27–29, 228–229, 233, 279–280) is indispensable. This should be supplemented by his companion work on the T'ang official system: Des Rotours (1947–1948:esp. 208–217, 339–344, 453–456).

79. *Chin shih* literally means "advanced scholar." The term "Doctor in Letters" is that used by Kracke (1953:61 passim), discussing the examination system in Sung times.

other examinations in astronomy and medicine were given by departments of government outside the civil service organization.

The creation of such a broad educational curriculum in the seventh century was an extraordinary achievement, but its practical value was severely curtailed by the very small number of candidates who chose to take their doctorate in anything but the generalized *chin shih* or Doctor in Letters. Statistics for the T'ang dynasty preserved in the *Wen-hsien t'ung-k'ao* encyclopedia (ca. 1319) indicate, for each year the examinations were given, the number of awarded *chin shih* degrees as compared with degrees in "the several [other] fields" *(chu k'o)*. Although no breakdown of the "several fields" is provided, they probably included the examinations on particular classics, mathematics, law, and writing, but not, for reasons just indicated, on astronomy and medicine. The statistics in Table 3, gathered for three random decades from early, middle, and late T'ang, demonstrate the great preponderance of *chin shih* degrees.

Table 3 **T'ang Dynasty Doctorate Degrees**

Decades	*Chin shih* degrees	Other doctorates
659–668	28	3
712–721	303	47
860–869	290	102
Total	621	152

SOURCE: *Wen-hsien t'ung-k'ao* 29/276c–280a

Reluctance to specialize continued and if anything increased during the Sung dynasty, as shown by the dropping of mathematics and writing from the examinations. In 973, a year for which statistics for each kind of examination happen to be available, only five persons passed the examination in law, as against twenty-six in letters (the *chin shih* degree), twenty-two in particular classics, twenty-nine in history, and forty-five in rituals. During the decade 1058–1067, 74 percent of all successful examination candidates took their degree in letters as against all the other more specialized fields combined (Kracke 1953:63–64).

Several centuries later (around 1600), near the end of the Ming, Matteo Ricci summed up the situation as it then applied to mathematics and medicine as follows:

> It is evident to everyone here that no one will labor to attain proficiency in mathematics or in medicine who has any hope of becoming prominent in the

> field of philosophy [i.e., of the classics as interpreted by Chu Hsi]. The result is that scarcely anyone devotes himself to these studies, unless he is deterred from the pursuit of what are considered to be the higher studies, either by reason of family affairs or by mediocrity of talent. The study of mathematics and that of medicine are held in low esteem, because they are not fostered by honors as is the study of philosophy, to which students are attracted by the hope of the glory and the rewards attached to it. (Gallagher 1953:32)

Of course this Chinese situation was by no means unique in 1600. As a Western parallel, let us consider the early seventeenth-century curriculum at Cambridge.[80] The leading studies at that time were classics, rhetoric, and divinity; mathematics was slighted and the various sciences practically ignored. During William Harvey's years at Cambridge (B.A. in 1597, M.D. in 1602), the so-called medical course was principally devoted to logic and divinity rather than "physick." And even as late as about 1630, the university statutes threatened Bachelors and Masters of Arts who failed to follow Aristotle faithfully with a fine of five shillings for every point of divergence from the *Organon*.

Nevertheless, at that very time a remarkable change in thinking was already well in the making. The young John Milton, when he delivered his third oratorical exercise in Latin at Cambridge sometime between 1625 and 1632, expressed the thought that though poetry, oratory, and history are all delightful, they are insufficiently useful; hence more study of such natural sciences as geography, astronomy, and natural history would be desirable. And by mid-century, mathematics was becoming such a prominent academic influence that in 1663 Isaac Barrow gave up his Greek professorship to accept the newly established Lucasian Professorship of Mathematics (in which he was Newton's predecessor). Before taking this step, he had complained, "I sit lonesome as an Attic owl, who has been thrust out of the companionship of all other birds; while classes in Natural Philosophy are full."

To return to China, the effects there of the persistent rejection of specialization are nowhere more apparent than in the field of administration. Such nonspecialization may in fact have constituted *the* major difference between China's traditional bureaucracy and the bureaucracies of modern societies, where sociologists from Weber onward have regarded functional specialization as a necessary feature of *all* bureaucracy.[81] The lack of it in China meant, to cite but one conspicuous example, that almost all the men who served as district magistrates

80. The information in this and the following paragraph comes from Merton (1970: 29–31, 95–96, 229–230).

81. C. K. Yang (1959:136, 163); also 136–146 and 161–164 for discussion of the whole topic of generalism in Chinese bureaucracy.

throughout the empire came to their posts with only the most amateurish knowledge of law and legal procedure—this despite the fact that the trying of law cases was one of their most demanding duties. A solution attempted by many Ch'ing magistrates (and no doubt those of earlier dynasties as well) was to engage the services of a private secretary who was well versed in legal matters. Such a secretary was in no way a part of the formal civil service. He might in fact be an unsuccessful examination candidate who had then taken up this specialization as a way of earning a living. Thus his services to the magistrate had to be made from behind the scenes, so to speak. He was the privately paid employee of the magistrate, whom he commonly accompanied on the latter's peregrinations from one post to another.[82]

Besides such private secretaries, the personally paid staff of the district magistrate commonly included both operatives (clerks, deputies, police, etc.) and servants, whose recruitment and employment took place entirely outside of the formal civil service. The separate existence of this legally unregulated substratum created tensions between it and the formalistically and impersonally oriented bureaucracy—tensions that remained unresolved down to the end of the empire. A possible solution might have been the extension of the formal bureaucratic system to include the staff organization, but this step was never taken—a clear testimony to the persisting strength of the dichotomy between the generalized *chün tzu* above and the specialized "small men" below (Yang 1959:161–163).

That the bureaucratic system nevertheless functioned as well as it did was due to several factors. One was the relative simplicity of the administrative machinery needed for governing a pre-industrial society, which meant that the entire civil service personnel of nineteenth-century China approximated only 40,000 officials. Even more important was that the formal bureaucracy of the central government was thinly superimposed upon a much denser network of informal social institutions and personal relationships, differing from community to community, through which most of the functioning of each community was actually effectuated. The glue that held these formal and informal systems of social control together was their common acceptance of Confucian moral values. It has even been suggested that such features in the traditional bureaucracy as nonspecialization, which to modern eyes seems like a weakness, may in fact have given it the necessary flexibility to coexist with a larger social order that to a considerable extent was self-

82. Yang (1959:160–162) and especially Ch'ü (1962:ch. 6), which discusses in detail the private secretaries of magistrates, some of whom might be specialists in other fields besides law.

functioning. This would in turn help explain why the traditional Chinese bureaucracy was able to maintain such extraordinary durability (Yang 1959:164).

However, the system could function effectively only as long as its social environment remained relatively unchanged. When external forces of the nineteenth century violently changed that environment, the system was unable to provide any adequate response. Then the traditional Chinese concept of the intellectual as a gifted amateur generalist proved incompatible with the ideas underlying modern science and technology. Furthermore, the concept placed China's literati at a hopeless disadvantage in their political and military confrontation with the industrializing West. It has been said of these literati, as they lived and thought on the eve of this confrontation, that they were "amateurs in the fullest sense of the word, genteel initiates in a humane culture, without interest in progress, leanings to science, sympathy for commerce, or prejudice in favor of utility. Amateurs in government."[83]

4b. The Role of the Merchant

The prevailing bureaucratic attitude toward the merchant from Han times until at least the middle of the T'ang (the An Lu-shan rebellion of 755) has been well summarized in the following words:

> The prime objective of state policy was a settled, stable and contented peasant population, carefully registered and controlled, which would provide regular and ample taxation in kind, and be readily available for labour service or military service when required. In such a society, the merchant was conceived of as a disturbing factor. . . . Not only did he convert to his own profit much of the peasant's surplus production. . . . Not only was he the advocate of a materialistic attitude . . . repugnant to the ethical precepts of Confucianism. . . . He also provided the population with a model of a possible means of social advancement based purely on the acquisition of wealth. . . . Moreover, he was an unstable element in society.[84]

The governmental regulations imposed on the merchant during the first nine centuries and more of the imperial age were of several kinds.

---

83. Levenson (1957:323), in his study of the amateur ideal among Ming and early Ch'ing intelligentsia, especially, but not exclusively, as exemplified in painting. To some readers the judgment may seem overly sweeping.

84. Twitchett (1968:65). The article provides a summary (1968:66–74) of the merchant's status until the middle of the T'ang and then (1968:74–95) of the growing subsequent changes. Much use of it has been made for the three paragraphs that follow. See also Twitchett (1963). Another useful survey along similar lines is that of Yang Lien-sheng (1970).

Socially, he was subjected to fluctuating but always onerous sumptuary regulations which, at various periods, forbade him to wear fine silks or furs, ride carriages or horses, carry arms, or make overly elaborate displays at weddings or other family observances. These sumptuary regulations, especially in post-T'ang centuries, were undoubtedly often violated or tacitly ignored. Yet they persisted in softened form down to the present century and could be invoked at any time if the government deemed it advisable.[85]

Economically, until the later T'ang, much of the commercial activity in large cities was confined to walled-in marketplaces where the hours of trading, quality of goods, freedom of exchange, accuracy of weights and measures, and much else were all subject to governmental inspection and regulation. The government also imposed controls on overseas trade and, within China, limited the freedom of merchants to travel from their own county to other regions. Fiscally, the system of tax payments in kind (primarily grain and cloth) gave the government an active role in the transport and distribution of these commodities, thus limiting the opportunities for private trading. Legally, during the T'ang, merchants and artisans alike were denied grants of government land unless a land surplus happened to exist in the region where they lived. Most important of all, both groups were excluded from the civil service examinations and thus from participation in the government.

From the mid-T'ang on, economic and social changes led to considerable improvement in merchant status, and many former restrictions, such as the segregation of trade within the government-controlled marketplaces, gradually diminished or disappeared. One of the results was the creation of new, large merchant fortunes, reminiscent of what had happened during the Han. The salt merchants who sold at retail to the public what they bought at wholesale from the government's salt monopoly are conspicuous examples. And again, as in the Han, the new wealth inspired numerous diatribes from the literati. Typical are the following lines of a ballad written by the poet Po Chü-yi around 800:

> The salt merchant's wife
> Has gold and silk in plenty.
> Yet she does not work in the fields or tend silkworms.[86]

---

85. Besides Twitchett (1968), see Ch'ü (1961:137–154) for a detailed discussion of sumptuary regulations of many sorts.

86. Compare the remarks made by Ch'ao Ts'o in 178 B.C, as quoted in sect. 3. Po's entire 29-line poem is translated in Twitchett (1968:85–86). The same article (1968: 81ff.) translates a series of other similarly biting comments by contemporaries of Po Chü-yi.

From Sung times onward, continued improvement in the status of the merchant led to a narrowing of the gap between him and the scholar-official. Some officials became involved in business enterprises, probably usually surreptiously, and some merchants, or at least members of their families, succeeded in entering officialdom. This they had to do by indirect means, because of the continued legal prohibition that prevented merchants and artisans alike from taking the examinations.[87] The most common device available to them was that of contributing grain in time of emergency with the expectation of receiving office in return.[88]

Yet despite the heightened influence exerted by many individual merchants during the last several centuries, the commercial class, considered as a class, never succeeded in effectively challenging the overall dominance of the scholar-official class. (Perhaps it would be more accurate to say that it never really tried.) Most important of all, it failed to produce an ethic that significantly differed from, and could compete with, the dominant Confucian ethic. These facts are strikingly illustrated by the great salt merchants of eighteenth-century Yang-chou (at the juncture of the Grand Canal with the Yangtze); they have been studied by Ho Ping-ti (1954) as *the* major example of Chinese indigenous quasi capitalism prior to the coming of Westerners in force in the nineteenth century. During the approximately fifty years of roughly the second half of the eighteenth century, thirty major merchant houses plus some two hundred lesser ones of the Yang-chou region together acquired fortunes aggregating about 250 million taels (ounces of silver). Many of them became millionaires or even multimillionaires.

Ho cites blatant cases of nouveau riche conspicuous consumption, but he also cites many instances in which merchants chose to spend lavishly in support of scholars and scholarship.[89] In other cases, members of merchant families themselves succeeded in becoming scholars. Thus between 1646 and 1802, somewhat fewer than 300 salt merchant families produced 130 holders of the *chin shih* (doctor in letters) degree, as well as 208 holders of the next lower degree, that of *chü jen* ("promoted man");

87. However, the ban was lifted under subsequent dynasties. See Ch'ü (1961:129 n. 2), which traces the ban's history from Han through Sung but makes no mention of it thereafter because it was tacitly omitted from subsequent law codes.

88. Kracke (1953:70, 76), which on the intervening pages also discusses other means of entering the civil service without taking the examinations.

89. Thereby imitating the example of the famous merchant Lü Pu-wei, allegedly the natural father of the Ch'in First Emperor, whose entourage of scholars produced (in 240 B.C.) the eclectic collection of philosophical essays known as the *Lü-shih ch'un-ch'iu* (Mr. Lü's Springs and Autumns). See Bodde (1940).

during the same period, 140 other members of the same group of families entered the civil service by purchase of degrees (a practice that, having been legally sanctioned during the Ch'ing, eventually became a serious abuse).

To the question, why did these families fail, despite their achievements, to produce a true commercial capitalism in the Western sense, Ho suggests several answers. One is the traditional absence of primogeniture in Chinese society, resulting in the rapid fissuring of family wealth. Another is the massive diversions of much merchant wealth from productive to nonproductive uses, of which he cites many instances. Yet I believe two other considerations were still more basic. The first is that, as late as the eighteenth century, the salt merchants of Yang-chou, like their professional forbears of a millennium earlier (see above), still remained subject to the controls of the imperial salt monopoly. Having given each merchant his concession to produce or transport salt, the government could at any time strip him of it again if it disapproved of his activities. The second consideration, already alluded to, is the merchant's failure, even at this late date, to create a distinctive ethic or mystique of his own as a viable alternative to the prevailing ethic. On the contrary, members of the merchant class, when they had the opportunity, often tried to make their way of life resemble that of the scholar-gentry. These two points underlie the following remarks by Ho Ping-ti:

> For all the power of capital, . . . the conditions congenial to the development of a full-fledged capitalism did not exist. In salt production, for example, however able and ruthless a factory merchant might be, he could not limitlessly expand his sphere of business at the expense of others, for each of them owed his position to government recognition. . . . The organization of the salt trade allowed little room for competition and rationalization. In fact, one may ask whether such ideas of competition and rationalization had ever occurred to the merchants. Individual fortunes might rise and fall owing to individual luck or folly, but the law ruled out the possibilities of ruthless competition among them. (Ho 1954:142)

> Small wonder, . . . in a society where the primary standard of prestige was not money but political and literary distinction, that the social mobility of the salt merchant families should have been along a single and almost fixed channel [directed toward the scholar-gentry]. (Ho 1954:166)

It is instructive to compare these remarks with those of another contemporary scholar, writing about the economic thinking of merchant and artisan in the classical Western world. "What was lacking," she writes, "was the modern form of the profit motive. The idea of making more money, not by buying more slaves and having more workshops, but by altering the method of production, simply did not occur to anyone" (Burford 1972:119). The same theme of noncompetition appears again in the account of medieval social theory quoted in section 2 of this chapter: "Peasants must not encroach on those above them. Lords must not despoil peasants. Craftsmen and merchants must receive what will maintain them in their calling, and no more" (Tawney 1926:23).

In Europe these attitudes were to change radically with the Renaissance. In China, despite the substantial advances achieved by the commercial-industrial class ever since the late eighth century, they were not enough to overcome the web of inhibition that had begun to be spun already at the creation of the empire.

Can these differing movements in China and Europe be explained? Although their possible causes are obviously multiple and complex, an anthropologist has pointed to one of them as worthy of special note. "A very important aspect of Indo-European society," he remarks, was "the relative weakness of kinship by comparison with China, and also with many pre-state societies" (Hallpike 1986:335). Because of this kinship weakness, the Indo-Europeans of Western (and particularly Northwestern) Europe began, already early, to create nonkinship sworn brotherhoods to provide fellowship and mutual assistance. From these associations eventually emerged the merchant guilds of medieval Europe:

> From the beginning, in early Germanic and Scandinavian society there had been the institutions of the libation and the oath by which non-kin could join together in a powerful union of mutual assistance under religious auspices, and this type of association was clearly appropriate in a society where the bonds of kinship were weak. Increasing social disorder only augmented the advantages of belonging to such groups. . . . The principle of contributions by members, initially only to defray the costs of feasts, could readily be applied to more purely economic ends. . . . Most significant of all, the later guilds were the basis of corporate, independent groups of merchants, bound by ties of mutual assistance in a confraternity, and this seems to have been of great significance in establishing the social influence of the merchant class. (Hallpike 1986:368–369)

The contributions of these guilds to later European socioeconomic changes, including the rise of capitalism, are well known. In China, on the other hand, kinship ties always retained their primacy down to recent times, while merchant guilds, although they existed, never achieved the power or independence of their European counterparts.[90]

90. Nevertheless, the Chinese guilds could, within their particular localities, play significant roles both as maintainers of discipline among their members and as protectors of the interests of these members in dealings with the outside world, especially officialdom. See Creel (1987:136–139).

4c. The Artisan and the Intellectual

Confucian criticism of the artisan, when expressed at all, was usually moralistic in tone. Generally speaking, however, especially in later times, the men of letters gave no particular thought to the matter, beyond accepting as an axiom Mencius's dictum that it is the natural duty of the manual worker to feed the brain worker. Regardless of which attitude prevailed, the result was the same: a polarization of mental and manual work which barred the literatus from any sort of manual activity other than painting and calligraphy, kept the hand worker in unlettered silence, and induced the former either to disregard the latter entirely in his writings or to describe his way of life in casual, patronizing, or idealized language.

Some of the consequences have been variously noted by Needham in his *Science and Civilisation in China.* Thus it is indicative of the bureaucratic orientation of Chinese society that the only technological treatise surviving from antiquity is the *K'ao kung chi* or *Artificers' Record,* which deals with the government workshops in the bureaucratic Utopia described by the *Chou li.*[91] Thereafter one has to wait until the Sung dynasty before major works on technology and engineering begin to appear.[92] By contrast, the first significant treatise on agriculture, the *Fan Sheng-chih shu* (Book of Fan Sheng-chih), was written a whole millennium earlier (ca. 10 B.C.). Though known today only in fragments, its great chronological priority possibly reflects the higher regard of the scholar-gentry for agriculture as against technology.[93]

No doubt the production of Sung technical works was encouraged by the increasing use of printing (Needham 1965a:170), as well as by other interrelated factors such as growing urbanization and expanding trade and industry. Yet the result was by no means an even flow of technological literature in all fields. In architecture, for example, output remained small, "presumably owing to the fact that architectural employment was not considered a very suitable employment for a Confucian scholar" (Needham 1971:80). In nautical technology, the distance be-

91. Needham (1962:12–17). The *K'ao kung chi* constitutes *chüan* 40–44 of the *Chou li;* tr. Biot (1841:2.456–611).

92. Needham (1965a:44ff., 165ff.; 1971:80ff., 323ff.).

93. The author of this text, Fan Sheng-chih, was an official under Emperor Ch'eng (r. 32–7 B.C.) and had taught agriculture in the area of the Former Han capital, Ch'ang-an. The surviving fragments of his treatise are translated and discussed in Shih Sheng-han 1959. Five centuries later (ca. 535) there appeared the first major agricultural treatise to have survived virtually intact, the *Ch'i-min yao-shu* (Essentials for the People's Welfare). Discussed in Shih (1958) and Bray (1984:55–59).

tween scholar and artisan was so great that systematic nautical treatises failed to appear at all. Shipwrights of the Ming dynasty may well have been "the most accomplished artisans of any age in any civilisation who were at the same time illiterate and unable to record all their skill."[94] But this was only part of a wider situation. Probably most Chinese craftsmen in all fields, down to recent times, remained illiterate. They learned their skills through rule of thumb and inherited tradition, rather than from written sources (Needham 1965a:47–48).

A concrete example of the artisan's difficulty in articulating his knowledge is that of Yü Hao (fl. 965–995), a master carpenter in government service but not an official. He probably had to dictate what he knew to a scribe in order to record his treatise on building techniques, the *Mu ching* (Timberwork Manual) (Needham 1954:153). The work failed to be included in the imperial bibliography and was eventually lost, very likely because of his nonofficial status. About a century later (1097), the scholar-official Li Chieh, who held the post of assistant in the Directorate of Buildings and Construction, built upon Yü's work to produce China's major architectural treatise, the *Ying-tsao fa-shih* (Treatise on Architectural Methods). In his preface he states that as preparation for his book, he had carefully studied the orally transmitted rules of master carpenters and other responsible artisans. Even so, one finds that he did not wholly succeed in fusing the scholarly and technical traditions. His earlier chapters are all filled with numerous quotations from earlier texts, but then he leaves them to describe the living practice of his own time and finally to lay down practical rules having very little relation to the earlier quotations (Needham 1971:81–84).

An even more vivid illustration of the differing statuses of official and artisan is Lei Fa-ta (1619–1693), the gifted carpenter who started his family on the road to success by leaving his ancestral home in Kiangsi for Peking, where he found employment in the workshop of the Ministry of Works. According to a story perhaps handed down in the Lei family, it once happened that a large beam broke in the T'ai Ho Tien (Hall of Grand Harmony), the great throne room still standing in Peking's Forbidden City. A replacement beam was hurriedly taken from one of the buildings at the imperial Ming tombs outside the city, and when it was raised to be fitted into place in the throne room, the K'ang-hsi Emperor himself (r. 1662–1722) was on hand to see the proceedings. At the crucial moment, however, the tenon of the raised beam failed to fit exactly into the mortise of the beam that was to receive it, so that the new beam was

94. Needham (1971:382). The great authority on traditional Chinese ships, the late G. R. G. Worcester, once observed that in the course of thirty years he found hardly any among the best shipwrights who could write (Needham 1971:380).

left hanging in mid-air. However, the top official immediately had Lei Fa-ta secretly put on the cap and robe of an official, and then, holding a hatchet in his sleeve, "climb up like a monkey" to the place of trouble. There, with a single blow of the hatchet, he caused the tenon to enter the mortise and the beam to fall into place. The emperor, greatly pleased, personally appointed Lei to be master carpenter in the workshop of the Ministry of Works, in which post he was followed by no fewer than six generations of descendants. The last of them died only in 1907, just before the end of the empire.[95]

Nothing could better illustrate the enormous distance between "white-collar" and "blue-collar" worker than this story of the carpenter (probably illiterate) whom the helpless officials had to smuggle in official garb into the presence of the emperor so that he could save the situation. Nor did Lei's imperial appointment as master carpenter do anything to lessen the distance, because this position was quite outside and below the civil service. This is why Lei's descendants could continue to hold it hereditarily. To the end of the empire, they remained artisans, not officials.[96]

As against the Confucian attitude toward manual work, one might expect Taoist writers to show more sympathy. And this they in fact do, as in the "knack-passages" occurring in the *Chuang-tzu, Lieh-tzu,* and other classical Taoist writings (discussed in Needham 1956:121–127). These passages all describe various extraordinary physical skills to illustrate the thesis that such skills cannot be taught or transferred, but are attainable by minute concentration on the Tao running through natural objects of all kinds. Among the skills thus described are those of Ting the butcher, musicians, cicada catchers, boatmen, swimmers, sword makers, carvers of bell stands, arrow makers, wheelwrights, animal tamers, mathematicians, and buckle makers.

---

95. The story, as well as a good deal more about the Lei family, is recounted in Chu Ch'i-ch'ien and Liang Ch'i-hsiung (1933:84–89). Unfortunately, the story's provenance is unstated except to say that it comes from a biographical study of the Lei family made by Chu Ch'i-ch'ien but apparently never published (unless it be the very text that is quoted in Chu and Liang). Chu could well have obtained his information from living descendants of the Lei family, perhaps supplemented by a family genealogical record, the existence of which seems to be suggested by the wording of Chu's account.

96. Needham (1962:31) remarks that very few of China's notable engineers are known to have attained high office in the Ministry of Works, very possibly because "the real work was always done by illiterate or semi-literate artisans and master-craftsmen, who could never rise across that sharp gap which separated them from the 'white-collar' literati in the office of the Ministry above."

As we look at this curious list, we see that some of the skills are non-productive (swimmers) or downright trivial (cicada catchers); others are useful but unrelated to craftsmanship (the butcher, the boatmen, etc.); and even among the craft skills, several (the sword makers, carvers of bell stands, etc.) merely serve the military, hunting, or ceremonial needs of the upper class. This leaves only one craft skill (that of the wheelwrights) or at the most two (if the rather peculiar "buckle makers" are included) that have broad usefulness for the ordinary man. It seems evident that although the authors of the "knack-passages" are in no way unsympathetic to manual workers, they also have no particular interest in the details of their crafts. Their anecdotes are written solely to illustrate their own theory of the Tao, not to throw light on technology. In order to find educated men who are at the same time acquainted at first hand with manual skills, one has to leave the "philosophical" Taoists entirely and look for the alchemists and other technologists or proto-technologists, usually of later date, who are loosely labeled "Taoists." But these were probably always few in number and mostly outside official circles.

On the basis of everything said so far, one would expect very little information to be recorded about individual craftsmen and their works in the written sources, especially those of official origin. By and large the expectation is justified, and yet, if one knows where to look, more information comes to light than might at first sight seem likely.[97] Some of the most fruitful sources are the dedicatory inscriptions on stone stelae prepared for various public buildings, especially Buddhist monasteries and temples. The inscription dedicating the Ta Hsiang-kuo Monastery of K'ai-feng, for example, describes this notable Buddhist structure (prior to its destruction in 1126) as having ten perfections. Six were paintings or groups of paintings whose artists are in each case identified by name. The remaining four consisted of a Maitreya figure cast by the priest so-and-so, a name tablet written by the T'ang Emperor Jui-tsung (r. 684) himself, the screens carved for the Buddha hall by so-and-so, and the pavilion reconstructued by the carpenter so-and-so.[98] The enumeration reminds us of church records in the West, although the importance of the data in the two civilizations is not the same: in Europe ecclesiastical architecture has dominated all other forms most of the time since the acceptance of Christianity; in China, aside from the

---

97. For occasional instances of named Han and T'ang artisans, see Needham (1965a:18–19). In what immediately follows, I am indebted to Professor Alexander C. Soper, of the Institute of Fine Arts, New York University, for several very helpful bibliographical suggestions.

98. Soper (1948:21–22); also recorded in Soper (1951:78–79).

period of Buddhist greatness, Buddhist architecture has played a lesser role.

A great many stele inscriptions are of course inaccessible or nonextant today, but the text of many others has been recorded in local gazetteers or in such art compilations (usually about painting) as the *Li-tai ming-hua chi* (Compendium of Famous Paintings of Successive Dynasties; A.D. 847). For example, the chapter in this work describing the Buddhist and Taoist temples in the two T'ang capitals of Ch'ang-an and Lo-yang naturally concentrates on the temple paintings and their makers. Yet every once in a while it notes the names of such other artists and craftsmen as the (master) carpenter so-and-so and (master) stonemason so-and-so, builders of a certain pagoda; or, in other temples, the men who modeled certain statues, the smith who made the metal decoration for a certain wooden screen, and the designer of a certain incense burner.[99]

One of the best as well as earliest of such inscriptions is that belonging to the famed Later Han funerary Wu Liang Shrine (A.D. 151). This reads:

> The able artisan *(liang chiang)* Wei Kai has engraved the inscription and carved the designs. He has arranged them in registers, giving full play to his artistic skill. The sinuous lines [of the designs] will provide a fine display for handing down to later generations, uninterruptedly for ten thousand years.[100]

In much later periods, if we turn to ceramics, it is occasionally possible to find pieces inscribed with the names of their makers. Examples include a few of the Tz'u-chou type pottery pillows of Sung time,[101] more of the *blanc de Chine* porcelains from Te-hua in Fukien,[102] and some of the products from the private (but not the imperial) kilns at Ching-te-chen, Kiangsi.[103] Relatively most numerous are the signed red teapots from Yi-hsing in Kiangsu, belonging to the last century and a half of the

---

99. *Li-tai ming-hua chi* 3.4; tr. Acker (1954:287, 309, 311, 324–325, etc.).

100. Tr. into French in Chavannes (1909–1915:1.106).

101. Such pillows, made by the Chang and Wang families and dated 1056 and 1119, are mentioned in Paine (1955:38).

102. Donnelly (1967:286) lists thirty-six individual potters or potter families for this ware, ranging in date very roughly from ca. 1660 to ca. 1900. "Nothing is known of the Tehua potters beyond their names. . . . Not a single dated piece of *blanc de Chine* is signed, and not a single signed piece bears a date" (Donnelly 1967:267).

103. Jenyns (1953:11) states that the names of two or three celebrated Ching-te-chen private potters are known in this way.

Ming, but more particularly to the reigns extending from 1573 to the fall of the dynasty in 1644.[104]

Turning to modern scholarship, a particularly valuable and systematic repertory of diverse artisans of all periods is the series of articles entitled *Che chiang lu* or *Collected Biographies of Master Craftsmen,* which was compiled in Peking in the early 1930s by members of the Society for Research in Chinese Architecture. The usually brief biographies, grouped according to occupations and then chronologically, are drawn from a wide variety of sources, including dynastic histories, encyclopedias, biographies of Buddhist monks, writings of private individuals, and—very important but ordinarily often disregarded—local gazetteers. The largest and most important group, to which the first five installments are devoted, consists of architecture (Chu and Liang 1932–1933). The following is the numerical breakdown of the architectural biographies by periods: beginnings through Chou: 15; Ch'in/Han: 13; Period of Disunity: 34; Sui/T'ang/Later Chou: 26; Sung/Liao/Chin: 28; Yuan: 8; Ming: 56; Ch'ing: 37; for a total of 217.

The other occupational categories and the number of biographies for each (undivided by periods) are: landscape architecture: 33 (Chu and Liang 1934); military technology: 38 (Chu, Liang, and Liu 1934); sculpture: 94 (Chu and Liu 1935); construction: 54 (Chu and Liu 1936); for a grand total of names for all categories of 436.[105]

The compilation suffers from evident weaknesses. Its biographies are usually unaccompanied by an explanation or evaluation, despite the uneven reliability of the sources. A great many names are not those of artisans at all, but of the patrons (including even emperors) who commissioned the construction of palaces, parks, and the like.[106] Still others are high officials having only nominal connections with the constructions attributed to them. A fair number of names in the early periods are those of legendary or semilegendary figures. The collection, as might be expected, leans heavily toward "official" as against private

---

104. See Jenyns (1953:II, 148) and especially Hedley (1936–1937:75–80), who lists thirty-six Yi-hsing potters extending from the early sixteenth century to 1644. I am indebted to Miss Jean Gordon Lee, East Asian curator emeritus, Philadelphia Museum of Art, for lending her copy of Hedley and informing me about the Paine and Donnelly studies.

105. The actual total is slightly greater because of the occasional sons, grandsons, and so forth, listed under the names of their fathers, whom I have not counted separately. The most notable such example, already discussed, is the six generations of the Lei family following the name of its founding figure, Lei Fa-ta.

106. This is true, for example, of almost all the persons listed under landscape architecture.

activities. There is a fair amount of occupational overlapping between "architecture" (*ying tsao;* understood in a broad sense as covering all aspects of the planning and construction of buildings) and "construction" (*ying chien;* city walls, dikes and other hydraulic works, bridges, palace complexes, etc., but sometimes also individual buildings). Finally, the collection omits many important categories.[107]

Despite all strictures, however, the collection presents a wide spectrum of valuable though undigested material that would well repay further exploration. Above all, it proves that a good deal more has been written in Chinese about individual craftsmen and technologists than most people would suppose.

How does all this compare with the status of the artisan in Europe? There are both similarities and differences. From the classical world, the supreme example of the intellectual who bridged the gap between men of his kind and manual workers was Socrates. Brought up himself as a stonemason, he is said to have engaged in endless conversations with various kinds of craftsmen and to have felt a special affinity for shoemakers.[108] But the intellectuals who followed Socrates sometimes wrote about manual workers (and merchants too) with a vehemence that reminds us of some of the extreme Chinese statements. "Their stunted natures," wrote Plato, "covet [philosophy], their minds being as cramped and crushed by their mechanical lives as their bodies are crushed by the manual crafts."[109] To which Aristotle echoed: "No man can practice virtue who is living the life of a mechanic."[110]

In some early Greek communities, the prejudice against *banausoi* (manual workers, both skilled and unskilled) prevented them from enjoying full citizenship rights. Even in fifth-century B.C. democratic Athens, a recognized way to attack a political opponent was to accuse him of being the son of a *banausos.* Sophocles, for example, was variously charged with having been the son of a carpenter, a smith, or a knife maker. This situation meant that even wealthy craftsmen could

---

107. The preface to the first installment mentions fourteen crafts as having been planned for inclusion, among them metalworking, ceramics, jade working, woodworking, and the like. But in the end, only five were actually covered. Moreover, the section on "construction" extends only through the Yuan dynasty; what would have been its final installment was apparently never published, no doubt because of the outbreak of the Sino-Japanese War in 1937.

108. Burford (1972:129–130). This study of craftsmen in Greece and Rome is the basis for what follows in this and the subsequent paragraph. See esp. Burford (1972:25–26, 29ff. 129–130, 156–157).

109. *Republic* 495D–E. Cited in Burford (1972:34).

110. *Politics* 1278A. Cited in Burford (1972:238 n. 334).

hardly hope to gain political office in their own name; they had to look to others who were better born to be their statesmen. In Rome, though prejudice was less outspoken and manual workers and merchants were never denied citizenship, it was yet possible for Cicero to describe "craftsmen and shopkeepers" as "the very sewage of the state."[111]

Remarkably, a fair number of *banausoi*—humble craftsmen as well as artists—succeeded, despite these prejudices, in making their names known to us as well as to their own age. This they did in several ways: by signing their products (sometimes with interesting added remarks), by attaching dedicatory signed inscriptions to the gifts they offered the gods, and by writing their own funeral epitaphs. The earliest known signature seems to be that written by the sculptor Euthykartides around 625 B.C. on the statue made and dedicated by him to Delian Apollo. Thereafter most dedicated statues were signed by their makers. In other fields, the signing of productions was also practiced in varying lesser degrees. It was widespread among Athenian vase painters and also current (both in Greece and Rome) among stonecutters, mosaic makers, metalworkers, gem engravers, and (in Rome but rarely in Greece) architects and painters. Funerary epitaphs further provide the names (and sometimes considerably more) of a variety of other artisans, some of them quite lowly: potters, carpenters, woodcutters, furniture makers, shipbuilders, silver smiths, blacksmiths, shoemakers, textile workers, miners. Additional information comes from the graphic representations of artisans at work, as produced in a variety of media. The artisans include cobblers, carpenters, potters, blacksmiths, founders, gold-leaf workers, minters, bricklayers, and sculptors, and among the media are paintings, grave stelae, terra-cotta figurines, wall paintings, and sculptured sarcophagi.[112]

The Middle Ages, or at least their latter centuries, reveal a rather similar situation despite their reputation for anonymity. The signing of finished works by painters, manuscript illuminators, sculptors, goldsmiths, and others goes back at least to the twelfth century, especially in Italy. Sometimes, as in the case of the classical craftsmen, the signatures are accompanied by interesting or amusing comments. The surviving 1292 tax lists of Paris indicate (with names) that of a total of 15,200 taxpayers, there were then 33 painters, 24 image makers, and 13 manuscript illuminators, as against some 350 each of shoemakers and tailors, plus other large artisan groups. Jean de Jandun, in his eulogistic treatise on Paris (1323), writes with approval of many kinds of artisans, among them sculptors and painters of figures, scribes, illuminators,

---

111. *For Flaccus* 18. Cited in Burford (1972:238 n. 334).

112. This paragraph is again based on Burford (1972), both texts and plates, passim.

bookbinders, parchment makers, and makers of clothing and ornaments, of metal vessels, and of instruments of war. For Gothic England, during the approximately four centuries from a little before 1180 to a little after 1560, the names are known of 51 masons (a term that in modern parlance would often embrace architects as well), 26 carpenters, 12 carvers, and 14 painters.

The professional activities of artists and artisans (in medieval as well as in classical times no clear-cut distinction was made between these categories) are portrayed both in painting (especially miniatures) and sculpture. Although the painter or illuminator is the artist most often shown, other depicted occupations include those of the ivory carver, tomb sculptor, statue carver (at Chartres), statue painter, mural painter (shown in one instance as being saved from a dangerous fall—a known historical episode), metalworker, Parisian goldsmith, and stonemason.[113]

In China, with all due respect for the data discussed earlier, it seems unlikely, taking the respective time spans and populations into account, that anywhere near as many individual artisan names are proportionately known as in Europe. The one great exception is that of Chinese scroll painters and calligraphers, whose names are legion. The reason of course is that they are members of the literati, not the artisanate, which is also the reason why I exclude them. It is likewise my subjective impression that artisans and technological processes are probably portrayed less frequently in Chinese art than in European art.[114] Curiously, there seem to be proportionately many more of them from the Han dynasty (tomb wall reliefs and paintings, molded bricks, pottery figurines, and so on) than from other dynasties.[115]

---

113. Most of the facts in these two paragraphs come from Martindale (1972:esp. 9–11, 66–69, 105–106). For drawing my attention to this work, I am indebted to Miss Evelyn A. Silber, research fellow at Clare Hall, Cambridge University. Thirty-one illustrations, in all of which artists and artisans are shown at work, are also contained in Egbert (1967); a number of them appear in Martindale (1972) as well. The statistics for known masons and such craftsmen in Gothic England are derived from the detailed table, with the names of such persons listed by categories and in chronological sequence, which occupies the endpapers of Harvey (1947). I am indebted to Mr. David R. Buxton, associate of Clare Hall, for showing me his copy of this work.

114. Reproductions of such Chinese art portrayals are scattered through the volumes of *Science and Civilisation in China*. See esp. Needham (1965a:Figs. 356, 358, 388, 396, 415, 507–509).

115. In making these East-West comparisons, only art portrayals are here taken into consideration, these being freely indicative of the thematic interests of their makers.

But as against the question of relative numbers, two other matters seem more basic, namely, individualism and literacy. On the Chinese side, the artisans rarely speak to us themselves—except, that is, through surviving physical examples of their work. What we know about them comes almost invariably from what scholars and officials have chosen to write in the dynastic histories and a variety of other works. Conversely, we have seen that in classical Europe and again in the late Middle Ages, it is the craftsmen themselves who in a variety of ways tell us their names and sometimes something about themselves.[116] This phenomenon has been pointed to as indicative of the greater individualism that seems to have characterized Western society during so much of its history. Concerning its beginnings in seventh-century B.C. Greece we are told: "The earliest signatures known are symptomatic of the new sense of individualism which manifested itself during the seventh century B.C., and which from then on was one of the characteristics which distinguished the Greeks most sharply from the peoples of the ancient Near East."[117]

To what extremes the assertion of the individual ego could go is amusingly shown by some of the statements that artists and ordinary artisans, both classical and medieval, have left about themselves:

> Mannes, son of Orymaios, who was the best of the Phrygians in the spacious land of Attica, lies in this fine tomb. And by Zeus I never saw a better woodman than myself.[118]

> I declare that at last the limits of *technē* [technical skill] have been reached by my hand.[119]

> No one rivalled me in *technē*.[120]

Excluded are the illustrations deliberately prepared for the *T'ien-kung k'ai-wu* and other technological manuals of Sung or post-Sung date, as well as those in similar manuals in Europe.

116. This statement is of course in no way intended to imply that our knowledge of European craftsmen is solely derived from what they themselves have recorded. To cite but a single famous literary source, the last five books of Pliny the Elder's *Natural History* (first century A.D.) contain much information not only about painters and sculptors, but also about mining and other technologies.

117. Burford (1972:212). The Confucian organicist view of man and its concomitant nonemphasis on the individual qua individual has already been alluded to (sect. 2). On individualism generally, see VI.3.

118. Epitaph of the fifth century B.C., quoted in Burford (1972:18).

119. Epigram by the painter Parrhasios (fl. 400 B.C.), quoted in Burford (1972:209).

120. Epitaph of Atotas the miner (fourth century B.C.), quoted in Burford (1972:177).

> I am the prince of writers [i.e., calligraphers]; neither my fame nor my praise will die quickly.[121]

> He would not know how to carve horrid or unseemly things even if he wishes to do so. . . . He is endowed above all others in the ordering of the pure art of sculpture.[122]

It is hard to imagine any Chinese artist or artisan praising himself quite so blatantly and naïvely. A near but very rare approximation is the unexpectedly vain but amusing statement by the influential Ming painter Tung Ch'i-ch'ang (1555–1636): "Not only do I regret that I cannot meet with the great masters of the past, but also I regret that they did not have the opportunity of meeting me!"[123]

Also indicative of differing attitudes toward self-expression is the considerable number of self-portraits to be found already in pre-Renaissance European art (and, of course, much more commonly thereafter).[124] Conversely, self-portraiture is exceedingly rare in Chinese art.[125]

---

121. Inscription written ca. A.D. 1149 by the calligrapher and illuminator Eadwine, monk of Canterbury; quoted in Martindale (1972:68).

122. Portion of a lengthy signed inscription carved on the pulpit (ca. 1302–1310) of the cathedral in Pisa by Nicola Pisana; quoted in Martindale (1972:105).

123. Quoted in Nelson Wu (1962:275). Equally boastful and untypical is a slightly later statement by the eccentric literary critic Chin Sheng-t'an (1610?–1661), expressed in his commentary (1641) on the novel *Shui hu chuan* (Water Margin). In his introductory remarks on the novel's twelfth chapter, Chin writes: "Since the invention of writing brush and ink there has never been finer writing [than this novel]. Since the existence of such writing, there has never been finer commentary [than mine.]" Quoted in John Wang (1972:30). The statement is particularly ironical in that when Chin prepared his edition of *Shui hu chuan,* he forged a new preface, which he attributed to the novel's author, as well as a new final chapter, which he substituted for the entire last fifty chapters of the novel.

124. Of the thirty-one portrayals of the medieval artist at work, as reproduced in Egbert (1967), seven (plates 3, 5–6, 11, 16, 29–30) are self-portraits. In antiquity, self-portraits are much rarer. See, however, the possible self-portrait of Phidias reproduced in Burford (1972:plate 78), as well as the self-portraits carved by two sculptors on their own grave stelae (plates 85–86). In a larger number of instances, sculptors were commissioned by other artisans (shipbuilders, shoemakers, potters, smiths, etc.) to portray the artisans in question on the grave stelae the sculptors made for them. See Burford (1972:plates 5–7, 9, 33, 36, 46, 48).

125. Quite exceptional, especially because it is the self-representation of a simple artisan, is the statue of a mason with chisel and uplifted hammer, surmounted by the inscription, "Tien the stone-carver," which is found in a Buddhist cave-shrine said to date just after A.D. 489. The cave-shrine is one of several on the grounds of

The second important consideration is literacy. The fact that even a small minority of European artisans used the written word for self-expression, whereas this happened very rarely in China, may conceivably reflect a higher degree of literacy within the European artisanate. It is possible that such literacy was helped by the availability of an alphabetic script as against the characters of the Chinese literary language, whose difficulties have been lengthily discussed in Chapter II. But I would further suggest that the city-based and commercially oriented civilization already characteristic of classical Europe may have been more conducive to the rise of a literate or semiliterate artisanate than was the rural-based and bureaucratically oriented civilization of China. At least it seems very probable that when, beginning in the Sung dynasty, a growth of trade and urbanization took place, it was accompanied by a rise in literacy—in the cities, not the countryside.

On the European side, it is evident that even some quite well known artists or artisans of antiquity had difficulty with the written word. An example is the Athenian potter and painter Exekias (late sixth-century B.C.), who managed to misspell his name on one of the two surviving pieces signed by him (Burford 1972:217). In contrast to him, the supreme example from European antiquity of someone who successfully combined extensive manual experience with book learning is the architect, engineer, and craftsman Vitruvius. His modern English translator characterizes Vitruvius as having "all the marks of one unused to composition, to whom writing is a painful task" (Morgan 1914:iv). Nevertheless, Vitruvius himself makes a moving plea in the introduction to his *On Architecture* (ca. 30 B.C.) for the joining of practice to theory, learning to technical skill:

> This knowledge [which, he writes, should be possessed by the architect] is the child of practice and theory. . . . It follows therefore that architects who have aimed at acquiring manual skill without scholarship have never been able to reach a position of authority to correspond to their pains, while those who relied only on theories and scholarship were obviously hunting the shadow, not the substance.

From this Vitruvius goes on to urge that the architect should "be educated, skillful with the pencil, instructed in geometry," and versed to varying degrees in history, philosophy, music, medicine, law and astronomy.[126] Had there been more Vitruviuses in Europe and

---

the Chi-hsia temple east of Nanking. See Needham (1965a:758). For further remarks on portraiture and self-portraiture in China and Europe, together with the parallel literary genres of biography and autobiography, see VI.3 below.

126. *On Architecture,* bk. 1, ch. 1; quoted in Burford (1972:104).

China, material progress would surely have moved much more rapidly.[127]

The Chinese craftsman-technologist, despite his lack of letters, succeeded superbly in achieving what was expected of him within the limits of traditional craftsmanship.[128] But his status remained relatively unchanged down to the end of the empire. Indeed, it apparently changed much less during the later dynasties than did that of the merchant, who nevertheless never quite succeeded, as we have seen, in breaking through the cake of custom.

In the West, the distant sequel to Vitruvius's eloquent plea was the marriage between craft practice and scholarly theory that actually took place in the seventeenth century; with it came the formulation of the Galilean experimental-mathematical method, and thus the birth of modern science (Needham 1959:150ff). Nowhere is the new social attitude better illustrated than in the statement of purpose formulated in 1663 by Robert Hooke for the newly founded British Royal Society. "The business of the Royal Society," he wrote, "is: To improve the knowledge of natural things, and all useful Arts, Manufactures, Mechanick practices, Engynes and Inventions by Experiment—(not meddling with Divinity, Metaphysics, Morals, Politics, Grammar, Rhetorick, or Logicks)" (Bernal 1954:317).

This theme is elaborated four years later by Bishop Thomas Sprat in his *History of the Royal Society of London* (1667), where he writes in defense of the Society's inclusion of men of all ranks and occupations:

> For now the Genius of *Experimenting* is so much dispers'd that even in this *Nation* . . . there could not be wanting able men enough, to carry them on. All places and corners are now busie, and warm about this Work: and we find many noble Rarities to be every day given in, not onely by the hands of Learned and profess'd Philosophers; but from the Shops of *Mechaniks;* from the Voyages of *Merchants;* from the Ploughs of *Husbandmen;* from the Sports,

127. What may at first sight look like a Chinese parallel to the idea underlying Vitruvius's plea is Wang Yang-ming's (1472–1529) doctrine of the "unity of knowledge and conduct" *(chih hsing ho yi).* Wang's doctrine stimulated continuing repercussions in later Chinese thought all the way down to such modern political figures as Sun Yat-sen and Mao Tse-tung. See Fung (1953:603–605) and, especially for the doctrine's later history, Nivison (1953). When Wang talked about "conduct," however, what he, like other Chinese thinkers, primarily had in mind was behavior in social situations. Thus his doctrine has little or no relevance to the question of the relationship between book learning and manual experience.

128. Needham (1965a:600), at the end of his discussion about mechanical engineering, points out that "the balance shows a clear technological superiority on the Chinese side [as against European technology] down to about + 1400."

the Fishponds, the Parks, the Gardens of *Gentlemen.* (quoted in Bernal 1954:320)

It was within a few years of these words that Lei Fa-ta, on the other side of Eurasia, was furtively dressing himself in the robes of a Mandarin so that he might be permitted, in the presence of the emperor, to do what none of the emperor's officials was able to do. His achievement earned for him and his descendants the position of master carpenter in the imperial workshops, but it did not narrow the gap between them and the learned men they served. Of course in seventeenth-century England too (and Western Europe generally) the new attitude was by no means total. Then and much later there continued to be craftsmen whose educational level and way of life were very similar to those of Lei Fa-ta. In 1668, for example, when the diarist John Evelyn witnessed the launching of the ship *Charles* at Deptford (below London on the Thames), he recorded her as "built by old Shish, a plain honest carpenter, master builder of this dock, but one who can give very little account of his art by discourse, and is hardly capable of reading, yet of great ability in his calling. The family have been ship carpenters in this yard about 300 years" (Naish 1957:491–492).

In China, we have noted (III.2) that the yin-yang and Five Elements theories probably had less negative effect on technology than on science and proto-science. The likely reason is the pragmatic orientation of much pre-modern technology generally, as against the more exclusively theoretical orientation of pre-modern science and proto-science. By 1668, when John Evelyn talked with old Shish, it is probable that the traditional technologies of Europe and China alike were both based more on practice than on theory and had both reached approximately the highest point possible for such technologies before the advent of modern science.

In Europe, however, the alliance of practice with theory had by then already begun to generate the revolution in science that was to transform society through industrialization. Seventeen years after Evelyn's encounter with Shish in Deptford, Newton's *Principia* was published in 1687. Less than a century after the *Principia,* steam was beginning to turn the wheels of Britain. In China, meanwhile, the gap between the intellectual and artisan or between theory and practice scarcely narrowed before the present century. Until 1907, as we have seen, Lei Fata's descendants continued to perform their traditional work in Peking's imperial workshops, apparently with little change.

VI. MORALS AND VALUES

1. The Moral View of Life

Almost every religion or philosophy moralizes to a greater or lesser extent; among most ancient Chinese thinkers, the tendency to do so was overwhelming. In some fields, such as art and literature, alternatives to the moralistic interpretation developed in the course of time, but this "secularization" of outlook usually emerged only slowly and rarely displaced the moralistic view completely. In the next several subsections we shall consider the ramifications of moralism in a number of fields, together with their possible consequences for science. As in earlier sections, Confucianism will be our primary concern, but again, as previously, we shall find from time to time that the Confucian outlook is shared to varying degrees by other schools.

1a. Moralism in the Arts

The classical Confucian approach to the arts—comprising literature (primarily poetry), the fine arts (primarily painting), and music (anciently regarded as closely related to poetry)—was in every instance strongly moralistic.[1] Excellent examples are provided by the two major ancient disquisitions on music: the twentieth chapter of the *Hsün-tzu*, "On Music," and the seventeenth chapter of the *Li chi* (Record of Ceremonial), "The Record of Music." The following are representative statements:

> Man cannot be without joy, and when there is joy, it must have physical embodiment. When this embodiment does not conform to the Tao, there will be disorder. The early kings hated this disorder, so they established the sounds of the *Ya* and the *Sung* to act as a guide.[2] They caused these sounds to be joyful and not degenerate, . . . to stir up the goodness in men's minds and to prevent evil feelings from gaining any foothold. This is the manner in which the early kings established music. (*Hsün-tzu* 20; tr. Dubs 1928:248)

---

1. See also De Woskin (1982:ch. 6) for the early Chinese moralistic approach to music.

2. The *Ya* and *Sung* are two sections in the *Shih ching* (Songs Classic). These and other poems in the *Shih* were originally sung and not merely recited; hence the association made in early times between poetry and music.

> The early kings, when they instituted the rites and music, did not do so to gain full satisfaction for the desires of the mouth, stomach, ears and eyes. They did it to teach the people how to regulate their likes and dislikes, and to turn them toward the correct Way of Man *(jen tao chih cheng)*. (*Li chi* 17; tr. Legge 1885:28.96)

> When the age is disordered, the rites become perverted and the music becomes licentious *(yin)*. (*Li chi* 17; tr. Legge 1885:28.109)

> Whenever depraved sounds *(chien sheng)* stir a man, his vital force in its unruly aspect *(pi ch'i)* responds to them, and as this acquires form, the licentious music arises within him. But whenever correct sounds *(cheng sheng)* stir a man, his vital force in its compliant aspect *(shun ch'i)* responds to them, and as this acquires form, the harmonious music arises within him. (*Li chi* 17; tr. Legge 1885:28.110)[3]

These passages not only illustrate early Confucian moralism but also express the idea that the two pillars of civilized society, the rites *(li)* and music *(yüeh)*, were created by the ancients once and for all in finished form; all that we later mortals have to do is faithfully practice them.[4]

No less moralistic is the early Confucian approach to poetry. A famous utterance occurs in the "Great Preface" *(Ta hsü)* of the *Shih ching* (Songs Classic). This essay has been traditionally ascribed to Tzu-hsia, a disciple of Confucius, but was almost certainly actually written by the Later Han scholar Wei Hung (fl. 25–57):

> Nothing approaches poetry[5] for establishing right and wrong, moving Heaven and Earth, and appealing to spirits and gods. The early kings used it to regulate the ties between husband and wife, to perfect filial piety and reverence, to deepen human relationships, to beautify moral instruction, and to transform manners and customs.[6]

The same sentiment reappears in Confucian remarks on painting. A good example is the statement attributed to the poet Ts'ao Chih (192–232):

> Of those who look at pictures, there is none who, beholding the Three Sovereigns and Five Emperors [the sage rulers of high antiquity], would not look

---

3. A person's vital force, *ch'i,* is discussed in *Meng-tzu* 2a.2, where it is pointed out that there are circumstances when a person is unable to keep this *ch'i* under control. On such occasions, as unruly *ch'i,* it causes the individual inadvertently to lurch, stumble, or otherwise misstep.

4. The concept of the Tao as created by and transmitted from the ancients was discussed in V.1.

5. *Shih,* "poetry," may here refer more narrowly to the *Shih ching* itself.

6. *Ta hsü* (Great Preface); tr. Legge (1960:4.Prolegomena 34). Also, more freely, tr. James J. Y. Liu (1962:66 and 1975:111–112).

up in reverence, or, beholding the degenerate rulers of the Three Decadences [the epochs of final decline of the Hsia, Shang, and Chou dynasties], would not be moved to sadness. Seeing a picture of usurping ministers stealing a throne, there is none who would not grind his teeth, nor any who, contemplating a fine scholar of high principles, would not "forget to eat."[7] At the sight of loyal vassals dying for their principles, who would not harden his own resolve? And who would not sigh at beholding banished ministers and persecuted sons? Who would not avert his eyes from the spectacle of a licentious husband or a jealous wife, and who, seeing a virtuous consort or an obedient queen, would not praise and value them? From this we may know that paintings serve to maintain the mirrored warnings [from the past against wrongdoing].[8]

Similar statements continue to be made down to recent times. Chang Yen-yüan (A.D. 847), for example, precedes his quotation of the foregoing statement by Ts'ao Chih with a comparable remark of his own: "To see the good serves to warn against evil; to see the bad serves to make one think of moral worth" (*Li-tai ming-hua chi* opening sect.; tr. Acker 1954:72–73). No less moralistic, though some nine centuries later, is Shen Te-ch'ien's statement in the first of two prefaces (1717) to his well-known anthology of T'ang poetry: "The task of the compiler of poetry is, by exercising discrimination in what he rejects and retains, to cause the mind of the reader to have a standard and be free from doubts." To which he adds in his second preface, some forty-five years later (1763): "Poetry's teaching can harmonize one's nature and emotions, strengthen human relationships, reform the government, and move the spirits, so that one may thus reach the prescient purpose of the makers of the poetry."[9]

Another eighteenth-century example is a passage from Hsü Ch'i's manual (1724) on the Chinese lute *(ch'in):*

When man is stirred by accomplished music, his nature reverts to rectitude. Then between ruler and subject the correct relations exist and between father

7. An allusion to *Lun yü* 7.18, where Confucius is described as "so intent upon enlightening the eager that he forgets to eat."

8. Quoted by Chang Yen-yüan in the opening section of his *Li-tai ming-hua chi* or *Compendium of Famous Paintings of Successive Dynasties* (A.D. 847); tr. Acker (1954:74–75). Inasmuch as quotation marks were unknown in Literary Chinese, it is possible that the final sentence does not itself belong to Ts'ao Chih's statement but is a concluding comment made by Chang upon that statement.

9. See Shen's *T'ang shih pieh-ts'ai* (Selections from T'ang Poetry), first and second prefaces. Some of the phrases in the second preface echo ones found in the "Great Preface" of the *Shih ching,* quoted earlier. A somewhat different rendition of the second preface is found in James J. Y. Liu (1975:115).

and son there is affection. Licentious desires are dissipated and there is a return to the Genuineness of Heaven *(t'ien chen).*[10]

Worth stressing is that this moralistic attitude persisted, at least in literature, down to the beginning of the present century.[11]

From moralism which thus judged a creative work according to the moral acceptability of its subject matter there gradually evolved another theory in which attention shifted to the moral qualities of the artistic creator himself. One of the earliest expressions of the new approach is the succinct statement by Wang Ch'ung in his *Lun heng* (A.D. 82/83): "The greater a man's virtue, the more refined is his literary work *(wen)*" (*Lun heng* 82(28/11); tr. Forke 1962:2.229). Thereafter the same idea recurs repeatedly. Hsi K'ang (223–262), for example, writes in his poetical essay on the Chinese lute *(ch'in):* "If the mind of the player is pure and serene, the music embodies the harmonious peace of perfect virtue. By its influence it can truly purify heart and spirit, and awake profound emotions."[12] And Chung Hung, in his evaluations of famous poets (early sixth century), writes about T'ao Ch'ien (365–427): "Whenever we look at his writings we see, in imagination, the virtue of the man."[13]

But it is in writings on painting that the theory seems to appear most frequently. Chang Yen-yüan says: "From ancient times those who have excelled in painting have all been men robed and capped [i.e., officials] and of honorable status, recluse scholars, and lofty-minded men who . . . have left behind them a fragrance that shall last a thousand years. This is not something that lowly and mean rustics from village lanes could ever do."[14] The same sentiments appear in the essays of the Ch'ing painting critic Shen Tsung-ch'ien (1781): "The high or low standard of the brush varies according to the character of the artist. . . .

---

10. Quoted in van Gulik (1940:80).

11. C. T. Hsia (1978:241), writing about the literary attitudes of Yen Fu (1853–1921) and Liang Ch'i-ch'ao (1873–1929) in the beginning years of the twentieth century, describes both of these noted scholars as "obsessed with the educational function of fiction."

12. Tr. van Gulik (1941:67). Elsewhere, however, Hsi K'ang boldly but perhaps inconsistently denies that music per se is either moral or immoral: "What is called music has nothing to do with immorality. Immorality and rectitude equally lie in the heart." Quoted in Holzman (1957:70).

13. Middle (second) *chüan* of Chung Hung's *Shih p'in* (Classification of Poets). Quoted in Cahill (1960:123).

14. *Li-tai ming-hua chi* 1.1; tr. Acker (1954:153). Kuo Jo-hsü, writing ca. 1075, makes very nearly the same remark. See Soper (1951:15).

The ancients made their character and status manifest in the art works [lit. "brush and ink"] produced by them for display to later generations; later generations, basing themselves on these material remains, admire the characters of these men as if they saw the men themselves."[15]

Without going into detail, it may be said that the same theory has dominated the sister art of calligraphy.[16]

Cahill has argued (1960:118–119) that the classical didactic approach to the arts in terms of subject matter was already becoming obsolete by 847, when Chang Yen-yüan made the first of the two statements quoted earlier. Although we have just seen that the same approach actually continued into the eighteenth century, I am quite ready to agree that already as early as Chang Yen-yüan, the moralistic preoccupation with subject matter may not have been the only or perhaps even the dominant approach to the arts.[17] This fact, however, by no means signalizes the abandonment of moralism per se, but only a shift from moralism of a cruder to a more sophisticated sort. This becomes apparent when Cahill, after noting the decline of the old theory, goes on to discuss the ideals of "literati painting" *(wen-jen hua),* the movement whose rise from the eleventh century onward was such an important development in Chinese painting.

"The fundamental contention of the *wen-jen hua* theorists," he writes, "was that a painting is (or at least should be) a revelation of the nature of the man who painted it. . . . A man of wide learning, refinement, and noble character will, if he adds to these attributes a moderate degree of acquired technical ability, produce paintings of a superior kind" (Cahill 1960:129). For the Sung *wen-jen hua* theorists, furthermore, painting is "a means other than descriptive by which it might communicate the ineffable thoughts, and transient feeling, the very nature, of

---

15. *Chieh-chou hsüeh-hua p'ien* (A Tiny Treatise on the Study of Painting) 3/6; tr. Lin Yutang (1967:205, 208).

16. In his book on Chinese calligraphy, Ch'en Chih-mai (1966:90) quotes the statement by the late eighth-century calligrapher Liu Kung-ch'üan: "When the heart is upright, the hand will be upright." He himself remarks similarly (Ch'en 1966:267): "All Chinese are in agreement that 'a person of low character cannot wield the brush properly.' "

17. Liu (1962:65–87) discusses four major approaches to poetry, as they were variously expressed at different times: (1) poetry as moral instruction (the didactic view we have been discussing); (2) poetry as self-expression (the individualistic view); (3) poetry as a literary exercise (the technical view); (4) poetry as contemplation (the intuitionalist view). Other formulations would also be possible. See Liu (1975: 106ff.).

an admirable man, and so contribute to the moral betterment of those who see it" (Cahill 1960:122–123).

Is this not a highly moralistic way of looking at the arts? Though Western parallels exist (see next note), it would seem that Western art criticism, at least since the Renaissance, was usually ready to maintain a reasonably clear distinction between the artistic merits of a creative work and the moral qualities of its maker.[18]

In pre-Sung China, as in pre-Renaissance Europe, many painters were professionals of nongentry background whose status was often little better than that of ordinary artisans. Only with the rise of the *wen-jen hua* tradition did painting gain full recognition in China as a cultural activity fully worthy of the scholar. The growth of this idea goes hand in hand with the idea, discussed earlier (V.4a), of the scholar as a nonspecialized but gifted amateur. By the same token, the moralism that measured the worth of a painting in terms of its creator's moral character could earily turn into a kind of "literary snobbery" that only served to perpetuate the traditional division between scholar and artisan.[19]

Examples of this attitude are easy to find among the *wen-jen hua* theorists, notably Su Tung-p'o (1037–1101), a major initiator of the movement. Thus concerning Wu Tao-tzu (d. 792), a professional artist generally regarded as the greatest of the T'ang painters, Su comments: "Although Master Wu was supreme in art, he can only be regarded as an artisan painter *(hua kung).*" By contrast, his high praise for Yen Su (d. 1040), an artist forgotten today, was probably inspired by Yen's scholar-official status: "Master Yen's brushwork is wholly divine and brilliantly fresh; it has left behind the calculations of the artisan painter *(hua kung)* and achieved the poet's purity and beauty." In more general terms Su also writes: "The artisans of the world *(shih chih kung jen)* may be able to create the forms perfectly, but when it comes to the principle *(li),* unless one is a lofty man of outstanding talent one cannot achieve

---

18. A well-known post-Renaissance exception is William Hogarth who, in preparation for his book *The Analysis of Beauty* (1753) initially drew up a table equating certain principles of beauty with moral qualities (e.g., "fitness excites a pleasure . . . similar to that of truth and Justice" and "uniformity and regularity are pleasures of contentment"). By the time his book was published, however, Hogarth had so changed his thinking that not only did he withhold the table from the printed version, but in its preface he sharply condemned those writers on aesthetic theory who might turn "into the broad, and more beaten path of moral beauty." See Hogarth (1955:170). For this reference I am indebted to Miss Renée Pigeaud, art historian, of the Netherlands.

19. See V.4.c. The term "literary snobbery" comes from Bush (1971:23).

it."[20] Su Tung-p'o's contemporary, Kuo Jo-hsü, expresses the same idea when, referring to what he calls "the work of ordinary artisans," he writes (ca. 1075): "Although it is termed painting, it isn't painting" (tr. Soper 1951:15).

The moralistic approach to the arts covers such a considerable sector of Chinese literati thinking that it deserves this sketchy presentation, even though in itself it is only tangential to the question of the development of science. More closely germane, I believe, is the discussion of historiography that follows.

1b. The Moralistic Interpretation of History

The belief that historical events are manifestations of divine purpose goes back in China to at least the eleventh century B.C., when the founder of the Chou dynasty justified his overthrow of the last Shang ruler on the grounds that by so doing he was carrying out the will of Heaven. This was the beginning of the theory of the Mandate of Heaven *(t'ien ming),* which later, through the efforts of Mencius, became a part of Confucian doctrine.[21]

Mencius himself then propounded the further idea that it is not only historical events themselves but also the recording of such events—historiography—that serves a moral purpose. Confucius, he stated, "made" *(tso)* the *Ch'un ch'iu* (Spring and Autumn Annals), with the result that "rebellious ministers and villainous sons were struck with terror" (*Meng-tzu* 3b.9).

Few scholars today believe either that Confucius composed the *Ch'un ch'iu* or that the terminology of that uninspiring chronicle embodies any didactic purpose. Mencius's statement, nevertheless, was of great importance because it marked the beginning of the "praise and blame" *(pao pien)* theory of history writing. Applied to the *Ch'un ch'iu* itself, the theory held that

> these annals were a book of judgements, in which praise or blame was allotted to those concerned in each historical event. Such moral judgements were not given explicitly but were expressed by a discriminating use of terminology. To give only one example: when a prince was killed, the personal name

---

20. The first two quotations are translated in Bush (1971:29–30) and the third in Bush (1971:42).

21. See Creel (1970:82–85, 93–97). Creel believes that the Duke of Chou, in particular, may have been responsible for formulating an "embryonic philosophy of history," in which the Mandate of Heaven was conceived as an entity, transmitted initially from the earliest Hsia dynasty to the Shang, and then from the Shang to the Chou.

> of the prince would be recorded if he had been bad, but the name of the murderer if *his* action was disapproved of. Not only were moral judgements included, but there was a purpose behind the terminology throughout the Spring-and-Autumn Annals. In this way a textbook of political ethics came into being, which needed a great deal of interpretation if the judgements were to be understood.[22]

In a very few instances, the principle of using selected terminology to express praise and blame of persons and events was deliberately carried out in later history writing. The most notable example is the *T'ung-chien kang-mu* (Outlines and Details Based on the *Comprehensive Mirror*), a chronicle history (403 B.C.–A.D. 959) completed by Chu Hsi and his disciples in 1172 as an abbreviation of Ssu-ma Kuang's famed *Tzu-chih t'ung-chien* (Comprehensive Mirror for Aid in Government). In his introductory section of explanation, Chu Hsi enunciates what are probably the most detailed rules ever devised for recording events in a Chinese history. For example, in order to define the procedures and terms to be used when recording "attacks and campaigns" *(cheng fa),* he formulates no less than fifteen articles embodying nearly ninety subarticles of details (Yang Lien-sheng 1961a:52).

This moralistic compilation, together with its supplements covering the Sung and post-Sung dynasties, became in turn the basis for the first detailed Western history of China, the massive thirteen-volume *Histoire générale de la Chine,* by the eighteenth-century Jesuit de Mailla.[23] And de Mailla's work in turn influenced in varying degrees some later Western histories of China, notably Cordier's four-volume work of the same name.[24] Thus a very attenuated but clearly discernible thread of moralistic interpretation leads all the way from Mencius's (false) attribution of an esoteric moral purpose to the *Ch'un ch'iu* down to Western twentieth-century history writing on China.

Probably more important than such deliberate use of selected moral terminology is the less self-conscious but all-pervasive moralism that underlies almost all Chinese history writing. This sees the figures of history as classifiable under one or another character type, the juxtaposition of which determines the pattern of events. The impersonal facts of geography, economics, and institutions, on the other hand, though often recorded in some detail, are rarely adequately analyzed for their

---

22. Van der Loon (1961:26). See also, among the many other accounts of the theory, Watson (1958:75–85).

23. Completed in Paris in 1737 but finally published only in 1777–1785.

24. Cordier (1920), which, it is claimed, "incorporated whole sections of de Mailla's work without acknowledgement." See Boxer (1961:315 n. 21), quoting a verbal statement made by the late O. B. Van der Sprenkel.

influence on events. This view of history could and did result in unwitting distortion of what was recorded. At the same time—and again often unwittingly—it diminished the effectiveness of another age-old tenet of Chinese historiography: the historian's duty to record events as accurately, objectively, and fully as possible, without introducing his own point of view (except, commonly, in a clearly labeled judgment at the end of each chapter). The tension between these two principles continued throughout Chinese historiography and was never properly resolved.

As a good example of historical moralism, one cannot do better than refer to what Wright (1960b) terms the "stereotype of the bad-last ruler." The first model of this stereotype was Chou Hsin, final ruler of the Shang dynasty, whose "crimes" were first enumerated by King Wu and other founders of the Chou dynasty, then recorded in the *Shu ching* (Documents Classic) and *Shih chi* (Historical Record), and from these became archetypes for the similar qualities subsequently attributed to the "bad-last rulers" of other dynasties. Wright's primary example of the latter is Sui Yang-ti (r. 605–616), the second and final ruler of the Sui dynasty.

From the relevant historical accounts, Wright has abstracted a general "characterology" of the "bad-last ruler" that includes the following qualities:

1. Tyranny (neglect of upright officials, favoritism toward sycophants, callousness toward the masses)
2. Self-indulgence (drunkenness, passion for expensive products, elaborate palaces and gardens, etc.)
3. Licentiousness (lust, sex orgies, sadism, etc.)
4. Lack of personal virtue (unfilial or otherwise improper behavior to family members, disrespect to Heaven and ancestors, etc.)

Stereotypes of this kind tended to be ascribed to any end-of-dynasty ruler simply because he had the misfortune to be the last of his line. As Wright explains (1960b:59), the historians who did this

> viewed the past as a rich repository of experience . . . [which] would serve the living as a guide to action; they viewed the past also as a continuum within which certain moral principles operated, . . . laid down by the sages. The past could only teach its lessons if the moral dynamics of history were . . . hammered down by argument and illustrated. One of the most persuasive ways to present such lessons was to exemplify in certain historical figures the virtues or vices which governed the course of human events. In doing this the tendency was to paint in black and white.[25]

---

25. A similar phenomenon was noted by Herbert Franke (1950), writing before Wright, in his study of the reign of Emperor Shun (1333–1368), final sovereign of the

Another moralism of a somewhat different "praise and blame" variety, decidedly relevant to science, has to do with the detailed recordings of eclipses, sunspots, comets, earthquakes, floods, droughts, fires, monstrous births, and other exceptional phenomena, which occupy many chapters in the dynastic histories. The belief that these phenomena came as responses of nature to human misconduct (particularly on the part of the government and perhaps even of the emperor) and the manner in which the "Phenomenalists" of Han times developed this belief into a "science" of omen interpretation has been described by Needham (1956:247, 378–382).

By comparing the records of solar eclipses contained in the Former Han histories (the *Ch'ien Han shu* and the relevant portions of the *Shih chi*) with the list of eclipses known by modern astronomy to have been actually visible in China during these two centuries, it has been possible for Bielenstein (1950) and Eberhard (1957) to demonstrate that some of the eclipses as recorded could not have actually occurred in China. More than this, these scholars have noted a correlation between the completeness and accuracy of the recorded eclipses and the recording of other categories of exceptional phenomena, whose number tends to fluctuate from one period to another.[26]

Both men therefore conclude that some, at least, of the portents recorded during the Former Han dynasty were the deliberate fabrications of the officials who reported them to the throne. Moreover, even among those portents that were recorded because they probably actually took place, political considerations may at times have influenced the decision of officials to reveal rather than suppress them. Eberhard further hypothesizes that sometimes the reporting of portents may have been used not only to criticize the government in general or the emperor, but also as a weapon in the struggle for power between one political faction within the government and another.

As against these conclusions, Noel Barnard (1973) points out that the accuracy of the recording of solar eclipses in the *Ch'ien Han shu* improves markedly from the early decades through the middle portion to the final decades of the Former Han period. He therefore doubts the existence of

---

Yuan or Mongol dynasty. Franke shows that whereas Shun is portrayed in later historical accounts as a weak and evil man, he appears in contemporary accounts (i.e., those written before it was known that he would bring the dynasty to an end) as a mild and benevolent ruler.

26. In Bielenstein (1950), the number of recorded exceptional phenomena ("portents") is presented in graphs according to reigns. Eberhard (1957) goes a step further by breaking the data down according to five-year periods. He also discusses at some length the general implications his findings have for both Chinese historiography and science.

continuing and purposeful manipulations in the records, in place of which he suggests, as "the simplest and most obvious explanation," the possibility that "the records were lost, or the data discarded during compilation." Yet he couples this hypothesis with the warning that the "unrealiability of portent records" is such as to be "now generally accepted"—a statement scarcely calculated to encourage modern quantitative research workers.[27]

Of the three scholars, Bielenstein is the most sweeping in his unfavorable conclusions. He believes that prolonged political manipulations have rendered the lengthy lists of naturalistic phenomena contained in the various dynastic histories largely useless for statistical purposes. Eberhard is more cautious, pointing out that his findings do not necessarily apply to periods other than the Former Han dynasty. Within this time span, however, he too refers forcefully to the unfavorable implications for science resulting from the alleged politicizing of the portent records:

> The reason why they [the official astronomers, astrologers, and meteorologists] did not develop their knowledge into a unified scientific system seems to be that they were not interested in pure science for science's sake. They did not spend time in developing abstract laws or in studying the process of thinking (logic). But they also were not interested in applied technical sciences, e.g., in developing theoretical tools which could be used to control the flight of a cannon shell or to direct ships safely across the sea. Their central interest was in politics, and all known "scientists" mentioned above were also personally deeply involved in politics. (Eberhard 1957:66)

Again he says:

> Just as the astronomers, astrologers, or meteorologists were motivated by an interest not so much in science as in politics, so the government sponsored those scientific fields only in so far as they were politically loyal, i.e., as long as they did not develop new ideas. New ideas were conceived as new tools for the political struggle and therefore suspect. Science, therefore, was hampered in its development. What was achieved, over and above applied "political science," was the result of the hobbies of certain individuals; it remained hidden in their occasional writings, as a curiosity, and it was not systematized, discussed, clarified, enlarged, or applied in nonpolitical fields. (Eberhard 1957:67–68)

Two vital questions come to mind, neither unfortunately of the kind that can be satisfactorily answered. The first is what percentage of the recorded portents in the Former Han histories was in fact influenced by politico-moralistic considerations, as against the percentage honestly

---

27. Barnard (1972:482–487 and [for the quotations] 486).

compiled with no thought of political manipulation. Regrettably, no quantitative answer seems possible.

The second question is the extent to which the situation perhaps characteristic of the Former Han continued to prevail under later dynasties. Here again no answer seems possible, at least without further laborious research. Yet the study of the last ruler of the Yuan dynasty that Herbert Franke (1950) has prepared independently of either Bielenstein or Eberhard suggests that continuity may in fact have been considerable. Thus this study reveals that two of the seven recorded eclipses for the decades 1346–1367 did not occur on the reported dates and three others were probably calculated, not actually observed. In all seven instances, moreover, the eclipse entry in the Yuan dynastic history is either immediately preceded or followed by other entries recording ominous or disastrous events, such as an earthquake or "a violent rain and hail, big as the head of a horse." These facts seem to indicate that the political doctoring of records of natural phenomena continued to be practiced as late as the early decades of the Ming dynasty, when the Yuan dynastic history was compiled. Further, it is known that emperors continued to acknowledge personal guilt for the occurrence of exceptional phenomena right down to the end of the empire, even though one may suspect that by then such expressions had become largely conventional.[28]

On the other hand, consideration should be given to certain contrary factors:

1. The interpretation of portents in Han times and thereafter was closely related to the yin-yang and Five Elements cosmological systems. Yet Chinese observations of natural phenomena had probably become an accepted practice long before the Han dynasty and before these cosmological systems had become generally accepted—in other words, before the idea of using portents for political purposes had probably arisen.

2. During the Han dynasty itself, when the belief in portents was probably at its height, it would have been difficult either to fabricate or to conceal reports of really massive occurrences such as earthquakes, floods, or droughts; this would have been far easier in the case of less conspicuous and more narrowly localized phenomena, such as unusual or fabulous animals, monstrous births, and other anomalies. Hence, before accepting the political use of recorded portents in Former Han times as a *uniform* possibility, it would be desirable to make a further

---

28. The decree of the T'ung-chih Emperor of 22 August 1862 was quoted above in IV.2 n. 27, in which the youthful sovereign accepted moral responsibility for the appearance of a flight of shooting stars. Similar statements continued to be made down to the early twentieth century.

quantitative analysis of the *kinds* of portents involved, to determine if they all occur with the same relative frequency.

3. As already noted, Barnard suggests that the faulty recording of Former Han eclipses may have resulted from the loss of records from the early part of the dynasty rather than from deliberate manipulation. Even so, however, the effectiveness of this argument is seriously reduced by Barnard's parallel admission that the portent records still extant are unreliable for quantitative research.

4. It is well known that the Han worldview, including the belief in portents, declined after the fall of the dynasty, owing to major political and social changes and the rise of Buddhism. Hence, despite the evidence cited above from the Yuan dynasty, one may suspect that after the Han, generally speaking, the belief in portents never returned to its Han level. It would be very helpful if spot checks could be applied to selected points in history to test this assumption.

In conclusion, although one may recognize with Bielenstein that the use of the records of natural phenomena for statistical purposes involves a certain risk, I am unwilling to conclude that this risk completely invalidates such use for all periods and all kinds of phenomena. As for Eberhard's strictures, they seem to contain an element of truth for the Han period itself and may also carry potentially serious implications for Chinese scientific development in general. The question that cannot readily be answered, however, is how extensive is that element of truth. The several points made above suggest, until further contrary evidence can be presented, that even in Han times the political manipulation of the records of natural phenomena, if it occurred, may not have been as extensive, and therefore as harmful, as Eberhard's interesting study initially suggests.

1c. Militarism and Expansionism

With the exception of the Legalists, all ancient Chinese schools of thought were strongly opposed to war and violence. Indeed, even the Legalists wanted a powerful army simply to build up a powerful state. Never, in Western fashion, did they glorify war as spiritually ennobling. Among classical Chinese thinkers, the most biting antiwar statements came from Mo Tzu, yet even he was criticized by the Confucians for basing his argument heavily on utilitarian considerations: the economic wastefulness of war and armaments. On this matter as all others, the Confucians were uncompromisingly moral. "Those who are good at war should suffer the highest punishment," said Mencius, and again: "There are those who say, 'I am good at marshalling troops, I am good at conducting battle.' These are great crimes. If the ruler of a state loves goodness *(jen)*, in all-under-Heaven he will have no enemies" (*Meng-tzu* 4a.14; 7b.4).

It has sometimes been argued that among the Chinese people as a whole, especially in later times, antimilitarism was less strong than traditionally supposed. Proponents of this argument cite the prevalence of warfare during many periods of Chinese history, the military expansion of the Chinese empire under major dynasties, the prominence of military events and figures in popular literature (fiction and drama), and the importance of military divinities in the pantheon (see esp. Fried 1952). The subject is involved and cannot be adequately discussed here. As on some other matters, one may readily admit the possibility of more than one level of ideas: the "great tradition" of the dominant scholar-gentry elite, which strongly disparaged militarism as an ideal (even though perhaps tacitly accepting it as a practical necessity) and the "little tradition" of the great mass of "small men," for whom militarism was probably much less a moral issue.[29] This bifurcation, however, does not really affect our discussion, because it is the thinking of the dominant elite that primarily interests us, this being the group that played the decisive role in Chinese scientific development. As one surveys the literary output of this group, there can be no doubt of its overwhelmingly antimilitarist tone, even in times either of political expansion or political breakdown.[30]

One such expansionist era was the Former Han dynasty, and it is therefore not surprising that the question of war and peace should crop up in the debate of 81 B.C. recorded in the *Yen t'ieh lun* (Discourses on Salt and Iron). At one point the Legalistically minded government spokesman alludes to the emperor's concern that China's fighting men receive adequate logistical support at the distant frontiers where they are stationed against their tribal opponents. The reply of the provincial Confucian literati is as might be expected:

> How can the Man or Mai tribesmen, with their barren lands, merit such trouble and worry, bringing with it the uncertainty of the Warring States period? And yet, if His Majesty wishes not to put them aside, let him but manifest his virtue to them and extend to them his favors. Then will the northern barbarians assuredly come of their own accord to pay tribute at the frontier passes. (*Yen t'ieh lun* 12; tr. Gale 1967:76)

---

29. The terms "great tradition" and "little tradition" are those of the late anthropologist Robert Redfield.

30. Fried himself admits (1952:347) that the scholar-gentry were, "as a class, firmly antimilitary," in contrast to the peasantry, who "regarded military roles as an avenue to high status."

To the argument that China enjoys peace only because of the sacrifices made by the frontier military men, who experience bitter cold and the constant threat of attack, the Confucians reply similarly:

> In antiquity the Son of Heaven stood at the center of all-under-Heaven. His internal domain did not exceed one thousand leagues, and the serried states of the [surrounding] nobles did not reach as far as the non-productive lands. . . . The "hundred surnames" enjoyed equilibrium and harmony, and the exactions of corvée labor were not strenuous. Today, however, the Hu and Yüeh tribes have been pushed back so that the highways and bypass [leading to them] have become circuitous and lengthy, and the troops are worn out. . . . This is why China faces imminent death and destruction, and why the "hundred surnames" clamor and clamor and refuse to be silent. (*Yen t'ieh lun* 16; tr. Gale 1967:100)

In similar vein the literati go on to criticize the achievements of Chang Ch'ien, China's first great explorer, who had returned to China around 126 B.C. from an expedition of twelve years, bringing with him the first authentic Chinese knowledge of Central Asia. "Chang Ch'ien," they say, "penetrated to strange and distant lands, but what he brought back had no use. The reserves in our treasuries are flowing away to foreign countries" (*Yen t'ieh lun* 16; tr. Gale 1967:102).

Striking is the similar tone of the Confucian officials some fifteen hundred years later when they criticize the voyages carried out by Cheng Ho on behalf of the government to the Indian Ocean and East Africa (1405–1433). In this case, their hostility was intensified by the fact that Cheng Ho was a eunuch. When, around 1479, the vice-president of the Ministry of War burned the documents that had been kept there regarding Cheng's seven voyages, he argued that they were "deceitful exaggerations of bizarre things far removed from the testimony of people's ears and eyes." And to his superior, the president of the ministry, he explained further: "The expeditions of the San-pao [i.e., Cheng Ho] to the Western Ocean wasted tens of myriads of money and grain, and moreover the people who met their deaths [on these expeditions] may be counted by the myriads. Although he returned with wonderful precious things, what benefit was it to the state?" (Duyvendak 1939: 395–396).

Duvyendak goes on to comment that the expeditions, in official eyes, were linked both with immoral extravagance and the baneful rise of eunuch power:

> This aversion helps to explain why in later times such expeditions have never been repeated. It also contributed to the formation of the Confucian mentality which frustrated fruitful intercourse with overseas barbarians in modern times. The desire for closer intercourse was felt to be unworthy of a Confu-

cian official; such a desire was identified with luxury and eunuch rule, it was extravagant and wasteful, and China, being economically sufficient unto herself, could very well do without the curiosities produced by foreign countries. (Duyvendak 1939:398)

One can applaud the Confucian attitude toward militarism and political expansion and yet recognize its probably negative implications for the creation of a modern science in China. This does not mean that Confucian antimilitarism should necessarily be regarded as a major inhibiting factor per se. It does mean, however, that antimilitarism constitutes one of the strands in a broader nexus of interrelated psychological factors that in their totality, it seems to me, have been profoundly unfavorable. Among them can be mentioned the subordination of the individual to the group in Chinese thinking, the constant preference for a middle way in matters of behavior and thought, the desire to avoid disagreement in interpersonal relations and intellectual issues at all costs, and the customary readiness to accept the status quo and established authority. These and other similar matters have either already been mentioned or will be discussed later (especially in the final three sections of this chapter).

More particularly, a psychological affinity is apparent between Confucian opposition to military expansion, attachment to one's native place (a strong Chinese characteristic), and reluctance to pursue massive exploration, settlement, trade, and exploitation abroad. The Chinese official view of Cheng Ho's voyages contrasts dramatically with the inexorable drive of Europeans, especially from the late fifteenth century onward, to explore, colonize, and ruthlessly exploit. It is difficult not to see a meaningful interplay between this drive and the struggle for religious freedom, the growth of individual self-consciousness, the rise of capitalist institutions, and the creation of a burgeoning science and industry, all characteristic of the modern West.[31]

On the Chinese side, John K. Fairbank, in his introduction to a symposium volume on Chinese ways of warfare, has perceptively outlined some of the ramifications of what he calls the Chinese traditional "pacifist bias":

War is not easy to glorify [in the Chinese tradition] because ideally it should never have occurred. The moral absolute is all on the side of peace. No economic interest sufficed to glorify warfare; no wealthy neighbors enticed Chinese freebooters across the border or over the sea. . . . Generals had few triumphs; and they lost their heads at least as often as anyone else. Chinese

---

31. For the profound psychological and behavioral differences between the Cheng Ho expeditions and the Portuguese voyages of exploration toward the end of the same century, see Needham (1971:514–516, 521–523).

> youth were given no youthful worship of heroism like that in the West. The most one could find was a Robin Hood figure like Chu-ko Liang (A.D. 222–280). Likewise, holy wars are not easy to find in the Chinese imperial records, just as an avenging God and the wrath of Jehovah are far to seek. Moral values are not handed down from a deity who is on one's side, ready to smite the infidel. The whole view of the world is less anthropomorphic and less bellicose than that of the Old Testament, or of Islam. (Fairbank 1974:7)

It would be misleading to end on this note. China after all *did* have its famed generals, its epochs of military glory, its intrepid travelers, its overseas navigators. The Chinese *oikoumene* expanded many times from its original nuclear area until by the first century B.C. it had come to embrace a territory greater than that of contemporary Rome at its greatest extent (Creel 1970:1–2). At the other end of the imperial spectrum—the past three or four centuries—Chinese migration spread like an expanding ink blob into surrounding areas, especially the lands and archipelagos of Southeast Asia.

All this is true, and yet Chinese militarism and expansionism functioned differently from those of Europe. Of course there were emperors—the Han Emperor Wu (140–87) was one of them—who felt a compulsive need for the personal glory associated with military expansion. But by and large, China's most serious external wars—those against a seemingly endless succession of tribal peoples from Inner Asia—stemmed from contradictions inherent in two different ways of life: that of the sedentary Chinese, based on intensive grain culture, and that of the Inner Asian tribesmen, based on intensive animal husbandry. The Great Wall—itself a Chinese defensive symbol—was an attempt to demarcate arbitrarily one way of life from the other. It failed because, geographically, such a precise division between settled agriculture and nomadic pastoralism was impossible and because both within and beyond the Wall lived peoples whose mixed economy was neither fully agrarian nor fully pastoral and who were ever ready to shift their political weight from one side to the other as conditions dictated.[32]

There is very little evidence that Chinese foreign wars were waged to gain quick loot in the manner of the Spanish conquistadores or to establish trading entrepôts in the manner of the several European East India companies. Probably, as the Confucian critics contended, these wars cost China far more than they ever brought back in the form of trade and tribute (between which there was sometimes not too clear a distinction). Some military campaigns, to be sure, especially those against southern tribal peoples, gained for the Chinese peasant new lands suitable for his way of life. For the most part, however, the Chinese south-

---

32. The classic analysis of this topic is that of Lattimore (1951).

ward spread within China itself, and from there to other lands in Southeast Asia, took place of itself as a gradual process of osmosis. Rarely was it directly encouraged by the government, and sometimes it was positively discouraged. Not infrequently, the spread consisted of Chinese ideas and institutions as much as of actual people.

Nor, as Fairbank has indicated, was religious fervor an important factor in Chinese warfare. The closest approach lay in the government's suppression of sectarian rebellions, but even then the dominant considerations were much more political than religious (see IV.2). On the other hand, perhaps it is significant that many of China's greatest travelers were not professional explorers at all, but religiously minded monks who traversed great distances to visit the Buddhist shrines of India and who recorded what they saw along the way.

In short, although Chinese militarism and expansionism cannot be denied, their expected consequences were blunted by a variety of ideological considerations. These included not only Chinese pacifism per se, but also disdain by the literati of the merchant and the artisan, suspicion of "useless" and costly technological constructions, and inculcation of the virtues of frugality and simplicity of living (see next subsection). How different all this was from the European world![33]

1d. Frugality and Technology

Frugality and simplicity of living, like antimilitarism, were virtues preached by almost all schools of classical Chinese thought. Here again the most famous exponent was Mo Tzu, who wanted men to have just enough, but nothing more, of food, clothing, houses, vehicles, and the

33. A more detailed and rather different account of the Chinese military tradition is that of Gawlikowski (1985), for a copy of which I am indebted to the author. In this article, Gawlikowski describes the major ancient Chinese writings on military matters, analyzes their prevailing attitudes toward military strategy (long influential among Chinese military figures all the way down to Mao Tse-tung), and shows how some of their ideas sometimes spilled over into such nonmilitary aspects of life as commerce, painting and calligraphy, games (especially *wei ch'i* or Chinese chess), sex relations (see also VI.2 below), and social behavior in general. Quite properly, Gawlikowski criticizes the frequent failure of writers on China to note the importance in Chinese civilization of the long-existent dichotomy between *wen* or the civilian way of life and *wu* or the military tradition (discussed in VI.4). All of this makes Gawlikowski's study well worth reading, but does not disprove, in my opinion, the contention here advanced that during most of China's imperial history, the civilian way of life was valued above the military by most members of the dominant elite, in contrast to the views often found among their counterparts in the Western world.

like. Besides this, he preached limited expenditures for funerals and religious ceremonies and the elimination of music.[34] But Mo Tzu's extreme utilitarianism earned him once more the criticism of Mencius and other Confucians, who distinguished sharply between the Confucian emphasis on *yi*, "what is right," and Mohist emphasis on *li*, "what is beneficial or profitable."[35] Of course the Taoists likewise urged simplicity in material life and ingenuousness in thinking, in order to turn men back from the corruptions of civilization to the genuineness of the child or "natural man."[36] Finally, the Legalists advocated frugality and hard work (to be reinforced by rewards and punishments) on the very practical grounds that these would build up the economic and military power of the state.[37]

Underlying these similar approaches was a basic consideration of which most thinkers were probably scarcely aware, and yet which probably subtly influenced the thinking of all of them—that preindustrial agrarian societies often have difficulty in meeting the economic needs of all their people, so that it becomes good policy to preach the doctrine of contentment with simplicity.

The usual Confucian moral approach, supplemented to some extent by economic considerations, appears in the twenty-ninth chapter of the *Yen t'ieh lun,* where the Confucian literati decry the growing elaboration and materialism of daily life since the simple days of antiquity. The theme is discussed in terms of food, clothing, horses and chariots, furs, eating utensils, sacrifices and music, coffins and tombs, domestic animals, and much more. To cite but one example: In antiquity the family rested on the relationship between one man and one woman. Later, however, members of the aristocracy commonly took as many as nine concubines for themselves besides a single wife. By "modern times" (the Former Han period), the nobles had hundreds of consorts, high officials, tens; and even men of middle rank enjoyed the services of several female attendants. The resulting sexual imbalance meant that many women confined to the households of the rich were forced to lead empty, unfulfilled lives, while many men of common status were doomed to die without ever having a mate. The summation reads:

---

34. *Mo-tzu,* esp. chs. 20–21 ("Limitations on Expenditures"), 25 ("Limitations on Funerals"), and 32 ("Condemnation of Music").

35. See, e.g., *Meng-tzu* 1a.1; 4a.14; 6b.4, and so on. The dichotomy goes back to Confucius. See *Lun yü* 4.16: "The superior man *(chün tzu)* is informed on what is right *(yi),* the small man *(hsiao jen)* is informed on what is profitable *(li).*"

36. See, e.g., *Lao-tzu* 3, 19–20, 29, 46, 59, 64, 80, and so on.

37. See, e.g., *Shang chün shu* 4 and 20; tr. Duyvendak (1928:201, 306). Also *Han Fei-tzu* 10 and 50; tr. Liao (1939:85–87; 1959:301).

> Luxurious unrestraint in the constructing of palaces and halls is the termite [i.e., consumer] of forest lumber. The decorating and carving of utensils and implements is the termite of natural materials. The embroidering and beautifying of clothes and garments is the termite of linen and silk. The eating of human food by dogs and horses is the termite of the five grains. Granting unrestrained license to mouth and belly is the termite of fish and meat. Nonlimitation of expenditures is the termite of the treasuries. Failure to check the loss of stored-up grain is the termite of fields and meadows. Lack of measure in funerals and sacrifices is the termite which harms the living. To pull down what has been built and change what is old is to inflict injury on the fruits of labor. To permit artisans and merchants to maintain contacts with their superiors is to inflict injury on agriculture. Thus a single bowl or cup may require the efforts of a hundred men, a single wind-screen may require the labors of a myriad men. Great indeed is the harm these cause! . . . Hence when the state suffers from inadequate reserves, its government becomes immobilized, and when the individual suffers from inadequate reserves, his person falls into danger. (*Yen t'ieh lun* 29 end; not tr. in Gale 1967)

It is evident here how the dislike of conspicuous and unnecessary consumption reinforced the Confucian suspicion of the merchant and the artisan. It also strengthened the opposition to political expansion and foreign trade discussed in the last subsection. Finally, it induced a hostility toward technology—at least some kinds of technology—by focusing attention on the artisan as the creator of unneeded luxury goods rather than as the innovator of useful technological improvements.

Suspicion of innovative technology underlies a Confucian story about Kung-shu P'an, the artisan of the fifth century B.C. who eventually became the deified "founder" of carpentry and related technologies. It seems that when the mother of Chi K'ang Tzu (a contemporary of Confucius who had usurped political power in the state of Lu) died, Kung-shu P'an, then still a boy, asked permission to facilitate the lowering of the coffin into the tomb by using a special mechanical device *(chi)* of his own devising. The request was rejected on grounds that it failed to conform to ancient practice, and Kung-shu P'an was told: "P'an, rather than try out your cleverness on another person's mother, why not try it out on your own mother?"[38]

Similar distrust of mechanical ingenuity appears in another story involving Kung-shu P'an. It is an anecdote in the *Mo-tzu* in which P'an

---

38. *Li chi* 2; tr. Legge (1885:27.184); Couvreur (1913:1.231). As Legge points out, the historicity of this story is doubtful because of the chronological gap between Chi K'ang Tzu (d. 468 B.C.) and Kung-shu P'an (who may have lived sometime between 470 and 380). Even if unhistorical, however, the story typifies the attitude toward mechanical devices on the part of the anonymous Confucian writer of the *Li chi* chapter in which it appears.

constructs a wooden bird (a kite), flies it for three days, and is very proud of his skill until Mo Tzu criticizes him by saying that his feat does not compare with the making of a linchpin. Such a pin, Mo Tzu continues, is a piece of wood only three inches long, yet it can carry a load of several tons. Only such a socially useful creation deserves to be called "skillful"; anything else should only be termed "clumsy."[39]

Wang Ch'ung, commenting in his *Lun heng* on this story, contributes a further revealing anecdote: "There is another report that Kung-shu P'an lost his own mother because of his skill. He constructed for her a wooden chariot and horses with a wooden driver, and when it was ready and she had taken her place inside, it sped away and never returned; thus was she lost to him" (*Lun heng* 26; tr. Forke 1962:1.498). It seems probable that different social classes viewed Kung-shu P'an with differing esteem. To artisans he was the patron saint of their crafts, but by some at least of the literati he was apparently suspected of creating useless and dangerously novel contrivances.

A Taoist example of distrust of manual cleverness is the story in the *Lieh-tzu* of the man of Sung who after three years of effort succeeded in carving a jade mulberry leaf indistinguishable from an actual leaf. He was rewarded with an official position, but Lieh Tzu commented bitingly: "If Heaven and Earth grew things so slowly that it took them three years to finish one leaf, there wouldn't be many things with leaves."[40]

But the most famous Taoist technological story is that told in the *Chuang-tzu* of the meeting between Confucius's disciple, Tzu-kung, and a farmer. Tzu-kung, seeing the farmer laboriously irrigating his garden with water brought by jar from a well, urges him to use a swape (a counterbalanced bailing bucket for raising water). But the farmer replies indignantly: "I have heard from my master that those who have cunning devices use cunning in their affairs, and that those who use cunning in their affairs have cunning hearts. . . . It isn't that I don't know [about the swape,] but I would be ashamed to use it" (*Chuang-tzu* 12; tr. Needham 1956:124, Watson 1968:134).

In his subsequent discussion, Needham (1956:125) suggests that the Taoist opposition to new technology may to some degree have been politically motivated. By this he means that it may reflect a fear on the part of the Taoists that the purpose of any new machine is simply to

39. *Mo-tzu* 49; tr. Mei (1929:256). For another version of the story, see *Han Fei-tzu* 32; tr. Liao (1939:34). There it is Mo Tzu himself who is said to have invented the kite, and it is a wooden ox-yoke peg, not a linchpin, with which the kite is compared.

40. *Lieh-tzu* 8; tr. Graham (1960:161). Also found in *Han Fei-tzu* 21 (one of the Taoistic chapters in this work); tr. Liao (1939:220).

uphold the interests of the Chou aristocracy against the peasants, either by cheating them of what was rightfully theirs or by chastising those who dared to rebel. While the possibility of such thinking cannot be wholly ruled out, we have seen from the present discussion that distrust of technology was by no means limited to the Taoists. The Confucians (and perhaps other schools as well) were likewise distrustful, though the basis for their attitude was moral and economic, not political.

A pejorative commonly used to criticize the techniques and products of distrusted artisans is *ch'iao,* translatable as "skill, skillful, cunning, cleverness."[41] Other similarly derogatory terms are *chi,* "art, dexterity," and *ch'i,* "implement, utensil" (meaning the product made by the artisans).[42] All three words are often reinforced by the adjective *ch'i,* "strange, extraordinary," or more rarely by another adjective, *yin,* "obscene, licentious." The following are examples:

The *Li chi,* in its chapter on "Royal Institutions," lists several categories of wrongdoers who deserve death, among them persons who "make licentious music, queer clothing, strange arts *(ch'i chi)* and strange implements *(ch'i ch'i),* thereby confusing the multitude" (*Li chi* 3; tr. Legge 1885:27.237; Couvreur 1913:1.308). Again, in its chapter on the "Monthly Ordinances," the same work says about artisans under the third month: "They are forbidden to carry out any licentious cleverness *(yin ch'iao)* such as will flood the minds of their superiors [with thoughts of dissipation and luxury]" (*Li chi* 4; tr. Legge 1885:27.266; Couvreur 1913:1.351).

The *Shang chün shu* (22; tr. Duyvendak 1928:313) complains that farmers, despite their hard work, are worse off than "merchants, shopkeepers, and artful *(chi)* and clever *(ch'iao)* men [i.e., artisans.]" In similar fashion, the *Kuan-tzu* (48[15/14b]) urges that "strange skills *(ch'i ch'iao)* be restricted and agricultural activities be furthered." On the Taoist side, Lao Tzu exhorts the ruler to "cut off cleverness *(ch'iao),* throw away profit, and thieves and robbers will not exist" (*Lao-tzu* 19). Later he remarks that "great cleverness *(ch'iao)* resembles clumsiness" (*Lao-tzu* 45).

In a document that belongs to the "new text" version of the *Shu ching* (Documents Classic) and therefore is of late origin, King Wu, in a harangue to his troops, says of his opponent Chou Hsin, tyrannical last ruler of the Shang dynasty: "He performs strange arts *(ch'i chi)* and licentious cleverness *(yin ch'iao)* for the delectation of his women."[43] An

---

41. In the story about Kung-shu P'an's mechanical device, *ch'iao* was the "cleverness" he was told to try out on his mother.

42. Confucius stated that "the *chün tzu* is not a utensil *(ch'i).*" See *Lun yü* 2.12 and V.3 above.

43. *Shu ching* 5.1.3; tr. Legge (1960:3.295). On Chou Hsin as the prototype of the "bad-last ruler," see VI.1b above.

edict issued by the Later Han Emperor Ho in A.D. 99 complains that "among the little people, some of the merchants forget the laws and prohibitions, and the extravagant products of bizarre cleverness *(ch'i ch'iao)* concentratedly pour forth and widely circulate" (*Hou Han shu* 4/4a). Apparently the situation failed to improve, because Emperor An began his reign in A.D. 107 with an edict in which he ordered the high officials to "prohibit extravagance and stop the making of frivolous and clever *(ch'iao)* objects" (*Hou Han shu* 5/1b).

The latest example of the phrase *ch'i ch'iao* known to me (though still later instances no doubt also occur) comes in the famous letter of 1793 from the Ch'ien-lung Emperor to King George III, made on the occasion of the Macartney embassy to China. In this the emperor curtly rejects the king's request for trading privileges. "I set no value on objects strange or ingenious *(ch'i ch'iao),*" he declares, "and have no use for your country's manufactures."[44]

Most artisan productions to which the literati objected on grounds of extravagance and frivolity probably had little significance for science. Some, however, may have been rejected simply because they seemed overly theoretical and of no immediate usefulness, and in such cases the loss to science may have been considerable. One thinks, for example, of Chang Heng's seismograph of A.D. 132 and the failure to develop it further. Plausible reasons, though the texts do not state them, may have been that Chang's invention seemed uncanny to his contemporaries and also did not seem to have much practical value. Or again, after Ma Chün invented a new kind of silk loom in the third century A.D., he failed to gain the support of a patron despite the eloquent arguments made on his behalf by a poet-friend. Though the reasons for the failure are not clearly specified, they too very likely included doubt about the practical value of Ma's invention.[45]

Finally, it should be noted that fear of technological unemployment seems to have had little or no unfavorable effect on the development of Chinese technology. On the contrary, as Needham has pointed out (1965a:28), pre-modern China, despite its abundant manpower, seemingly offers almost no known historical instances of suppression or

---

44. Bland and Backhouse (1914:325) and Teng and Fairbank (1954:19). The term is rendered identically in both translations.

45. On the inventions of Chang Heng and Ma Chün, see, respectively, Needham (1959:626–635 and 1965a:39–42). As Needham indicates, the dynastic histories mention two sixth-century writings (now long lost) which may have discussed Chang's invention. These references, however, are too brief and vague, and the time between them and Chang Heng is too great, for them to constitute acceptable evidence of any real follow-up to Chang's invention.

objection to labor-saving devices because of fear of unemployment. This contrasts sharply with the situation in Europe, where several such instances are known. However, as should also be stressed, the Chinese attitude changed markedly in the nineteenth century as a result of the introduction, actual or planned, of Western-type machinery and factories into China. Illustrative are the following words by the official Chu Yi-hsin, written in 1892:[46]

> In Western countries the territory is large, the population is thin, therefore they even use machines for agriculture. If China employs them, one person's agricultural work may take away the livelihood of ten others. These ten others, if they do not want to sit down and wait for death by starvation, could choose a risky road [i.e., they might become bandits].

Even stronger is the statement by Chang Tzu-mu which, though undated, was probably made around 1880:

> The farmers in the south, the miners in the northern mountains, and those who are pulling carts and manning boats: all these several thousand or million people have been working all their lives at hard labor and if this is suddenly replaced by machines, they will lose their way of making a living. Would they not gather together to cause disorder or start a rebellion? This is probably the reason that, following widespread adoption of machines in Europe less than a hundred years ago, great revolutions there have repeatedly occurred.

The contrast between such statements and the earlier Chinese silence about technological unemployment suggests that prior to the nineteenth century, changes in Chinese technology came too gradually to have any marked immediate social and economic effect. In other words, the Chinese preindustrial indifference to the possibility of technological unemployment may simply reflect the fact that such a possibility was at that time hardly imaginable.[47]

---

46. This and the following quotation both come from Teng and Fairbank (1954:186).

47. As against the theme of frugality discussed in this section, a very few Chinese can be found who have advocated increased government spending, especially on public works, as a means for providing employment and alleviating suffering during times of famine or other calamity. Yang Lien-sheng (1957) cites relevant passages from *Kuan-tzu, Hsün-tzu, Shih chi,* and *Yen t'ieh lun* in antiquity, as well as from Shen Kua's *Meng-ch'i pi-t'an* of 1086, followed by an essay of the mid-sixteenth century. Yet as Professor Yang himself points out, any such justification for spending constitutes "an uncommon idea in traditional China." Along the same line, he also writes in the second paragraph of his study: "Traditional Chinese thought in general has been in favor of saving and frugality and against spending and lavishness."

1e. The Moralistic View of Nature

A good part of the following chapter will be devoted to the general topic of the several ways in which the Chinese viewed nature. Here we are interested only in one of these possible views: what might be called the moralistic view, meaning by this the attribution of human characteristics to the forces and objects of nature. This view is primarily but not exclusively Confucian. The eclectic *Huai-nan-tzu* compilation (ca. 140 B.C.), for example, contains a passage (3/28b) in which various possible sequences of the Five Elements are enumerated. In each sequence, any element that generates another element is termed a "Generator" but more literally a "Mother"; any element thus generated is the "Offspring"; and the interactions between the two are variously termed "righteousness," "fostering," "special effort," "control," and so on. Such personifications of natural phenomena and things are common enough in every early society, but in China, what Needham (1956:261 n. a) calls "this weakness in ancient . . . science" continued to be a weakness much later as well. The following are a few representative examples, both early and late.

Water, as is well known, is an important symbol in Taoism.[48] In the writings of other schools, however, it also appears frequently, so that there is no surprise in finding it ascribed various human moral qualities in three classical texts, two of them variants of the same basic statement.[49] For the example presented here, we go to the *Ta Tai Li chi* version, in which Confucius's disciple, Tzu-kung, begins by asking the Master why the *chün tzu* (the morally superior man) should be particularly observant when he gazes at a large river. Confucius's reply is:

> Water, to the *chün tzu*, is the counterpart of various qualities. It gives itself to things without partiality: in this it is like virtue *(te)*. What it touches lives, what it does not touch dies: in this it is like benevolence *(jen)*. Its flow moves downward to every nook and cranny, conforming to their patterns: in this it is like righteousness *(yi)*. It plunges unhesitatingly into an abyss of a hundred fathoms: in this it is like courage *(yung)*. Its flow as a shallow stream becomes an immeasurable depth: in this it is like knowledge *(chih)*. It is soft and yield-

48. Waley (1934:56). Examples are *Lao-tzu* 8: "The highest good is like water, whose goodness benefits all things without ever striving"; *Lao-tzu* 78: "Nothing under Heaven is softer or weaker than water, yet for attacking what is hard and strong, nothing surpasses it."

49. See *Hsün-tzu* 28(20/3b–4a), not tr. in Dubs (1928), and *Ta Tai Li chi* (Record of Ceremonial of the Elder Tai) 64 end; tr. Wilhelm (1930:147). The third water passage, similar in general idea but textually different, occupies all of *Kuan-tzu* 39 (14/1a–4b); tr. T'an and Wen (1954:86–88) and Needham (1956:42–45).

> ing yet subtly penetrating: in this it is like discrimination *(ch'a)*. It accepts bad things without rejection: in this it is like forebearance *(pao meng)*. Whatever enters it dirty emerges clean: in this it is like the transforming power of goodness *(shan hua)*. By collecting in whatever [hole or hollow] is to be measured, it ensures levelness: in this it is like rectitude *(cheng)*. When it fills [a vessel], it needs no cover: in this it is like measuredness *(tu)*. It may have a myriad windings, yet always proceeds eastward: in this it is like determination *(yi)*.[50] These are the reasons why the *chün tzu*, when he sees a great river, is sure to contemplate it.

Another personification, this time of jade, likewise occurs in three slightly differing versions.[51] In the most elaborate of the conceits (that in the *Li chi*), jade is likened to eleven different qualities and things: benevolence, knowledge, righteousness, propriety, music, loyalty, good faith, heaven, earth, virtue, and the Tao.

Some nine centuries or more later (in 803), we find the famed poet Po Chü-yi (772–846) writing an essay, "On Cultivating Bamboo" *(Yang chu chi)*, in which he ascribes certain qualities to the bamboo and indicates how these each provide the *chün tzu* with reminders of proper conduct:

> Why is it that the bamboo suggests the man of moral worth *(hsien)*? The shoots of the bamboo are firm *(ku)*, and through firmness virtue is implanted. So the *chün tzu*, on seeing its shoots, thinks about planting well and not pulling up. The nature of the bamboo is to be upright *(chih)*, and through uprightness the character is established. So the *chün tzu*, on seeing its nature, thinks about taking a central stand without sideward leanings.[52] The heart of the bamboo is empty *(k'ung)*, and through emptiness the Tao is embodied. So the *chün tzu*, on seeing its [empty] heart, thinks about responding to the needs [of the world] while yet vacuously accepting [all that comes].[53] The joints of the bamboo represent integrity *(chen)*, and through integrity the will is established.[54] So the *chün tzu*, on seeing its joints, thinks about polishing his name and conduct, leveling his unevennesses, and following a single undeviating course. It is for these reasons that the *chün tzu* often plants bamboos in his

50. Almost all major Chinese rivers, despite their turnings, flow in general from west to east. The word *yi* (lit. "thought") should be taken here as equivalent to *chih*, "determination," which is the reading in the *Hsün-tzu*.

51. (1) *Hsün-tzu* 30(20/9a–b); not tr. in Dubs (1928); (2) *Li chi* 45; tr. Legge (1885: 28.464) and Couvreur (1913:2.697–698); (3) *Kuan-tzu* 39(14/2a); omitted in T'an and Wen (1954), tr. Needham (1956:43).

52. The strong Chinese preference for centrality has been discussed in III.2.

53. The sentiment in this and the preceding sentence is Taoist or Taoist/Buddhist.

54. *Chieh*, the "joint" or "node" of a bamboo, also means, in an extended nonmaterial sense, "limit" or "restraint." This is why it should suggest to the *chün tzu* the idea of *chen*, "integrity," a word also translatable as "chaste, pure, virtuous, uncorrupted," and so on.

courtyard. . . . Thus it is that the bamboo, as compared with other plants and trees, is like a man of moral worth *(hsien)* as compared with the common herd *(chung shu)*.[55]

Po Chü-yi's essay probably gave initial impulse to what, in painting, became a popular practice following the rise in Sung times of "literati painting" (*wen-jen hua;* see VI.1a). This practice consisted of identifying certain plants and flowers as symbolizing qualities thought to be particularly appropriate to the scholar-gentleman. Besides the bamboo, other plants thus identified and hence often depicted in painting included the orchid, chrysanthemum, plum blossom, and pine tree. Sometimes, too, such symbolism was extended to the animal kingdom. Horses wearing halters, for example, could symbolize high-minded would-be recluses chaffing under the restrictions imposed on them by government service.[56]

Our final example is once more a description of water, this time written around 1710 by the painter Shih-t'ao (also known as Tao-chi) who, as a scion of the Ming imperial family, became a Buddhist monk in order to isolate himself from the conquering Manchus:

Water, as a wide expanse, broadly distributes its benefits: this is because of its virtue. By seeking lowly and humble places it conforms to propriety: this is because of its righteousness. As the tidal ebbing and flowing it never ceases its movements: this is because of the Tao. Having cut a channel for itself, it leaps forward as a wave: this is because of its courage. Swirling about, it reaches an even level: this is because of the Dharma *(fa)*. Filling up what is far away, it penetrates everywhere: this is because of its capacity to discriminate *(ch'a)*. Despite its twistings and turnings, it keeps toward the east: this is because of its determination.[57]

Of the seven qualities and abstractions here symbolized by water, five are identical with those in the *Ta Tai Li chi* passage with which we began. The only new ones are the Tao (which, however, appeared both

---

55. Po Chü-yi, *Yang chu chi,* in *Ch'üan T'ang wen* (Complete T'ang Prose) 676/5b–6b.

56. These symbols are discussed in Bush (1971:101–105). Concerning the bamboo, she writes (pp. 102–103): "Bamboo stood for the character of the scholar, upright and enduring despite hardships and old age. Like bamboo he could yield to circumstances without breaking, and his open mind was compared to its hollow stem." This sounds very much like an echo of Po Chü-yi's essay, even though, as Bush remarks, Su Tung-p'o (1037–1101), who was influential in the *wen-jen hua* movement, was the person more immediately responsible for making bamboo an important subject for painting.

57. *Shih-t'ao hua yü-lu* (Shih-t'ao's Talks on Painting) 18.67; tr. Lin Yutang (1967:153 [greatly modified]).

in the *Li chi* passages on jade and in Po Chü-yi's essay on bamboo) and the Buddhist Dharma or *fa* (a reminder that Shih-t'ao was a monk). Shih-t'ao precedes his disquisition on water with similar remarks on mountains: in their manner of seeming to make obeisance, he says, they are like propriety; in their grouping together, like caution; in their massiveness, like generosity, and so on. Probably this juxtaposition of mountains and water was inspired by Confucius's enigmatic and rather Taoistic remark: "The man of wisdom delights in water. The man of virtue delights in mountains" (*Lun yü* 6.21).

Persons who think of bamboos as "men of moral worth" and of the downward flow of water as an example of the Confucian virtue of propriety *(li)* are unlikely to have much interest either in the growth of bamboo or the movements of water as scientific phenomena. Or even if they have, they are not going to know the methodology necessary to satisfy this interest. The most famous example is Wang Yang-ming's attempt around 1492, when he was twenty, to investigate the principles *(li)* of the bamboos growing in front of his pavilion. After seven days of constant study he fell ill, as had his friend after a mere three days (Fung 1953:597; Chan 1976b:2.1408–1409). The intellectual background for this fruitless attempt will be discussed in the next subsection.

1f. Moralism and Science

In classical Confucian writings it is rare to find any expressed interest in the investigation of what would today be termed scientific phenomena. The *Meng-tzu* contains a single passage which stands out from the rest of the work because it is so different in tone. It is also beset by semantic problems that make it particularly obscure. Without going into these, I quote only the two final sentences to indicate the general flavor. These read: "Consider the heavens so high and the stars so distant. If we seek out the causes, we may, while at our seats, bring forth [i.e., calculate or determine] the solstice days of a thousand years."[58]

As against this lonely passage, the statements by Mencius and other early Confucian writers prescribing the proper interests of the *chün tzu* are almost always strongly ethical. Thus Hsün Tzu writes:

---

58. *Meng-tzu* 4b.26; tr. Legge (1960:2.331–332). The translation here of the key second sentence differs from those of Legge and others, who place the solstice in the past: "go back to the solstice of a thousand years ago." But the Chinese text gives no indication as to whether the action refers to past or present. Hence I make the translation indefinite, believing the statement to signify that the solstices of either a millennium past or a millennium in the future may be calculated on the basis of currently made observations.

> As to anything that does not serve to separate right from wrong, to range the crooked apart from the straight, to differentiate good government from misrule, or to bring order to the ways of mankind: skill in such matters will be of no benefit to men, and lack of such skill will do them no harm. . . . For these are the doctrines of unruly persons living in a degenerate age. (*Hsün-tzu* 21(15/10b; tr. Dubs 1928:277–278)

And the literati in the *Yen t'ieh lun* comment even more pointedly apropos Tsou Yen, the "father" of Chinese naturalistic speculation: "Therefore the *chün tzu* should have nothing to do with things that are of no practical use. What is not concerned with governmental matters he should not investigate" (*Yen t'ieh lun* 53[9/9a]; tr. Needham 1956:252).

Yet in the *Ta hsüeh* or *Great Learning*—a short treatise traditionally attributed to a disciple of Confucius or to his grandson, but possibly written by a disciple of Mencius—there occurs the key paragraph which, many centuries later with the rise of Neo-Confucianism, was to become a focus of controversy. Despite its great familiarity, its quotation here is desirable:

> The ancients who wished to radiate radiant virtue throughout the world, first administered well their own states. Wishing to administer well their states, they first brought accord to their families. Wishing to bring accord to their families, they first cultivated themselves. Wishing to cultivate themselves, they first corrected their minds. Wishing to correct their minds, they first sought to make their intentions utterly true. Wishing to make their intentions utterly true, they first extended their knowledge. This extension of knowledge lay in the investigation of things. (tr. Legge 1960:1.35–358; Gardner 1986:19)

Each link in this chain sequence is moralistic in usual fashion until, in the very last sentence, the key term *ko wu* 格物, "investigation of things," appears for the one and only time in early Chinese literature. This term, as it came to be interpreted by the Chu Hsi wing of Neo-Confucianism, has been hailed by modern scholars as a close Chinese approach to the inductive method,[59] or again as "an anticipation of the scientific method which unfortunately failed to mature" (Graham 1958: 79). It is indicative of the enormous difficulties faced by any proponent of the existence of a scientific methodology in Neo-Confucian China that what little support he or she can find in Confucian literature should center on a two-word phrase of very uncertain meaning, embedded in a small moralistic text of dubious authorship that had been almost totally

59. Hu Shih (1922:4): "It is clear that the interpretation which the Cheng [Ch'eng] brothers and Chu Hsi gave to the phrase *kueh wuh* [*ko wu*] comes very near to the inductive method. . . . But it is an inductive method without the requisite details of procedure."

ignored for more than a millennium (until Han Yü called attention to it around 804).[60]

The semantic ambiguity of *ko wu* becomes apparent as soon as we examine its components. *Ko,* first of all, is a word with a wide range of meanings. In different contexts it can mean "to correct," "to arrive at," or "to oppose" (Graham 1958:74). Cheng Hsüan (127–200), the noted Han commentator on the classics, interprets it as meaning "to come," or rather "to cause to come," and therefore "to attract." Typically, he then attaches a moral meaning to *ko wu* by writing: "He who is deep in the knowledge of the good will attract *(ko)* good things; he who is deep in the knowledge of evil will attract *(ko)* evil things."[61]

Ch'eng Yi (1033–1108) and Chu Hsi (1130–1200) come close to Cheng Hsüan by interpreting *ko* as "to reach" (i.e., to come to).[62] Other scholars, however, glossed it very differently. For the historian Ssu-ma Kuang (1019–1086) it meant "to guard against" (the excitation of desires by external things) (Graham 1958:74). For Wang Yang-ming (1472–1529) it meant "to rectify"; Wang's disciple Wang Ken (1483–1540) similarly explained it as meaning "to apply the correct pattern" of oneself to external things; and the pragmatically inclined Ch'ing thinker Li Kung (1659–1746) explained it as meaning "to practice a thing with one's own hand."[63] D. C. Lau's careful study of *ko wu* leads him to conclude (Lau 1967:357) that the term originally meant something like "to get to understand things." Therefore, he believes, Chu Hsi (and before him Ch'eng Yi) was essentially correct in interpreting *ko wu* as "to reach into things" (from which comes Legge's "investigation of things"). Chu Hsi showed excellent insight here, even though he did not bother to marshal evidence to support his interpretation.

Of greater interest, however, is how Chinese thinkers have interpreted the second word, *wu,* translated above as "things." "Things" is a conveniently ambiguous word in English because it covers not only concrete objects but also abstract affairs, occurrences, and the like. The primary meaning of *wu,* however, is "creature," meaning a member of

---

60. In his *Yüan tao* (Origin of the Tao). See V.1 beginning.

61. Cited in Lau (1967:353), which is a study of the original meaning of *ko wu.* Among other relevant discussions, see Chan (1957:324 and 1963:561–562, 611–612). Above all, see Gardner (1986:passim). Gardner's monograph focuses on Chu Hsi's work upon the *Ta hsüeh,* translates it with the text, and discusses the history of the term *ko wu,* though without stressing its possible significance for science (our primary concern here).

62. Fung (1953:529) and Lau (1967:353).

63. See, respectively, Fung (1953:602, 628, 634). For Wang Yang-ming and the *ko wu* controversy, see also esp. Ching (1976a:ch. 3).

the animal or (less commonly) vegetable kingdom; from this the word can readily be further extended to cover all kinds of physical objects, inorganic as well as organic. Only very rarely, however, does it occur in the early texts in the yet more extended sense of intangible, abstract matters. Therefore it is deeply indicative of the persistent human-centeredness of Confucian thinking that Cheng Hsüan nevertheless interprets *wu* in *ko wu* as equivalent to *shih* 事, "affairs, occurrences."[64] This definition is accepted not only by Chu Hsi and his school but, I believe, by almost all other Chinese thinkers who have analyzed the term. Moreover, Chu Hsi proceeds still further to remove *ko wu* from the field of concrete entities by interpreting the phrase not merely as meaning "to reach into things" (i.e., affairs), but, from this, "to reach into the *principles (li)* of things/affairs" (Lau 1967:353; Gardner 1986:54).

The result of interpreting *wu* as "affairs" is what might be expected. To be sure, a very few passages can be found in which Ch'eng Yi and Chu Hsi acknowledge that objects of the physical world (or rather, the *li* or principles of these objects) deserve investigation. Ch'eng Yi says, for example: "The scholar should understand everything, at one extreme the height of heaven and thickness of earth, at the other why a single thing should be as it is." Or again: "A single grass and a single tree both have principles which must be investigated."[65] And Chu Hsi says along the same lines: "From 'the Ultimateless that is also the Supreme Ultimate' above,[66] down to small things like a blade of grass, a plant or an insect below, each has its principle *(li)*. . . . If we do not investigate a certain thing, we shall miss the principle of that thing. We must understand them one by one."[67]

Yet it is abundantly evident from other statements that the interests of both men (and, it might be added, of almost everyone else who discussed *ko wu*) are overwhelmingly on the side of human activities and relationships, considered from an ethical point of view. Ch'eng Yi puts it this way:

> Each individual thing *(wu)* has its own principle *(li)*, which we should exhaustively study. There are many ways for so doing: one may read books

64. Lau (1967:353); Gardner (1986:23). What Cheng actually says is that "*wu* is like *(yü) shih.*" By this he means that in the particular phrase *ko wu, wu* is to be understood as equivalent to *shih;* in other contexts the two words would not necessarily be equivalents.

65. Both quotations are cited in Graham (1958:77).

66. I.e., the cosmological beginning of things. The phrase comes from the opening words of Chou Tun-yi's famous essay on cosmological evolution. See Fung (1953:435).

67. Quoted in Chan (1967:93–94).

and expound their meaning; one may discourse on personages of the past and present, discriminating the right and wrong in them; or one may take a proper stand in one's reactions to and contacts with [present day] affairs *(shih)* and personages. All these are means of studying principle exhaustively.[68]

The one thing one may not do, the modern scientist feels tempted to add, is to study Nature! Chu Hsi, as usual, follows Ch'eng's views closely:

In *ko wu,* nothing comes before the five categories [*wu p'in,* i.e., the five human relationships].

Ko wu consists in plumbing the fact that a given matter should be like this and another matter like that. For example, he who rules should rest in benevolence, and he who serves as subject should rest in reverence.

[An example of *ko wu* is,] for example, when the people of Han greatly explored the question of why the Ch'in had failed and the Han had succeeded.[69]

Ko wu, as interpreted in Neo-Confucianism (and very likely as intended by its original author), really had nothing to do with the procedures of the physical sciences despite what some modern writers have said about it. In the first place, its "investigation" rested more on intuition than on any fixed methodology. In the words of Graham (1958:78):

The investigation of a thing consists of thinking followed by a sudden insight into its principle. This insight reminds one a little of the "satori," the sudden and permanent mystical illumination of Zen Buddhism, but it is really quite different, a purely intellectual illumination in which a previously meaningless fact, as we say, "falls into place." . . . Of the thinking which precedes it, the only kind which is specified is inference from principles already known.

Wang Yang-ming, lacking any formulated methodology for studying the *li* of bamboo, could only hope that his lengthy meditative contemplation of the courtyard bamboo might eventually result in an intuitive "falling into place" of things. It is not surprising that when this did not happen after seven days and nights of steady concentration, he experienced a temporary mental breakdown.

In the second place, the orientation of *ko wu* from the very start was, as we have seen, ethical rather than scientific. Again in the words of Graham (1958:79):

The whole purpose of the Investigation of Things is moral self-development; the principles which really matter are moral principles, and investigation is mainly concerned with uncovering them in human affairs. There is no idea

---

68. Quoted in Fung (1953:529) and (partially only) in Gardner (1986:25).

69. *Chu-tzu yü-lei* (Classified Conversations of Chu Hsi) 15.514 (for the first two quotations) and 15.522 (for the third).

> of adding to a common stock of knowledge; the object of investigation is to discover how to live, a discovery which has already been made once and for all by the sages[70] and which each individual must make over again for himself.

Chu Hsi too, it would seem, regarded *ko wu* as a technique directed primarily toward man rather than nature, toward internal understanding rather than external knowledge:

> *Ko wu,* as Chu understood it, served as a Confucian corrective to the meditation of the Buddhists . . . ; like Buddhist meditation, *ko wu* led to a realization of the self, to an enlightened state; but *ko wu* was a reality-affirming process that underscored the centrality of the relationships between self and society. (Gardner 1986:58)

Toward the end of the Ming dynasty, however, a few scholars reacted against the Chu Hsi or, more often, the Wang Yang-ming wing of Neo-Confucianism, holding that one or the other was overly metaphysical, abstract, intuitional, or removed from early Confucianism. A good example, particularly with reference to the "investigation of things," is the scholar and official Fang Yi-chih (1611–1671), author of the *Wu li hsiao shih* or *Small Encyclopedia of the Principles of Things* (1644).[71] In this work, Fang insists that *wu,* "things," comprise not only *shih,* "affairs," but concrete physical objects as well. Indeed, he maintains, even heaven and earth (meaning the cosmos) constitute such "things." Fang's keen interest in natural phenomena is evidenced by the frequency with which his encyclopedia discusses astronomy, meteorology, geography, human physiology, and other naturalistic topics. In such passages Fang frequently mentions, quotes, or even refutes (especially on teleological matters) the scientific writings in Chinese of Ricci, Aleni, Schall, and other Jesuits who were then in China. It seems reasonable to conclude that Fang's knowledge of these Western writings (a knowledge already possessed by his father, a high official) must have materially strengthened, if it did not initially inspire, his interest in *wu* as physical "things."[72]

Fang's opposition to introspection and his emphasis on objective facts as prerequisites for scholarly research resemble those of the somewhat later Ch'ing scholars who, because of their insistence on factual evi-

---

70. See V.1, beginning, on the creation and transmission of the Tao of the sages.

71. For what follows on Fang, see Peterson (1975 and 1979).

72. Fang's knowledge of the Jesuit writings was shared by some other scholars of his time who, like him, tended to be more receptive to Western scientific ideas than to Western theology. See Gernet (1980 and 1985:20–22, 57–63, 261 n. 201). As pointed out below, the interest in science later declined.

dence, came to be known as the "School of Evidential Research" *(k'ao cheng hsüeh)*.[73] A noteworthy difference, however, is that only a few such scholars shared Fang's scientific interests to the point of writing seriously themselves on these topics. On the contrary, the great majority focused their energies on classical, historical, and textual studies.[74] This continuing emphasis on *shih,* "affairs," even by scholars who were otherwise often critics of Neo-Confucianism, demonstrates the continuing power of the Confucian tradition in late imperial China. The situation contrasts sharply with the secularization of natural philosophy taking place at about the same time in Europe.

Several centuries earlier, however, during the rebirth of Confucianism in Sung times, its power had seemed a good deal less certain. Then its urgent task, faced as it was by Taoist naturalism and Buddhist metaphysics, had seemed one of rescuing the ethical teachings of the classics from threatened oblivion by integrating them into a reasoned theory of the universe. Using great insight and imagination, the Neo-Confucians, culminating but not ending with Chu Hsi, succeeded in giving human ethical values a proper place within the vast framework of a total moral universe. What they did was a remarkable achievement, but it would be wrong to see it as in any way connected with the formulation of a methodology for modern science. The Neo-Confucians would not have understood what we today speak of as "science," and their own orientation did not and could not lead them in the direction of that concept.[75]

---

73. Surveys of the development of this movement from Sung times onward are contained in Elman (1983 and especially 1984).

74. "European science in eighteenth-century China was not . . . developed past the limits we have just sketched partly because of faulty Jesuit transmission that failed to challenge native classicism or provide an academic alternative. The preeminent position of classical studies in *k'ao-cheng* scholarship, along with its historical focus, remained intact" (Elman 1984:84). During Fang's early life and the decades immediately preceding, several major works on natural history, agriculture, and technology had been compiled, among them the *Pen-ts'ao kang-mu* (Great Pharmacopoeia) (1596), *Wu pei chih* (Treatise on Armament Technology) (1628), *T'ien-kung k'ai-wu* (Exploitation of the Works of Nature) (1637); and *Nung-cheng ch'üan-shu* (Complete Treatise on Agriculture) (1639). After Fang's time, however, such compendia became much less frequent, although the *T'ao shuo* (Description of Pottery) (1774) deserves mention.

75. I admire the study by De Bary (1975a) of Neo-Confucian methods of cultivation but find it difficult to accept its statements that Neo-Confucianism was open "even to science" (p. 166); that it enjoined, "with a religious sanction," the "systematic, rational inquiry into what we call the natural sciences" (p. 167); and that it "was not inherently antipathetic to scientific study" (p. 205). See also pp. 187 and 204.

2. The Question of Sex

In view of the universal importance of sex, it is striking, but probably not surprising, to find how few of the original Chinese writings on the subject have survived and how scant until fairly recently has been the attention paid to it by modern scholars.[76] The first serious study in a Western language is that of van Gulik (1951), reproducing and discussing erotic color prints of the Ming dynasty. This was joined ten years later by the much more comprehensive, as well as more readily available, further study by van Gulik (1961), which still remains the standard work on the subject. In recent years there has been an increasing flow of additional publications, some of them serious,[77] others popular and uncritical attempts to capitalize on the current interest in sexual matters. These latter lean heavily on the translations made by predecessors, notably van Gulik, sometimes without giving due acknowledgment.

Among classical Chinese thinkers, only one seems to have had the boldness to recognize the sex drive as a dominant element in the human psyche. "Food, sex: that's human nature." So asserted Kao Tzu in his famous debate with Mencius on the subject of human nature (*Meng-tzu* 6a.4; tr. Legge 1960:2.397). Mencius, curiously, offers no direct response to this assertion, which makes one wonder if the record of the debate is complete at this point.[78] But there can be no doubt that this reduction of human nature to two instinctual drives, devoid of any metaphysical component, would be thoroughly distasteful to virtually all Confucians. The only conceivable exception might have been Hsün Tzu, in view of his denial of any spirituality in Heaven and his consequent assertion that human nature in its raw state is "evil," by which he really means uncivilized.

As is true of many Chinese key terms, the word *se* 色, translated above as "sex," is ambiguous in that it embraces several meanings,

---

76. This section has already been published, with very slight changes, as Bodde (1985).

77. For example, Beurdeley (1969) (important for its pictures despite its often loose scholarship), Ishihara and Levy (1970) (an annotated translation of a Chinese sex manual preserved in the *Ishimpō,* a Japanese medical work of 982), and Needham (1983:184–218) (on sexuality as one aspect of Chinese physiological alchemy). On female footbinding—an important aspect of Chinese sexual mores—see Levy (1966).

78. The comment on the subject by Richards (1932:50) is no doubt too caustic: "The initial inarticulation of Kao Tzu's statement, 'Food, Sex, Nature,' need not detain us once we have realized how foreign to our own ways of thinking any thought which is satisfied with it must be."

among which "sex" is by no means primary. The basic meaning appears to be "color," from which are successively derived other important meanings, including "countenance," "appearance" (facial), "beauty" (primarily feminine), and hence "sex" (in the sense of feminine allurement). This sequence reveals at once the disapproval usually attached to *se* when it carries a sexual connotation. Confucius, for example, remarks of the *chün tzu* or true gentlemen that "in his time of youth, before his blood and breath have yet become stable, he guards himself against [the temptations of] sex *(se)*" (*Lun yü* 16.7). Legge (1960: 1.312–313) and Waley (1938:205–206), in order fully to bring out the word's unfavorable connotations, both translate it here as "lust," and by and large it is not a word in good repute in the Confucian literature.

Very occasionally, however, *se* is used in a favorable or at least a neutral sense. When King Hsüan of Ch'i, for example, confesses to Mencius that he suffers from a weakness, namely that he likes *se,* Mencius replies by citing the example of an early ancestor of the Chou ruling house who also liked *se* and loved his wife, but saw to it that among his people there would be no dissatisfied women and unmarried men. Mencius assures King Hsüan that if he too were to give his people the conditions for satisfying the liking for *se,* universal rulership could become his (*Meng-tzu* 1b.5). It can hardly be doubted that *se* signifies sexual satisfaction here, even though Legge (1960:2.163–164) translates it as "beauty," presumably in the sense of "feminine beauty."

There are many uncertainties and no clear consensus in attempts to evaluate the sexual mores of traditional China. In the first place, there is the question of how significant the differences really were between Confucianism, with its strongly patriarchal orientation, and Taoism, with its traditionally greater emphasis on the female side of life. Van Gulik initially maintains that "the difference between the two Schools in their attitude to this subject was only a matter of emphasis, the Confucianists stressing eugenics and the obtaining of offspring, the Taoists stressing the sexual disciplines for prolonging life and for obtaining the Elixir of Immortality" (van Gulik 1961:78). Yet later he seems to contradict himself by saying that "Taoism has been on the whole much more considerate to woman, and has given much more thought to her physical and emotional needs than Confucianism ever did" (van Gulik 1961: 84). The latter statement seems to me closer to reality.

Yet granted this greater consideration for women, there still remains room for argument as to whether the Taoist attitude, at least from a modern point of view, should really be considered "progressive." The major purpose of the Taoist sex manuals was less to ensure pleasure for the participants or engender progeny than to secure long life for the male. This the male could achieve by always inducing the sexual satisfaction of his female partner (or partners), in order thereby to absorb into himself the vital power believed to be emitted by the woman at the moment of female orgasm. In order to receive maximum benefit, it was essential for him not to lose his own vital force, as embodied in his semen. His aim, therefore, was to prolong the sex act as much as possible in order to gain maximum benefit from the woman, but at the same time to experience actual ejaculation himself as rarely as possible. It might be argued that frequent stimulation of the male in this way with only infrequent ultimate satisfaction could be neither physiologically nor psychologically healthy.[79]

Although from the woman's point of view the Taoist insistence on her satisfaction is certainly praiseworthy,[80] doubt arises as soon as we consider its motivation. For as we have seen, it is the man's (not the woman's) vital energy that is to be built up through the prescribed sex hygiene, the woman's that is to be transferred to the male for this purpose. Thus looked at, the procedure can be, and has been, condemned as a typical example of male selfishness. Thus we are told by Ishihara and Levy (1970:227): "The generalization that the Taoists accepted 'the equality of women with men' . . . is not borne out by the specific texts that we have translated. Our texts show that the woman was manipulated and made to serve man's longevity-salvation objectives." And the specialist on religious Taoism, K. M. Schipper, writes still more forcefully (1969:24) on one of the Taoist texts:

> This is only one example of all the methods of extracting the breath and essential female force. The remarkable thing is its absolute selfishness. The woman is without exception considered to be an enemy. Sexual union does not lead to creation by the other [i.e., to joint procreation by the couple]. The semen must be withheld to fortify one's own body, and to create in it the immortal embryo.

Though the Taoist view of sex is thus really far from truly egalitarian, it nevertheless seems miles ahead when compared with the Confucian attitude, which approves of sex primarily for procreation and regards overt expressions of affection between the sexes as immoral. The puri-

---

79. This situation could be bettered to some extent by the Taoist-sanctioned practice of "making the *ching* [the semen] return" (i.e., at the moment of ejaculation, pressing the urethra so as to divert the seminal excretion into the bladder, from which it would be later voided with excreted urine). The Taoists supposed that in this way the seminal essence was caused to ascend and rejuvenate the upper part of the body. See Needham (1983:197–201).

80. This opinion is also expressed in van Gulik (1961) (e.g., at p. 84) in contrast to van Gulik (1951), where the estimate of Taoist theories is less favorable. See also the strongly pro-Taoist view of Needham (1983:217–218).

tanical attitude of early Confucianism could be demonstrated by many passages. The following are two taken from the chapter in the *Li chi* entitled "Rules for the Household":[81]

> Except at sacrifices and funeral rites, they [men and women] do not hand objects to one another. Or if there is such a handing of something, the woman receives it in a basket. And if she have no basket, they both sit down [on the ground], where he puts the thing on the ground and she then picks it up. Whether outside or inside [the house], they do not both go to the same well or same bath. They do not share the same sleeping mat, do not request or borrow things from one another, and do not wear the same kind of clothing.
>
> Men and women do not share the same clothes rack. She [the wife] dare not hang [her clothes] on the husband's rack nor store [her belongings] in her husband's boxes or baskets. She dare not bathe together with him.

Lest we think these and similar passages are perhaps exaggerated, coming as they do from a ritual text, they should be compared with a passage in *Meng-tzu* (4a.17; tr. Legge 1960:2.307) in which someone asks Mencius: "Is it the proper practice that a man and a woman do not allow their hands to touch when giving or receiving something?" Mencius replies that it is. The interlocutor then asks whether, in that case, a man should use his hand to rescue his sister-in-law from drowning. Mencius replies that he would be a beast if he did not, for though it is ordinarily proper for man and woman not to touch hands, the use of the hand to rescue a woman is necessitated in this particular case by the emergency. The important point here is not that Mencius granted the possibility of touching hands in an emergency but rather that he accepted the taboo on hand touching as the accepted mode of behavior under normal circumstances.

One might argue that the Confucian injunctions about relationships between the sexes were either only theoretical and never really expected to be carried out; or that, if originally taken seriously, they failed to be perpetuated; or yet again, that even if they were perpetuated, they were taken seriously only by the small top minority constituting the "scholar-gentry." Without going into the only too familiar details of filial piety, arranged marriage, subordination of wife to husband, domination by her parents-in-law, and so on, I think that each of these assumptions can be disproved by citing a few representative passages from the systematic community studies made in recent times by well-known Chinese anthropologists. In general, the studies deal with rural villages or small towns in which most of the people interviewed are of nongentry origin. The first passages come from the account by Martin C. Yang (1945) of

---

81. *Li chi* 10; tr. Legge (1885:27.454–455 [first passage] and 470 [second passage]).

his native village of about 750 people on the Shantung coast in North China:[82]

> The meeting of the engaged boy and girl before marriage is definitely improper.
>
> Unmarried people are not given information about sex. When a daughter is ready for marriage a mother may tell her everything about being a wife except the sexual aspect. A girl may learn something from her brother's wife, but this is unusual and is strictly forbidden in a decent family. A boy is similarly handicapped. . . . His father, brothers, or uncles do not speak to him of such matters. . . . In general, every couple has to go through a period of trial and error.
>
> Although the couple often achieve a genuine affection for each other after a brief period of living together, they must not let their love be apparent. . . . A young husband must not mention his wife too often. . . . A young wife must also keep from showing that she loves her husband.
>
> In the last thirty or forty years, there has been only one case of divorce. The wife had become pregnant [through someone else] before her marriage.
>
> In Taitou there have been only two cases in which men married, or simply took widows.

The next passages come from Fei Hsiao-t'ung's famous study, made in the 1930s, of a town of slightly fewer than 1,500 persons in the lower Yangtze valley, not far south of the T'ai Lake in Kiangsu:

> Before the birth of the child, her husband, at least overtly, is indifferent to her. He will not mention her in conversation. Even in the house, in everyone's presence, if he shows an intimate feeling for his wife it will be considered improper and consequently will become a topic for gossiping. Husband and wife do not sit near each other and very seldom talk to each other in that situation. Rather they talk through a third party. . . . But when a child is born, the husband can refer to his wife as the mother of his child. Thereafter, they can converse freely and behave naturally towards each other. . . .
>
> The people are prejudiced against any intimate relations of a woman with a man outside wedlock. To prevent such a possibility prenuptial chastity is strictly maintained by social disapproval of any intimate association between grown-up girls and boys. . . . Adultery of married women is still more grave. Husbands, in theory, can murder the adulterers with impunity. But in practice it is seldom done. (Fei 1939:47)

Our final examples come from the field study (1941–1943) made by Francis L. K. Hsü of a fair-sized town (some 8,000 persons) in western Yunnan, not far from the wartime Burma Road, but two to three weeks by foot from Yunnan's capital, Kunming:

---

82. Yang (1945:109, 114, 54, 116, 118, respectively).

> If a girl were known to have been mixed up in any clandestine affair, her only chance of marriage would be to go to some distant village or to Kunming, where no one would know her history. (Hsü 1948:98)

> In her second marriage a woman is a dishonored object. The marriage may touch off the wrath of the gods as well as the spirit of her departed husband. (Hsü 1948:104)

> The development of a close relationship between the sexes is considered detrimental to the supremacy of filial piety. . . . Women must not be attractive to men and must not display their personal charms. . . . Sex is considered unclean, and women carry the burden of this uncleanliness. That is why there are various taboos on mothers who have just given birth and on women during menstruation. . . . Women are inferior to men and must assume a strictly subordinate role. (Hsü 1948:207)

> The culture says to the male: No upright man shows signs of intimacy in public with any woman, not even his wife; your primary duty is toward your parents. . . . It says to the female: To be attractive to men is unnatural. . . . It is shameful to be sexually attractive, even to your husband. . . . Your main duty is toward your parents-in-law. . . . While the gratification of sex in the physiological sense is not barred, all possible awareness of it is to be eliminated, and all secondary expressions, such as tenderness of feeling and mutual possessiveness, are to be banned. (Hsü 1948:246)

The values and practices here described can hardly be regarded as other than constituting a very repressive Confucian brand of sexual puritanism.[83] The cited passages, which could readily be multiplied from other sources, demonstrate the persistence down to recent years in rural China of a view of life that, despite its general acceptance by almost all members of the population, must have created untold suffering over the ages, both physiologically and psychologically.[84] In saying this, I wish in no way to assume an attitude of superiority. As I fully recognize, sexual repression, cruelty, ignorance, and inequality have been widespread in most or all of the so-called higher civilizations during the greater part of their history. They have by no means disappeared from modern Western society, despite its recent sexual permissiveness.

---

83. How fanatical this puritanism could be is shown by the 1943 episode reported by Hsü (1948:27, 225) of the young man who had returned to his town in Yunnan from Hong Kong (where he had obviously picked up new ideas), and who, together with his bride, had a bucket of human excrement thrown over his head because they were walking down the street holding hands.

84. It should be noted, however, that what Hsü calls the "estrangement of the sexes" could, as he points out (1948:247–248), be more rigidly adhered to in wealthy homes, where all necessary work was performed by servants, than in the homes of the poor, where man and wife were commonly obliged to do hard work in cooperation, and where therefore the relations between them were necessarily less formal.

About Confucianism it has often been stated, and is no doubt true, that it accepts sex as entirely "natural" and morally permissible as long as it is performed within the bonds of matrimony and primarily for purposes of procreation. Hsü's insistence, therefore, that the group he studied regarded sex as unclean seems at first sight puzzling.[85] But I believe that the idea of uncleanliness has a Buddhist origin and that its prominence in Hsü's Yunnan town reflects the amalgamation of Confucian, Buddhist, and Taoist ethics that has dominated the thinking of ordinary Chinese during the past several centuries. This interpretation is confirmed by Eberhard's study of Chinese concepts of guilt and sin (Eberhard 1967:esp. 64–65, 80–81). For this Eberhard analyzed a considerable number of popular moral tracts (*shan shu,* lit. "good writings"), most of them modern (nineteenth and twentieth centuries), but in one case going back to a Buddhist tract of the sixth century. The influence of Buddhism in these writings is of course very strong, but Taoist and Confucian ideas are also conspicuous, the latter increasingly so in the more recent tracts. The tracts stress the uncleanliness of body functions and of sex and prescribe so many taboos on the latter that according to Eberhard's calculation, only some hundred days in the year were recognized as auspicious for sex relations. Eberhard's conclusion is:

> If one had to keep all these rules constantly in mind and had to be afraid that in any case of violation either his health would be destroyed, or miscarriages would occur, or children with bad character qualities would be born, or the deities would be offended and mete out punishments, Chinese sex life could not have been very "natural" and sexual inhibitions and fears must have been quite strong from fairly early medieval times down to the present. (Eberhard 1967:80–81)

The great unknown quantity is the extent to which these prohibitions were really heeded. Hsü (1948:151–152) noted similar moral tracts in circulation in his Yunnan town and found that some twenty-one birthdays of gods and goddesses were annually taboo to sexual intercourse. However, most of his informants confessed to having failed to observe or forgotten one or another of the tabooed dates at one time or another. Nonetheless it can hardly be gainsaid that the Buddhist view of sex must have reinforced Confucian puritanism, especially among the less well educated sectors of the population for whom the moral tracts were especially designed.

Before we leave this topic, note should be taken of the insistence with

---

85. Besides his statement in the next-to-last of the passages quoted above, he elsewhere writes (1948:151): "Sexual intercourse, as well as everything that is connected with it, is unclean."

which three otherwise disparate groups of writings (two Chinese, one European) all proclaim the need for sexual taboos. On the Chinese side, the writings include not only the just-mentioned moral tracts but also the sex manuals of early imperial centuries;[86] on the other side of Eurasia, they comprise several penitential pronouncements of the Christian church in medieval Ireland. Perhaps it is unsurprising that such religiously oriented penitentials, like the moral tracts of China (also religiously oriented), should view sex as sinful or potentially sinful and hence attempt limitations on its practice. More remarkable, however, are the similar restraints demanded by the Chinese sex manuals, despite their nonreligious character and strongly favorable views of sex. No doubt the motivations of the three groups differed (for example, correct sex hygiene was a major concern of the manuals but not of the other writings), as did the particular days they specified. Yet the net result in each case was the same: the imposition on married couples of sexual prohibitions variously ranging from somewhat under two-thirds to well over two-thirds of the days in each year.[87]

Let us turn now to the question of possible historical changes in Chinese sexual attitudes. It is a major thesis in van Gulik's 1961 study of Chinese sexual life that Confucian puritanism assumed prominence only from the thirteenth century (late Sung dynasty) onward and became an obsession only during the Ch'ing dynasty (1644–1911), when the Chinese "showed a nearly frantic desire to keep their sexual life secret from all outsiders." Before the thirteenth century, he maintains, the ancient sex manuals continued to be widely studied, the separation of the sexes was not strictly enforced, and sexual relations were freely talked and written about. Van Gulik even goes so far as to attribute the

---

86. These sex manuals are now traceable to the second century B.C. and probably go back still earlier. See Harper (1987), which translates and comments on the introductory section of a text on sexual techniques found at Ma-wang-tui, near Ch'ang-sha, Hunan, in a tomb dated at 168 B.C.

87. The Chinese moral tracts accepted only some hundred days annually as auspicious for sex relations, which means that they regarded the remaining 265 days or so as inauspicious. There is a striking numerical agreement here with the days of abstention enunciated in some of the Irish penitentials: approximately 260, for example, in one of the oldest, the Penitential of Cummean (mid-seventh century), whose prohibited days included Wednesday, Friday, Saturday, and Sunday of each week, as well as a forty-day period preceding each Easter and Christmas and following each Whitsunday. See Bieler (1963:117), translating the Penitential of Cummean from Latin (at ch. II, para. 30). In the Chinese sex manuals, the tabooed days are somewhat fewer: 200 and more annually, including the first and last days of each lunar month, the days of full and quarter moon, days in the recurring Chinese cycle of 60, and so on. See Maspero (1981:531–533).

change in attitude to a very specific historical experience: the hiding by Chinese householders of their women from the Mongol soldiers who were billeted among them when the Mongols were conquering the country in the mid-thirteenth century. "One suspects," he writes, "that it was during this period that the germs of Chinese prudery came into existence."[88]

I am very skeptical that a change in thinking of this magnitude, if indeed it was as great as van Gulik indicates, could have been sparked by such a relatively short-lived situation. Rather I believe that it was part of a much longer and slower process of social and intellectual change that had begun already in late T'ang (618–906) or early Sung (960–1279) times and continued into and beyond the Mongol period. Van Gulik himself points to indications of change in sexual attitudes beginning long before the Mongol invasion. One was the growing practice of female footbinding (tenth century onward), which brought dancing by women to an end.[89] Another was the change in feminine fashion that began during the Southern Sung, resulting in the formerly bare throats and lowcut bosoms of the ladies being covered by a short jacket worn under the outer robe and buttoned in front with a high tight collar (van Gulik 1961:222, 237).

Intellectually, the growth of prudishness can hardly be divorced from the rise of Neo-Confucianism (eleventh century onward). Specifically, Ch'eng Yi (1033–1108), the man who more than any other started the school that Chu Hsi brought to completion, was apparently a single-minded and humorless man, an ascetic and a true puritan. In the words of a modern Chinese scholar, Ch'eng "was stern and arrogant; unblessed with an aesthetic sense, he self-righteously refused to join any gathering devoted to the enjoyment of art"; on one occasion, when invited by a colleague to drink tea and see paintings, "he declined, saying that he never drank tea and did not know anything about paintings" (Li 1972:57).

Ch'eng Yi was more responsible than anyone else for strengthening the social convention (it was not a law) denying remarriage to widows (but not of course to widowers). The following is his famous interchange on the subject with an interlocutor, as included (and therefore approved) by Chu Hsi in the *Chin ssu lu* (A Record for Reflection):[90]

---

88. For this paragraph, see van Gulik (1961:esp. xi–xii, 22, 237, 245–246). The two quotations appear on xi and 246, respectively.

89. See V.1 above, where the rise of footbinding was contrasted with the earlier practice of court ladies riding horseback and even playing polo.

90. *Chin ssu lu* 6.13; tr. Chan (1967:177 and [for Chan's remark immediately below] xxv).

Question: "According to principle *(li),* it seems that one should not marry a widow. What do you think?"

Answer: "Correct. Marriage is a match. If one takes someone who has lost her moral integrity *(chieh)* to be his own match, it means he himself has lost his integrity."

Further question: "In some cases the widows are all alone, poor, and with no one to depend on. May they remarry?"

Answer: "This theory has come about only because people of later generations are afraid of freezing or starving to death. But to starve to death is an extremely small matter. To lose one's moral integrity . . . is an extremely serious matter."

It is not surprising that Wing-tsit Chan, translator of the *Chin ssu lu,* comments here: "This is perhaps the most extreme statement in the *Chin ssu lu,* the most controversial, and, in the twentieth century, the most condemned."

Despite the evidence of growing sexual repression from Sung times onward, however, I believe it would be a mistake to underestimate the considerable degree of repression that must have existed previously. First of all, we have seen that the sexual views of classical Confucianism were essentially just as repressive as those of Neo-Confucianism; the one real difference was the greater effectiveness with which the Confucian ethic was universalized during the Neo-Confucian period (though, as indicated above, there was a considerable admixture of Buddhist and Taoist ideas on the popular level). Even before Neo-Confucianism, however, the influence of Confucian values was always great, especially in fields like law and administration, where the government exercised immediate control.

A striking example is the provision in the T'ang penal code of 653 imposing one year of penal servitude on married couples who conceived a child during the standard twenty-seven months of mourning required upon the death of a parent of the husband. Perhaps this article was not consistently enforced, and there is some evidence that it was intended primarily for members of officialdom rather than the populace at large. Even so it was an extraordinary attempt to legislate sexual morality, and it remained on the statute books of successive dynasties until its abolition in 1374 by the founder of the Ming dynasty, who said that it ran counter to the dictates of human nature.[91]

---

91. See Bodde and Morris (1967:39). The law itself is art. 7 in bk. 12 of the T'ang Code, and is also referred to in art. 2 of bk. 3 (the context of which is the basis for the supposition that the law may have been applicable primarily to officialdom). The Ming founder, when he abolished the law, could very well have cited the dictum of Kao Tzu as quoted at the beginning of this section, but of course he did not do so.

In the second place, any weaknesses in pre-Sung Confucianism were compensated for by the corresponding strength of Buddhism, which, as indicated above, regarded sex as unclean. Buddhist asceticism and Confucian prudery, for example, were equally scandalized by the communal sex practices current among certain Taoist groups around A.D. 400. The result was a countermovement which had made great inroads into this kind of Taoism by the middle of the sixth century and led to its disappearance by the seventh (Maspero 1981:533–541).

More significant are two long-lasting phenomena that, I believe, sharply distinguish China from most other high civilizations. The first is the virtual absence of any tradition of dancing between the sexes. The dancing noted by van Gulik as having been halted by the rise of foot-binding was not heterosexual at all, but a solo accomplishment performed by courtesans for the delectation of their male admirers. As such, it was part of the catering to male needs and had nothing to do with egalitarianism between the sexes. Equally solo were the dances sometimes performed by female characters in stage plays (where, however, during much of the Ch'ing dynasty, female roles were always performed by men). On the other hand, the group dances performed in Confucian temples on the occasion of sacrifices to the sage or those performed in conjunction with other ceremonies belonging to the state cult were apparently always restricted to males. Only by going back to Chou and pre-Chou times and to the peasant mating festivals that Granet (1932) postulated for that epoch does the possibility arise that dances may then have occurred involving both men and women. But even this is speculation, and if such dances did exist, they are not recorded and had probably virtually disappeared by the creation of the empire in 221 B.C. All of this is puzzling in view of the existence down to modern times of heterosexual dancing among the many tribespeople—the Miao, Lolo, and others—scattered through south and southwest China. The significance of the dance in Chinese culture is an unexplored subject that would repay study.[92]

Another major feature that distinguishes China sharply from the Indo-European (but not the Semitic) civilizations is the almost total absence of any tradition of nudity in art. Wellnigh no indication of it

---

92. The brief and sparsely documented study by Giles (1914a) is only an introduction to the subject. More scholarly are Kaltenmark (1963) and Kuchera (1977), the former on sacred dances in China, the latter on a particular kind of dance (the *ch'i p'an wu* or "dance of the seven plates") performed in Han and post-Han times for entertainment purposes by a few highly skilled professionals, probably mostly women. Neither article controverts the point here being made, namely, that dancing between the sexes was virtually unknown in traditional Chinese civilization.

can be found in early Chinese religious art; and when, in late imperial times (especially during the Ming), nude male and female figures are portrayed in various acts of love in the secular erotic paintings and blockprint illustrations known as "spring pictures" *(ch'un hua),* one cannot but be struck by their aesthetic crudity, painful literalness, anatomical clumsiness, and general unattractiveness.[93] Phallicism existed in prehistoric and early historic China; for example, the upright pointed ancestral tablets were, it is believed by some, of phallic origin. Yet already by late Chou times the original symbolism had probably been largely forgotten, and though traces of phallicism (and a corresponding cult of the female sex organ) sometimes occur much later, their manifestations were generally kept sub rosa.[94]

Our overall conclusion, then, is that the prevailing Chinese attitude toward sex during the greater part of imperial history was repressive, puritanical, and masculine. Perhaps the sexual conventions were somewhat looser for the unlettered masses than for the gentry, yet Confucian prudery, supplemented by Buddhist asceticism, seems to have been remarkably widespread among all social classes. Even the Taoist manuals and practices, enlightened though they seem by comparison, reflect a society in which the male always held first place. Even at their peak they probably reached only a tiny fraction of the total population, and long before the coming of Neo-Confucianism they had been largely suppressed and destroyed.

During the latter part of the Ming there seems to have been something of a reaction against the rigors of Neo-Confucian puritanism. This is evidenced by the appearance of erotic literature of which the *Chin P'ing Mei* novel is the supreme example; also the graphically frank "spring pictures." Yet these were probably all produced for, and reflected the lives of, only a small, well-to-do, predominantly mercantile urban class. This is certainly true of the *Chin P'ing Mei* (whose male protagonist is a merchant), and the "spring pictures" equally unmistakably portray a life of great luxury within large urban households containing numerous concubines and servants. This fact is made especially plain when, as in some cases, the pictures show the master of the household extending his favors to two or even three women simultaneously.

Finally, what is the relevance of all this to science, especially Chinese science? Unfortunately, it is very difficult to give a firm answer. I myself would be happy if it could be demonstrated with assurance that sexual

---

93. See the many illustrations in van Gulik (1951) and Beurdeley (1969).

94. See Karlgren (1930, 1942) and Ling Shun-sheng (1959a, 1959b). Nudity was occasionally practiced for magical or religious reasons (e.g., the exposure of naked shamannesses to the sun to induce rainfall). The subject is well treated in Schafer (1951).

emancipation (but *not* unbridled license), together with religious diversity, political democracy, and the rise of modern science, are *all* equally parts of a single spectrum labeled "freedom." Unfortunately, at least as far as sex is concerned, such assurance is apparently not yet possible.

A large and often repetitive book was published in 1934 to demonstrate that sexual repression is a prerequisite for the cultural flowering of a tribe, people, or nation. The author (Unwin 1934) based his findings on the study of eighty "uncivilized" societies, to which he added the Sumerians, Babylonians, Hellenes, Romans, Anglo-Saxons, and English as "civilized" peoples. For each group he measured on a numerical scale its degree of sexual freedom, using as his criterion the extent to which prenuptial chastity and postnuptial sexual denial (outside the legitimate marital relationship) were insisted upon by the society in question for its women. (Significantly, he did not apply this criterion to men.) Against these measurements he then attempted to measure the degree of cultural creativeness of the given society. His conclusion was that imposed sexual limitations result in sublimation which in turn generates what he called "social energy." In his own words: "A society which displays *productive social energy* [emphasis in original] develops the resources of its habitat and by increasing its knowledge of the material universe bends nature to its will" (Unwin 1934:314).

From this he went on to say that the continuation and intensification of sexual checks for at least three generations leads to a further unparalleled outpouring of energy (exemplified in only three of his selected societies—the Athenians, Romans, and English):

> Under the influence of still greater sexual checks, the society bursts its boundaries, conquers, slays, subdues, and explores; but, if this intense continence remains part of the inherited tradition for two generations, the energy increases abundantly, changes its form. . . . The society . . . exhibits a terrific mental energy that is manifest in the arts and sciences, refines its craftsmanship . . . exerts considerable power over its environment. (Unwin 1934:431)

One regrets that the author did not reduce the number of obscure tribes in his study and correspondingly increase the number of higher civilizations. Had he included at least China, he would have found that this country perfectly fitted his criterion of strict prenuptial chastity and limited postnuptial sexual opportunity outside marriage, and did so over a period ten or twenty times as long as his specified three generations. Would he have found that it too therefore "conquered, slayed, subdued, explored" and "bent nature to its will"?

Quite the contrary thesis was advanced more recently in a study specifically on science (Feuer 1963), which one reviewer has described as "simple-minded, one-sided, frequently ignorant, and stimulating."[95] The author's general thesis is that sexual freedom is a primary ingredient in the movement for general intellectual freedom out of which, he believes, modern science arose. To support this thesis, he makes a frontal attack on the theory of the interrelation between English Puritanism and science, especially as formulated in Merton (1970; originally 1938).[96] In so doing, however, he rather unexpectedly admits that Puritanism *was,* after all, helpful to science—not, however, because of its religious asceticism, but because its spirit of revolt won freedom from institutions, ideas, and persons. "We might state the matter concisely by saying that it was the Protestant politic, not the Protestant ethic, which lent a helping hand to the scientific movement."[97]

It is a pity that Feuer did not include Victorian England in his discussion, because this would seem to have been an age of intense sexual repression that nevertheless somehow achieved great scientific and technological progress. Concerning China and Japan, which he does discuss, his conclusion is that "it is in this repression of the emotional energies, this thwarting of the natural direction of sexuality, that the primary cause for the failure of Asian civilization [i.e., the civilizations of China and Japan] to produce a sustained scientific method will be found" (Feuer 1963:253).

Some of Feuer's remarks in support of this thesis reveal insight,[98] others only ignorance or misunderstanding.[99] Worst of all are certain comments on Chinese psychological attitudes. For example, he asserts that "all the classical schools of Chinese philosophy were affected by self-defeat and renunciation; all looked upon nature through the masochistic mode of resignation"; he finds that a "feeling of profound doubt and despair" has been a persistent theme in Confucian writings: and, for

---

95. This comment is the first sentence in the review by Fleming (1965).

96. This theory was discussed in IV.2.

97. Feuer (1963:228). Merton's new preface to the 1970 edition of his book (originally published in 1938) contains several pages (Merton 1970:xxiv–xxvii) of rebuttal, in which, among other things, Merton finds Feuer guilty in one passage of having misquoted him (Merton) slightly but significantly.

98. As on individualism or the incompatability of Taoism with science, for which see Feuer (1963:248, 256).

99. As when he compares the Chinese attitude to law with the English preference for common law (ignoring the fact that the Chinese compiled massive law codes for each major dynasty) or suggests that the mechanism of the Legalists might have been as conducive to science as the mechanism of seventeenth-century European philosophy (ignoring the Legalists' usual disinterest in the natural world; see III.4 at n. 7). See Feuer (1963:251, 253).

him, cyclical conceptions of history are "always a projection of pessimism."[100]

Thus we reach the end of this section, unhappily unable to achieve a clear-cut conclusion. In China, where the attitude toward sex was generally repressive, a modern science failed to arise. In India, where the social life of all periods was characterized by "vigorous sexuality" (Basham 1959:170), a modern science likewise failed to arise. In Europe, where sexual habits changed greatly from one age to another, a modern science did arise. It almost seems as if sex, despite its overwhelming importance for the individual, may after all not be such a decisive factor in the development of societies, including their scientific development.

Yet of this we cannot be really certain as long as the historical aspects of the subject still await adequate study, not only in China but elsewhere. Meanwhile, though one may not agree with Kao Tzu's apparent belief that food and sex make up the whole of human nature, there is no question that they constitute an extremely important part—so much so that despite our concluding uncertainty, we could not possibly have left sex out of our story.

3. Individualism and Self-expression

From time to time we have touched on topics closely connected with individualism: the Confucian emphasis on man as a social being, the Confucian and Taoist emphasis on humility and yielding, the significance of individualism for modern science, the relative anonymity of Chinese artisans compared with those of Europe, the relationship of individualism to certain traits of character.[101] But so central is this topic to our whole inquiry that we turn to it again, this time together with the cognate theme of self-expression.

The *Oxford English Dictionary* defines *individualism* as "self-centred feeling or conduct as a principle; a mode of life in which the individual pursues his own ends or follows out his own ideas; free and independent individual action or thought; egoism." Surprisingly, considering how strong these notions have been in Western civilization, the first occur-

---

100. Feuer (1963:257–258). We have seen (III.3 above) that cyclical conceptions of history, while prominent in Chinese historiography and philosophy, are by no means the only Chinese ways of viewing time.

101. Very relevant to this section, but published after it was written, is the volume edited by Munro (1985), in which fifteen scholars discuss various aspects of individualism in China from the standpoints of Confucianism and Taoism.

rence of the word in English, according to the *Oxford Dictionary,* came only in 1835.[102]

The modern Chinese equivalent for "individualism" is a compound term, *ko-jen chu-yi,* which means something like "single-man-ism." The term is a modern neologism, borrowed from Japan, where it was coined under Western influence, probably early in this century. Were it not for the lateness of "individualism" as a word in English, it would be tempting to conclude that the absence of a lexical counterpart in pre-modern China reflects the absence there of the idea as well. As we shall see, this is not wholly true. However, as we shall also see, ideas and behavioral standards that in modern Western society would be termed "individualistic" and therefore be often (not invariably) approved, would in Chinese society (past and present alike) usually be viewed unfavorably.[103] This is very much the case with *ssu,* possibly the nearest indigenous Chinese lexical approach to "individualism," whose connotations—"private, personal, secret, selfish, partial, illicit," and so on—are all strongly unfavorable in Chinese eyes.

Another word that on occasion—by no means always—shares some of the derogatory connotations of *ssu* is the first person pronoun *wo.*[104] Used pejoratively, *wo* comes semantically close to "ego" or "egotism" in English. Confucius, for example, is said to have "had no obstinacy, no 'I' *(wo)*" (*Lun yü* 9.4).[105] And Mencius fiercely criticized Yang Chu on grounds that Yang "acted for the 'I' *(wei wo)*" (*Meng-tzu* 7a.26). Until recent decades it was customary for educated Chinese, in formal conversation, to avoid the first and second person pronouns by using such circumlocutions as "humble person" *(pi jen),* meaning the self, and "gentleman" (*hsien sheng,* lit. "earlier born"), meaning the other person. In Literary Chinese, as noted earlier (II.4), the "rule of economy"

---

102. It appears in the English translation by Henry Reeve (London, 1835–1840) of Alexis de Tocqueville, *Democracy in America,* pt. 2, bk. 2, ch. 2: "Individualism is a novel expression, to which a novel idea has given birth." The French equivalent word, *individualisme,* was first used in 1820 by J. de Maistre, according to Lukes (1973:4).

103. The term "*modern* Western society" is used here advisedly, because individualistic behavior, although no doubt fairly frequent in pre-Renaissance times, was probably often viewed then more ambivalently than in recent centuries, when individualism came to be exalted by many Western thinkers.

104. Other words signifying the first person pronoun also exist in Chinese, but for the sake of simplicity I restrict the discussion to *wo.*

105. This passage is cited by Bauer (1964:12) in his study of the rise of self-consciousness and of autobiographical writing in Chinese literature. Also relevant, but published after the present section was written, is Bauer (1979).

allowed writers often to leave the subject or object of a sentence unspecified if they wished, thus obliging the reader to deduce them from the context.

In poetry, where this practice was particularly prevalent, the result was good to the extent that ambiguity may be considered a desirable feature in poetry, but it was also dubious, at least to the Western reader, in that it gave to Chinese poetry its frequent air of impersonality. Of course not all poets were necessarily impersonal. Ch'ü Yüan, for example, uses the first person pronoun no fewer than 83 times in the 187 lines of his celebrated *Li sao* poem (ca. 300 B.C.). This, however, is decidedly exceptional. Generally speaking, the Chinese poet's attitude to the reader is: "This is the situation, it's up to you to react to it," not "Here I am, see how strongly I feel."[106]

Biography and autobiography, when considered as literary genres, can be useful indicators of changing attitudes toward individualism and self-expression. By definition, both are commonly thought of as media for the portrayal of distinctive personal character. Yet it may not always be thus. Greek biographical writing, which apparently began in the fifth century B.C.,[107] included such masterpieces as Plato's portrayal of Socrates, but most of it consisted of conventional funeral elegies and rhetorical exercises depicting ideal types of character. The first century A.D. was marked by the extraordinary biographical creations of the four evangelists, as well as Plutarch's *Lives,* but what followed consisted for many centuries of highly stylized hagiography (with the single major exception of Saint Augustine's *Confessions,* ca. 400). It was only from the Renaissance onward that personalized biography really came into its own as a significant literary genre.

In China the situation is reversed. The earliest biographies, those inserted by the historian Ssu-ma Ch'ien into his *Shih chi* (ca. 90 B.C.), are also the most vivid and personal, though sometimes, for that very reason, less than fully reliable. Since that time, the Chinese have produced more biographies than any other pre-modern civilization: many thousands in the standard or dynastic histories alone and a great many more in the form of funerary inscriptions (not unlike the Greek funeral

---

106. Both the quotation and the reference to Ch'ü Yüan come from Hightower (1961:160, 162–163). Another study of how creative personality was viewed in traditional Chinese literary criticism is that of Gálik (1980), whose opinion about Western-style individualism in China is very definite (1980:191): "The Chinese cultural realm . . . never knew individualism of the kind current in romantic Europe."

107. When the poet Ion of Chios wrote brief sketches of such contemporaries as Pericles and Sophocles. For this and the rest of this paragraph, see Kendall (1974: 1010–1011).

elegies) and other literary media. But following Ssu-ma Ch'ien, the historical biographies became increasingly depersonalized and stereotyped according to set formulas. For the most part, they tended to be quite short (most of those in the *Ming shih* or *Ming History,* for example, do not exceed 1,000 characters), and centered their attention almost entirely on the formal aspects of a man's public life. As Twitchett has pointed out (1962:34),[108] Plutarch's kind of writing

> was not what the Chinese reader expected of a biography. Just as the readers of a medieval saint's life were primarily interested in the record of his sanctity and holy deeds—even if these were portrayed in conventional set forms—so the scholar-bureaucrats for whom the dynastic histories were designed as manuals of precedent were satisfied with the bare account of a subject's performance of his limited role as a member of the bureaucratic apparatus.

Again, we are told by Nivison (1962:458)[109] that the

> later historical biographies . . . gave essential information about a man (essential from the point of view of the bureaucrat-historian) in a highly formal way—his family background (but usually not his date of birth),[110] his official career (if he had one) in outline, quotations from his writings (if he wrote),[111] and anecdotes, often stereotyped and quite false, intended to indicate his character. This portrait of character conceives of the man as falling into a type, at most realizing potentialities present from birth, but never exhibiting a dynamic and changing personality.

In the West, the existence since early times of the epic and the tragedy did much to shape the writing of biography along lines of conflict and changing personality. In China both literary forms were absent.[112]

On the other hand, autobiography developed much earlier in China than in the West.[113] Ssu-ma Ch'ien's *tzu hsü* or "self-narration," appended as the final chapter to his *Shih chi,* preceded by half a millennium the first real autobiography in the West, Augustine's *Confes-*

---

108. See also the parallel but somewhat more extended discussion of Chinese biography in Twitchett (1961).

109. Nivison (1962) has also supplied the information above about the lengths of biographies in the Ming history.

110. See also III.4.

111. See also II.9 at n. 93.

112. Twitchett (1961:110) and James J. Y. Liu (1962:152–155).

113. On Chinese autobiography, in addition to Bauer (1964), there is the detailed study by Hervouet (1976), which discusses not only various genres of Chinese autobiography but numerous individual works as well, ranging chronologically from Ssu-ma Ch'ien to the twentieth century.

sions.[114] In autobiography, even more than in biography, one would expect a high degree of individualism and self-consciousness, and this is in fact achieved to a considerable extent in at least two early Chinese autobiographies besides that of Ssu-ma Ch'ien.[115] By and large, however, Chinese autobiography shares the qualities of Chinese biography in general: shortness, impersonal form (most autobiographies, including that of Ssu-ma Ch'ien, are written in the third person), and omission of details of personal life.

Many Chinese autobiographies, in fact, were not written purely and simply as autobiographies at all, but, like that of Ssu-ma Ch'ien, were appended as *tzu hsü* to longer literary works. Their primary purpose was to explain the circumstances under which such a work was composed rather than to give a systematic account of the author's life. Further, their personal quality might be diluted by the author's attention to biographical facts about his father and earlier forebears as well as himself. All these features characterize Ssu-ma Ch'ien's own autobiography, only about one-fourth of whose respectable length of some 9,000 characters is strictly autobiographical (Bauer 1964:19–20).

Perhaps sensing some of these weaknesses, Sung and post-Sung writers increasingly made use of a more extended biographical form, the *nien p'u* or "yearly register," in which the facts of the subject's life were recorded under year-by-year entries. This form permitted greater length and detail and became so popular that from early Sung down to the beginning of the twentieth century, some 1,934 *nien p'u* had been compiled by 1,208 persons, of whom more than 230 wrote autobiographically.[116] This kind of biography, which continued to be produced in large numbers during the Republic, most nearly resembles the Western scholarly biography. Yet it shares the weaknesses commonly found in other kinds of Chinese annalistic writing:

> Like the annalistic historical style upon which it was modelled, the *nien p'u* is usually a succession of carefully dated discrete facts, with no attempt to con-

---

114. On the Western side, following the appearance of the *Confessions,* there was a hiatus of many centuries (as in biographical writing) before the outpouring of autobiography that came with the Renaissance, especially from the sixteenth century onward: the autobiographies of Benvenuto Cellini (1558–1563) and Jerome Cardan (1570–1576) and Montaigne's *Essays* (1571–1589). See Clark (1935:26–31).

115. Those of Wang Ch'ung in his *Lun heng* (A.D. 82/83) and of Ko Hung in his *Pao p'u tzu* outer ch. 52 (ca. 320). See respectively tr. Forke (1962:1.244–262) and Ware (1966:6–21).

116. These are the figures cited by Hervouet (1976:130–131) from an unnamed bibliography. However, Howard (1962:468 n. 12–13) indicates much lower figures: some 645 datable *nien p'u,* of which 111 are autobiographical.

> nect them in any meaningful causal pattern; it makes no attempt to provide any explicit interpretation of its subject, and is best described as materials for a biography rather than biography itself. (Twitchett 1962:37)

Before leaving the topic of self-expression, a bare mention should be made of Chinese portraiture and self-portraiture as artistic parallels to biography and autobiography. Self-portraiture has always been excessively rare in China (see V.4c) and ordinary portraiture, though much more common, seems on the whole to have been a good deal less important than in the West. It reached peak popularity during the T'ang dynasty, especially in court circles of the eighth century (Sickman and Soper 1956:90), but thereafter declined. One cause is probably the rise in Sung times of *wen-jen hua* (literati painting), with its emphasis on naturalistic subjects (landscapes, birds and flowers, etc.), and its tendency to look down on the T'ang painters as artisans.[117] More broadly, however, the decline of portraiture may perhaps be seen as another indication of the posited shift from the allegedly greater individual freedom of T'ang and pre-T'ang China to the tightening social controls of Neo-Confucian China.

Returning now to the central topic of individualism, there is nothing surprising in the fact that its manifestations in thought and behavior are most clearly evident during times of political and social strain and uncertainty. Such conditions marked the Warring States period, when many rival schools of thought flourished, and they appeared again in the third century A.D., after the Han empire had disintegrated and its holistic worldview faded. The unconventional behavior of the so-called Seven Sages of the Bamboo Grove and other similarly uninhibited persons of this period has induced one scholar to comment: "A situation in which Chinese gentlemen surrendered themselves to drink, did not mourn for their parents, took no thought of funerals, neglected etiquette, and went naked, indicates a collapse of the Chinese moral structure."[118]

During the late Ming dynasty (sixteenth and early seventeenth centuries) a recrudescence of individualistic thinking occurred. It came in part as a reaction against the by then frozen conventions of orthodox Neo-Confucianism, but more immediately was stimulated by the growing weakness, ineffectiveness, and corruption of the Ming government. In his lengthy study of the subject, De Bary (1970) has traced the mani-

---

117. See VI.1a end and Lawton (1973:ix).

118. Shyrock (1937:16). Although Holzman (1957:32–34) adduces evidence that throws doubt on the historicity of some of the specific anecdotes about the Seven Sages, there can be no question that their age was generally one of revolt against the old Confucian standards.

festations of this individualism in a line of late Ming thinkers culminating with Li Chih (1527–1602). Li was a man of many lives and contradictions. Born into a Fukien mercantile family whose forebears included men and women of Muslim faith, he himself became a scholar and served as a minor Confucian official for several decades. Then, when in his fifties (1585), he turned to Buddhism, sent his family home, and himself entered a monastery where, however, he refused to follow the monastic discipline. During his subsequent years as a seeming monk, he was accused, probably validly but exaggeratedly, of sexual and other immorality. In 1600, because of the hostility aroused by his unconventional writings, an officially instigated mob burned his temple residence. Finally, in 1602, he was arrested and sent to prison in Peking where, while awaiting formal charges, he committed suicide by cutting his throat.

Li Chih was generally opposed to conventional morality but had little to offer in its place. Among his iconoclastic ideas were the assertions that selfishness is innate in the human mind, that standards of right and wrong change with the times, that morality cannot be taught, that the scholar-officials are morally inferior to the merchants, and that human passions are no obstacle to the achievement of Buddahood. Although contemptuous of ordinary authority, he came, in a manner similar to Jonathan Swift, to look upon ordinary mankind as no better than beasts. Against these creatures, he extolled the Ch'in First Emperor (r. 245–210) as the "greatest emperor of all time" and Emperor Wu of Han (r. 140–87) as "the greatest sage of all time." Thus his neuroses and hatreds led him, in a manner only too familiar today, from a critic of authority to an upholder of total power. He died a martyr more to his own contradictory convictions than to any ideal of universal intellectual freedom.[119]

Space has been given to Li Chih precisely because his kind of individualism is so thoroughly familiar to modern Western society and so completely un-Chinese. Indeed, Western-style individualism is so hard to find in traditional China that it may be misleading to apply the term there. If we really wish to describe the Confucian concept of distinctive personality, the word to use is probably not "individualism" at all but

---

119. For the foregoing account of Li, see especially De Bary (1970:188–222), supplemented by Hsiao Kung-ch'üan (1976) and Hok-lam Chan (1979–1980 and 1980). The chapter on Li Chih in Ray Huang (1981) appeared only after these paragraphs had been written, but I am glad to note that his assessment of Li's thinking is similar to mine in important respects (see esp. 208–213, 215–216). A survey of Chinese interpretations of Li Chih from 1907 through the 1970s will be found in Cheng Pei-kai (1983–1984).

"individuation," meaning the fullest development by the individual, within his or her social milieu, of his or her creative potentialities. Here is the way Carl Jung distinguishes between the two conceptions:

> Individualism consists in deliberately giving prominence to and emphasizing presumed originality, as opposed to collective considerations and responsibilities. Individuation on the other hand means a better and more complete fulfilment of man's collective responsibilities, in that, by making adequate allowance for what is peculiar to an individual, better fulfilment of his social aptitudes may be expected than when these characteristics are neglected or suppressed. . . . Individuation can therefore only represent a process of psychological development leading to the fulfilment of given individual conditions or, in other words, to the evolution of the particular predetermined entity that is man as we know him. He does not thereby become "selfish" in the generally accepted sense of the term, but simply fulfills his own specific nature. As we have already noted, there is a world of difference between this and egoism or individualism.[120]

Near the end of his study, De Bary acknowledges that the examples of Ming "individualistic" thinking that he has discussed are very far from representing a norm for China, and he succinctly indicates some of the factors that prevented individualism of the Western type from ever becoming popular:

> Individualism of the Western type is a product of a different historical development. In China the extreme weakness of the middle class, the nondevelopment of a vigorous capitalism, the absence of a church which fought for its rights against the state, or of competing religions which sought to defend the freedom of conscience against arbitrary authority; the lack of university centers of academic freedom, deriving from their original function as monastic sanctuaries; the want of a free press supported by an educated middle class—the list, of course, is almost endless of the elements lacking in China which contributed in some way to the rise of Western types of individualism. (De Bary 1970:220)

I believe that all these points are closely relevant to our central question of why a modern science did not appear in China. However, I also agree with De Bary that the future will probably be more receptive to the Chinese kind of individualism (what we would call individuation, though he does not use this word) than to the extreme Western individualism which in the past has functioned so creatively and also—especially in recent times—with such appalling destructiveness:

> The possibility must be allowed for that any type of individualism which develops in the future may tend more to keep within the allowable limits of

---

120. Jung (1928:185–186); quoted and tr. in Abegg (1952:186).

Chinese tradition than to expand to the outer limits of Western forms of individualism. It is not inconceivable, for instance, that a Confucian type of individualism or personalism could eventually prove itself more adaptable to a socialist society than modern Western types of individualism. (De Bary 1970: 224–225)

4. Competition: The Example of Sports

"The *chün tzu* [superior man] competes in nothing" (*Lun yü* 3.7). This statement, uttered by Confucius apropos the archery contest, provides a good motto for this section. He explains it further by saying that the *chün tzu,* even when at the archery ground, always bows and makes way for his opponent when ascending or descending the ground, as well as at the drinking assembly that follows (*Lun yü* 3.7). The theme of noncompetition is closely linked with the Chinese view of individualism discussed in the preceding section. Though the theme has been touched on still earlier (see VI.1c), its importance is such as to deserve further examination in terms of a concrete illustration: the relative roles played by competitive athletic sports in China and the West.

No history of Chinese sports has ever been attempted, perhaps because so little has been written about them by Confucian-minded scholars, and also because organized sports seem in fact to have played quite a minor role in total Chinese life. The remarks that follow are necessarily scattered and sketchy. In pre-imperial China, when fighting between feudal noble families was endemic and Confucianism had not yet become influential, we would expect a correspondingly greater acceptance of the idea of individual physical competition. And indeed, recorded events very occasionally support this expectation. In 307 B.C., for example, King Wu of the state of Ch'in died of an injury incurred when competing with another person in lifting a bronze tripod vessel. (The person in question was promptly executed together with his relatives.)[121] Yet even in that feudal age such an event was exceptional, and during imperial times, beginning in the Han dynasty, it become almost unheard of.[122]

Although chariot driving is listed in the ritualistic texts as one of the

121. *Shih chi* 5; tr. Chavannes (1895–1905:2.76).

122. Nevertheless, a single similar instance occurred during the Former Han dynasty: that of Liu Hsü (128–54 B.C.), the eldest son of Emperor Wu, who as a youth was known for his lifting of bronze tripod vessels and bare-handed fighting with bears and other wild animals. Perhaps for this reason, he was passed over for succession to the throne and, many years later, ended his life by suicide after having committed murder. See *Ch'ien Han shu* 63/4a–b; summarized in Dubs (1938–1955: 2.181).

"six accomplishments" *(liu yi)* expected of the aristocracy,[123] it would seem that in Western (i.e, early) Chou times, the chariots of the nobles were driven by menials, and only later, during the Spring and Autumn period, did this accomplishment acquire prestige (Creel 1970:274). There is very little evidence at any time for the existence of organized chariot racing such as was so common in Greece and Rome.[124] Another recreation much more popular among the aristocracy all through history was hunting, but I leave it out of consideration because it was not competitive in the sense of being a game played by competing individuals or groups of individuals according to fixed rules.[125]

Archery is the one sport of feudal China that was competitive (despite Confucius's dictum at the beginning of this section) and about which significant information exists. Its importance derived, of course, from its hunting and military associations, and the earliest Chinese school of which there is record was apparently an archery academy under the personal patronage of the Western Chou kings. Contests in archery, sometimes attended by the king himself, were held at the school in conjunction with feasts and sacrifices. According to the much later ritual literature, whose accounts may be idealized to some extent, the contests were highly stylized, with the participants carrying out their successive movements to the accompaniment of music, almost as if taking part in a ballet. Elegance was apparently judged to be as important as the actual skill in hitting the mark. A significant feature was that the competition was between teams, not individuals. The shooting was done in alternate succession, first by a member of one team and then by a corresponding member of the other, until all had had their chance.[126]

---

123. The others were ceremonial, music, archery, writing, and arithmetic. See *Chou li* 13; tr. Biot (1841:1.297–298).

124. The "competitive games" of Han times (see below) may have included charioteering, though how this was organized and whether it actually embraced racing is unknown. Other than this, textual references that may be relevant are extremely few and ambiguous. Typical is the succinct statement in *Han Fei-tzu* 21; tr. Liao (1939:222). This says that Viscount Hsiang of Chao (r. 457–425; this was probably before he acceded to the title) was studying driving from Wang Liang, an almost legendary master of chariot driving. "All at once he started racing with him [Wang Liang]. He changed horses three times but thrice he lagged behind." Textual emendations are necessary to arrive at the names of the two protagonists here.

125. On Chinese hunting see, inter alia, Schafer (1968) and Bodde (1975:351–352, 381–386).

126. Creel (1970:273–274, 393–394). Of the accounts in the ritualistic literature, the most detailed are those in the *Yi li* (Observances and Ceremonies) 8–10, 13–14; tr. Couvreur (1916:101–176, 212–283) and Steele (1917:1.74–121, 150–188).

Even in early imperial times, archery continued to be a polite accomplishment of the *shih* or "gentleman/scholar." During the Later Han dynasty, for example, students who were enrolled at the national academy outside the capital, whenever they periodically took written examinations on the classics, shot arrows at targets as a necessary preliminary; the questions they answered were those written on the targets they succeeded in hitting (Creel 1970:407–409).

Apparently the closest Chinese approach to massive spectator sports of the sort, say, known so well in Rome, was the "competitive games" first referred to in 208 B.C. as being witnessed at the Ch'in court by the Ch'in Second Emperor. As we shall see below in connection with wrestling, the term *chiao ti,* here somewhat loosely rendered as "competitive games," later came to be used (and possibly was so used earlier as well) to signify wrestlinglike contests between paired individuals. However, if one takes *chiao* in its original concrete meaning of "horn" instead of secondary meaning of "competition," the term *chiao ti* may be understood in the sense of "horn-butting." And indeed, statements by later commentators suggest that although the games, in a broad sense, probably included more than one competitive event of a military nature, such as archery or charioteering, their core may have consisted of contests between paired men wearing horns with which they butted one another. It is possible that these contests were "humanizations" of a popular bull cult, known until modern times in certain parts of eastern and southeastern coastal China, in which bulls fought with one another and then were ritually slaughtered to fructify the fields.

The "competitive games" themselves continued during the early part of the Han dynasty, when there may have been an arena to accommodate spectators. In the third month of 108 B.C., for example, it is stated in the Former Han history that when the *chiao ti* games were performed, "within the space of three hundred leagues *(li),* everyone came to observe." Unfortunately, we are not told who "everyone" was. But in 44 B.C. the games were abolished, quite possibly (although this is not stated) because of combined moralistic and economic objections. Thereafter they became largely forgotten.[127]

---

127. On the "competitive games," see esp. Dubs (1938–1955:2.92, 129–131, 314) as well as Diény (1968:58–60), who discusses the games more as choreographic spectacles than as martial competitions. For the bull cult, see Bodde (1975:206). *Added note:* See Loewe (1990) for a detailed study raising the possibility that the "competitive games" may have been reenactments of a mythological battle between the Yellow Lord (Huang Ti) and Ch'ih-yu (an almost satanic divinity associated with metallurgy/weapons and with rain). As the author himself points out, the evidence linking the games with the battle rests primarily on a late (mid-T'ang) text and so is far

Another much longer lasting game, probably of folk origin and going back at least to the Warring States period, is *ts'u chü*[128] or "football," more literally "kick-ball."[129] Many casual references to it appear in the Han histories—enough to show that it was particularly popular among military men.[130] However, as so often happens, the texts give no detailed account of what they mention so casually.[131] From them it appears that the game was played by kicking a leather ball filled with hair, but as to how many persons took part and how they competed, we do not know.[132]

The game continued into post-Han centuries and became particularly popular during the T'ang, when it was even played at court. This is why more is known about its rules at that time. Apparently the goal consisted of a round hole made in a cloth stretched between two standing bamboos some thirty Chinese feet high. Players belonging to the two sides would try alternately, in a manner similar to the old archery contests, to kick the ball through the goal.[133] Thus the T'ang game defi-

---

from conclusive. I am much indebted to Dr. Loewe for letting me read his article prior to publication.

128. The name can be written with more than one combination of slightly different characters.

129. In a speech allegedly addressed by the itinerant diplomat Su Ch'in to King Hsüan of Ch'i (r. 319–301), the diplomat mentions football among the recreations popular during his time in the Ch'i state capital. See *Chan-kuo ts'e* (Strategems of the Warring States), Ch'i no. 126; tr. Crump (1970:157–158); repeated in *Shih chi* 69/3b.

130. The military references include: (1) The Han physician Shun-yü Yi kept a record of twenty-five of his medical cases which has been preserved in his *Shih chi* biography. In the last of the cases (datable as 154 B.C. or earlier), he warned a military man not to overexert himself; the man nevertheless "played football" and quickly died. See *Shih chi* 105/6a; tr. Bridgman (1952–1953:44 [with further discussion and references at p. 97 n. 240]). (2) The catalogue of the Han imperial library, under its section on military writings, lists a book in twenty-five chapters entitled *Ts'u chü* (Football). See *Ch'ien Han shu* 30/8b. (3) The noted general Huo Ch'ü-ping (d. 116 B.C.), in order to raise morale among his troops when they were suffering from food shortages in Inner Asia, had an area dug out for them to play football. See *Ch'ien Han shu* 55/4a.

131. Other references occur in *Ch'ien Han shu* 65/3b and 97A/1b. A modern discussion appears in Eberhard (1942:1.46–47).

132. A funerary wall relief of A.D. 123 shows a man kicking a ball, while near him are two other men who sit and talk. See Chavannes (1909–1915 [vol. 1:55, 1913; Plate XV, fig. 25, 1909]). But the military background of football makes it probable that it was played by teams, not single individuals.

133. See Giles (1914b), which includes illustrations from Chinese treatises but is deplorably incomplete or defective in its bibliographical citations. A new scholarly study of Chinese football is badly needed.

nitely was not a contact sport. Evidence from the Ming novel *Shui hu chuan* (Water Margin) indicates that football continued to be popular in high circles until as late as 1100—perhaps in fact until the early Ming, when the novel was probably first put together.[134]

Another very different kind of competition, whose popular origins are lost in antiquity but which never gained official popularity, is that of the midsummer dragon-boat races, held until recent years on the fifth day of the fifth lunar month. In Central and South China it was the custom on this day for long and narrow "dragon boats," each propelled by some fifty paddlers, to race each other on lakes and rivers. Upsets and drownings sometimes occurred, and on occasion the races almost took on the nature of a naval battle, with the competing boats pursuing or grappling with one another. In the words of Eberhard (1968:395-396): "It was a ritual sacrifice of humans to the river in order to ensure fertility. . . . It is perfectly clear that the boat race is a fight between two parties and the losing party is sacrificed."[135]

Yet another sport, also fertilistic in its original purpose but known to us primarily as one more amusement at the T'ang court, is tug-of-war *(pa-ho).*[136] Variations of the sport occur in many countries, among them Korea, Burma, Assam, Ceylon, and even Morocco.[137] Its magical aim is to secure rain and thus good crops, which is why, in China, it took place in spring. During the successive reigns of the T'ang Emperors Chung-tsung (705-710) and Hsüan-tsung (712-756), tug-of-war was performed in the famous imperial Pear Garden in the presence of the emperor. A huge main cable was used, four hundred or five hundred Chinese feet long, to which dozens of lesser cables were spliced, thus enabling more than five hundred persons to pull on either side. In one

134. The beginning of the novel describes how a fellow who was skilled in football, though otherwise useless, gained the favor of the prince who was to become the Sung Emperor Hui-tsung. This alleged event would have happened in 1100, the year before the prince ascended the throne. See Irwin (1953:202 n. 3).

135. The most detailed study of the festival is that of Aijmer (1964). Among many other accounts of this extremely well known festival, see Bodde (1975:314).

136. Lit. "to pull [into] the river." The term suggests (and there is some support from accounts of the practice in other countries) that a stream separated the competing sides and the contest continued until one side had pulled the other into the water. See Des Rotours (1933), which is the basis for all except the reference to Korea in this paragraph.

137. For a detailed account of how, in present South Korea, tug-of-war contests are still occasionally conducted in the spring between all available residents of rival villages, lasting sometimes as long as three or four days for each contest, see Choe Sang-su (1960:43-45).

of the contests witnessed by Emperor Chung-tsung, three high ministers and five generals competed on the western team, against which the eastern team included another high minister and two sons-in-law of the emperor.

A further notable Chinese sport is polo, again a game exceedingly popular at the T'ang court, but unlike tug-of-war, uncompromisingly aristocratic. Of Iranian origin, the game was introduced to China early in the seventh century by Tibetan diplomatic envoys and thereafter, despite Confucian protests, was played by officials, princes, emperors and even court ladies (the latter riding donkeys). Like football, it retained popularity into the Sung, but unlike football it could be quite a violent and dangerous game.[138]

Equally violent was our seventh Chinese sport, wrestling. Here, as frequently in these texts, the problem arises of how to deal with scattered references that almost never say more about the sport than that it took place at such-and-such a place and time and was witnessed by so-and-so. Chinese wrestling has apparently not yet received the attention from scholars it deserves.[139] Of its two most common designations, the first is that same term, *chiao ti,* whose several meanings were mentioned earlier in our discussion of the Ch'in and Han *chiao ti* or "competitive games." The other and later designation is *hsiang p'u,* which can be translated both as "mutual striking" and "mutual overturning." That *chiao ti* (in the sense of wrestling) and *hsiang p'u* are virtual synonyms is attested in the Chinese texts and further confirmed for the Japanese language by the dictionaries.[140]

The earliest apparent evidence for wrestling in China is a small bronze plaque, perhaps of the fifth century B.C., on which the intertwined figures in high relief seem to be wrestling with one another.[141] Here the unresolved question is whether they are really wrestling in the sense of playing a sport according to fixed rules or are simply combatants in an unorganized free-for-all. Much less uncertain but also much

---

138. See Giles (1914b), Laufer (1932), Schafer (1963:66–67), Goodrich (1938, 1975) and especially, for many interesting details, James T. C. Liu (1985). I am grateful to the late Professor Goodrich for providing me with copies of his and Laufer's articles.

139. The popular two-page account by Yi Shui (1980) is unscholarly and inadequate.

140. Unequivocal is the statement in the thirteenth century *Meng liang lu* 20/13b: *"Chiao ti* is another name for *hsiang p'u."* In Japanese dictionaries, *chiao ti* and *hsiang p'u* are both defined as wrestling, though the former is pronounced *kakutei,* the latter *sumō.*

141. The plaque was recovered from a tomb southwest of Sian during excavations of 1955–1957. See the Feng-hsi Archaeological Team (1959:527–528).

later is an Eastern Han tomb mural of the second century A.D. in which two stalwarts are portrayed confronting each other in wrestling pose. The two are clearly differentiated by the fact that one wears a brown loin cloth and short scarlet skirt whereas the identical garments of the other are reversed in color.[142]

In later centuries, archaeology largely gives way to textual references, of which the following are a few representative examples: In about A.D. 606 a straitlaced official at the Sui court protested in a throne memorial against the frivolities of popular amusements then widely practiced on the day of the Lantern Festival (fifteenth of first lunar month); among them he mentions *chiao ti.*[143] The protest, however, had no apparent effect, and during the following T'ang dynasty several emperors are known to have attended army-sponsored *chiao ti* wrestling contests. Three late T'ang examples occurred on dates corresponding to 16 February 820, 7 August 826, and 20 March 839. Starkly indicative of how violent and unrestrained the sport could be, even when viewed by the emperor, is the brief account of the 826 performance. This states that the wrestling, accompanied by considerable drinking, continued for many hours until the first or second watch (7–9 and 9–11 P.M.), by which time some of the military participants had received battered heads *(sui shou)* or broken arms *(che pei).*[144] More than three centuries later, nonetheless, wrestling (by then called *hsiang p'u*) continued at the Southern Sung capital (modern Hangchou), where it consisted of contests between divisions of the Imperial Guards held on the emperor's birthday and at court banquets. Also, at pleasure grounds for the populace elsewhere in the city, wrestling by professionals periodically occurred.[145]

---

142. The mural comes from tomb no. 2, excavated in 1960–1961 at Ta-hu-t'ing, Mi-hsien, Honan province. See Anon. (1974:plate 38 and facing page of Chinese text as well as p. 8 of English-language insert). The same plate is reproduced as plate 1 in Dolby (1983), where it is also briefly discussed on p. 12. For tracing these references, I am greatly indebted to Dr. Michael Loewe of the University of Cambridge; Miss Diane Perushek, curator of the Gest Oriental Library, Princeton University; and especially Professor Susan Blader of Dartmouth College.

143. *Sui shu* 62/3b–4a and *Pei shih* 77/4a; tr. De Groot (1886:1.140).

144. See, respectively, *Chiu T'ang shu* 16/1a, 17A/4b, 17B/12a.

145. See Gernet (1970:224–226). Professor Gernet's interpretation of the *hsiang p'u* contenders as "boxers" (*boxeurs* in the original French edition) seems to me questionable. Although striking with the palm of the hand may conceivably have been one of the acts permitted under the rules for *chiao ti* and *hsiang p'u* contenders, I believe that their struggles, in large measure, were what in the West would be called wrestling.

It would be misleading to end our account of Chinese sports at this point. Other athletic competitions of a somewhat different sort do in fact occasionally appear in the texts, notably horse and dog racing[146] and cock and drake fighting.[147] At least two of them (dog racing and cock fighting) go back to Chou times, and probably all four persisted for many centuries during the empire. Unfortunately, here again the texts rarely provide any information other than the names of the sports and, often, of the individuals who played them.[148] Among the four, cock fighting is mentioned earliest and seems to have been the most popular. Its first reference—an item more informative than most—states for the year 517 B.C. that when the cocks of two aristocratic families fought one another, the owner of one bird protected its head with a sheath while the owner of the other equipped his bird's feet with metal spurs.[149]

---

146. *Tsou ma* (running horses) and *tsou kou* or *tsou ch'üan* (running dogs or running hounds). Assuming that *tsou* functions here as a transitive verb rather than adjective, the phrases can also be rendered as "running the horses" and "running the dogs or hounds" (i.e., making them run).

147. *Tou chi* (fighting cocks) and *tou mu* (fighting drakes). Here again the prefixed word may be verbal: "fighting the cocks" and "fighing the drakes" (making them fight).

148. The following is a chronological list—no doubt incomplete—of classical and immediately postclassical references: (1) *Tso chuan* Chao 25 (517 B.C.); tr. Legge (1960:5.710) and Couvreur (1914:3.387); (2) *Chan-kuo ts'e,* conversation with King Hsüan of Ch'i (r. 319–301), Ch'i no. 126; tr. Crump (1970:157–158); repeated in *Shih chi* 69/3b; (3) *Chuang-tzu* 19 (third century B.C.); tr. Watson (1968:104); repeated in *Lieh-tzu* 2 (perhaps ca. A.D. 300); tr. Graham (1960:56); (4) *Shih chi* 101/2a (soon after 154 B.C.); tr. Watson (1961:1.526); (5) *Yen t'ieh lun* 9 (soon after 81 B.C.); tr. Gale (1967: 56); (6) *Ch'ien Han shu* 8 (shortly before 73 B.C.); tr. Dubs (1938–1955:2.202); (7) *Ch'ien Han shu* 97A/7a (shortly before 73 B.C.); (8) Ts'ao Chih (A.D. 192–232), poem *Tou chi* (Cock Fighting) in *Ts'ao Tzu-chien chi* (Collected Works of Ts'ao Tzu-chien = Ts'ao Chih) 5/1b–2a; tr. Waley (1918:61); (9) Ts'ao Chih, poem in ibid. 6/3b; (10) Ko Hung, latter part of his autobiography (ca. A.D. 335), *Pao-p'u-tzu* outer ch. 52; tr. Ware (1966:18). Horse racing is mentioned in nos. 6, 9, and 10 of these entries, dog or hound racing in nos. 2, 4, and 10, cock fighting in all except 10, and drake fighting in 10.

149. Since this was written, Cutter (1989) has published an excellent comparative study of cock fighting in China and England, with particular emphasis on how Chinese and English poets have described the sport. He finds the main cultural difference between the two societies lies in the graphic emphasis of the English poets on the blood and violence of the sport, whereas the Chinese poets are much more metaphorical in their language and focus their attention on the beauty rather than the aggressiveness of the cocks. Cutter attributes this Chinese tendency to the general predominance, even in early Chinese history, of *wen* (the civilian aspect of life) over *wu* (the martial aspect). (For more on the *wen* and *wu* dichotomy, see below.)

Despite our scanty knowledge about these sports, three generalizations seem possible:

1. To judge from the references, the sports were primarily the recreations of aristocrats (including kings and emperors) and of persons close to the aristocrats; there is little evidence that they enjoyed much popularity among ordinary people.[150]

2. None of the four was a team sport and none, probably, attracted massive crowds of spectators, even though each surely had its circles of supporters and participants.[151]

3. Above all, although each was a physically competitive sport, the actual competitors were in each case animals, not humans. Horse racing was of course a partial exception, yet I suspect that even in its case the human riders were more often hired commoners than the aristocrats who patronized the sport.

On the whole it would appear that these four sports were less prominent, less physically demanding in human terms, and less significant for our topic of prevailing Chinese attitudes toward sport than were the seven described earlier. For these reasons, these four animal sports will not be discussed further in these pages.

Now, for the sake of comparison, let us leave the Chinese scene and look at some of the highlights of European sports. The subject has

---

150. Of the ten references listed in note 148, the first concerns Chou aristocrats, the fourth Yüan Ang, adviser of Han emperors, the fifth "privileged families" of Han times, the sixth and seventh the Han Emperor Hsüan (r. 73–49) and his father-in-law, while the eighth and ninth were written by the third son of Ts'ao Ts'ao, the man who controlled China during the final decades of the Han. Of the three remaining references, the second belongs to a general statement about the prosperity of Ch'i's capital around 300 B.C., the tenth is Ko Hung's autobiography in which he denies that he ever participated in such sports, and the third, from the *Chuang-tzu/Lieh-tzu,* is an anecdote about an imaginary Taoist worthy and therefore historically irrelevant. Evidence that kings and emperors of much later times continued to enjoy the sports is found in the account of the king who ruled the Ch'i dynasty for a few months in A.D. 493, described as a dissolute gambler and dog racer who "loved fighting cocks, which he bought for prices reaching several thousands"; also in that of Emperor Hsi-tsung (r. 874–888) at the end of the T'ang dynasty, who allegedly "loved football and took pleasure in cock fighting." See respectively *Nan shih* (History of the Southern Dynasties) 5/2a and Wang Tang, *T'ang yü lin* (A Forest of Talks About the T'ang; ca. 1105) 7/256. Further similar references could no doubt be found.

151. The eighth reference in n. 148 above describes how the guests (probably aristocrats) at a party, after tiring of watching the dancing girls, departed in the evening for an amusement place where, in a cleared room, they sat on long floor mats (Waley translates loosely "long benches") to watch the cock fights.

apparently been studied much more critically for the classical world than for the complex ages following the fall of Rome.[152]

Although the first Olympiad occurred in 776 B.C., references in Homer and other evidence make it probable that athletic contests go back at least five hundred years earlier to Mycenaean times. The Olympics, of course, were only one of several series of games held at various centers of the Greek world. All were associated with religious festivals, it being believed that the gods enjoyed watching the contests. The events consisted of boxing, wrestling, the pankration (a wrestling in which hitting was allowed), four kinds of footraces, and the pentathlon (a sequence of long jumping, discus throwing, javelin throwing, wrestling, and a footrace). Chariot racing was formally introduced at Olympia in 680 B.C. but is already recorded in the *Iliad.* Mule-cart races were added in 500.

Competitors and spectators alike at the games were male only, and from about 700 onward the performers performed naked (in earlier times they wore shorts). During the early centuries the competitors, with only a handful of known exceptions, were all from relatively well-to-do families, and amateurism was stressed. Only one possible case of professionalism is known before the end of the fourth century B.C. But from Alexander's time onward, professionalism and commercialism became rife almost everywhere except at Olympia itself, and with them came corruption.

In Rome, on the contrary, the prime purpose of organized sports was the amusement of the spectator rather than the physical and moral well-being of the participant. Beginning in early times, the Romans provided chariot racing, athletic meetings, gladiatorial combats, and fights with wild animals as public entertainment. In 41 or 33 B.C., mention is first made of the *xystus* or trade union of athletes, an organization concerned with improving the pay of its members, some of whom enjoyed enormous popular esteem. The Roman Circus Maximus, where the chariot races were held, could, around the middle of the first century A.D., accommodate something like a quarter of a million spectators. In short, the development of sports in Rome, as in Greece after Alexander, followed a direction depressingly similar to that of sports in the world today.

Besides the above sports, the Greeks and Romans also had ball

---

152. See esp. Harris (1964), which concentrates on Greek athletics, and Harris (1972), which broadens the scope to include sports of all kinds in both Greece and Rome. On the classical Olympic games, the most recent contribution I have seen is Finley and Pleket (1976). What is said in the next four paragraphs is largely based on the two Harris works.

games. These were played either with a small ball consisting of feathers or hair, covered with panels perhaps of leather, or with a larger inflated ball made of ox or pig bladder. These ball games, in contrast to those other games just described, were informal and unstructured, of plebian origin, and never became popular spectator attractions. One of them involved throwing a ball against a wall and is probably the ancestor of fives, tennis, rackets, and squash. Most references are to the throwing and catching of the ball, but there is also some slight evidence of the use of a bat or, in one instance, of what look like hockey sticks. On the other hand, there is no evidence in Greek literature, and very little in Roman, for any kicking of the ball. The fact that Plato denies that men can be either left- or right-footed points to the absence of anything like football in his time.[153]

The history of postclassical sports is scattered and obscure. In the late Middle Ages and the Renaissance, a rugbylike game, in which the ball was both kicked and carried, was played under various names in a variety of places, including the city of Florence, places in Brittany and Normandy, and small English towns, especially in Derbyshire. In England, at least, the game could be violent and dangerous, involving hundreds of people on opposing sides struggling throughout the day to move the ball through streets or fields to one or the other end of a given community. Fatal accidents are recorded in 1280, 1322, and 1303 (when the unpopular servant of a Cistercian abbey was murdered and his head used as a football). By the end of the fifteenth century, the game had become better organized with a marked-off field, goals, and rules. Yet even down to the beginning of the present century, primitive mass football continued to be played in certain towns (such as Workington on the Cumberland coast), traditionally on Shrove Tuesday. It has been suggested that the ball symbolized fecundity and would bring magical powers to the person or persons who gained its possession (Young 1968: chs. 1–2).

Of other ball games, tennis, according to the *Oxford English Dictionary,* is mentioned around 1400 in England and probably has an earlier history in France. Cricket ("small stick"), under the Saxon name of *creag* ("stick or staff"), is mentioned in 1300 as being played by Prince Edward, grandfather of the Black Prince (Altham 1962:1.19–21). Athletic contests such as boxing, wrestling, and footraces were no doubt performed sporadically ever since the fall of Rome. Henry VIII and François I, for example, are known to have wrestled one another at Calais in

---

153. Plato, *Laws,* 794d: "The right and left hand are supposed to be by nature differently suited for our various uses of them; whereas no difference is found in the use of the feet and lower limbs"; tr. Jowett (1953:361).

1520. But such sports became popular and better organized only from the seventeenth century onward (and in many cases much later). The British were pioneers in keeping records of sporting events, for which the invention of the stopwatch in 1731 was a helpful development. Much earlier, and of course completely aristocratic, was knightly jousting, which reached its prime in France before the Hundred Years War (i.e., before 1338). It came to an end in France in 1559 after Henri II was killed in a tournament (Umminger 1963:175–180, 191ff.).

Let us summarize now the Western attitude to sports, first as found in the classical world. The most important point is the tremendous intensity of the competitive spirit, already evident in the earliest Greek times and by no means confined to physical contests alone. "Greek (and Roman) society," a modern study tells us,

> was shot through with the competitive spirit. The normal Greek word for an athletic contest was *agon,* which could be used for any contest or struggle (hence our word "agony"), such as a battle or a lawsuit, as well as for games. Choruses and playwrights competed, as did runners or wrestlers. Ambition, in Greek *philotimia* (literally "love of honour"), was always praise-worthy, in public affairs as in private. (Finley and Pleket 1976:21)

In the second place, the competitive spirit was initially very closely associated with amateur status, and this in turn with good birth. This is why the competitions included music, drama, and the like, but did not extend to creations by the *banausoi* (the artisans). In later classical times, when sports had become professionalized, the competitive spirit remained strong, but for economic rather than idealistic reasons. Betting by spectators was widespread.

Third, the competitive spirit showed itself above all in contests between individuals, not groups. We have seen that the ball games of the Greeks and Romans were largely unstructured and uncompetitive, as well as probably of plebian origin.

When we look at sport in the postclassical world, we see the same characteristics supremely represented in the medieval knightly tournaments. These had originated as mock battles between masses of fighters, but as the rules of the tournament were worked out, it was inevitable that the focus should shift to the individual: "Given the individualistic character of the Frankish warriors, the clash dissolved into a mass of single combats" (Umminger 1963:176).

At the other end of the spectrum, the mass football games, best known in primitive form in Britain, were anything but aristocratic, involving as they did large, nondescript masses of humanity. Though nonindividualistic, they nevertheless remained intensely competitive, no doubt because of the communal solidarity and primitive religious beliefs on which they were founded. English football, in its organized later form, has been hailed as a democratizing influence. With reference to the Tudor period, Young writes that "the football game was a meeting point, for the artisan and the artist, for the peasant and the scholar, for the modest and the mighty" (Young 1968:81). And again, referring to the seventeenth and eighteenth centuries, he places football with the music club and the theater as institutions that broke down class barriers (Young 1968:35).

Turning now to the Chinese side, let us first remember the evaluations we made of the four animal sports discussed above (horse and dog racing, cock and drake fighting) and the resulting decision to exclude these sports from further consideration. This means that the remarks that follow are based solely on the seven other Chinese sports covered earlier, namely archery, the "competitive games," football, the dragon-boat races, tug-of-war, polo, and wrestling.

1. Important to Chinese civilization has been the persistent interplay through the ages of two opposing traditions or ways of life: the civil or civilian tradition, called *wen,* and the military tradition, called *wu.* (They are excellently discussed in Fairbank 1974:2–9.) Ideas, activities, and things have all commonly been classified as pertaining in part or in whole to one or the other of these traditions. Of the seven Chinese sports here being considered, for example, it might be said that because they are physically competitive, they therefore all belong more to the *wu* than the *wen* side of life. Nevertheless, gradations between them exist. Thus at the extreme *wu* end we find wrestling, certainly the most martial and individually competitive of all the Chinese sports we have examined. Yet against this, at the *wen* antipode, stands the already cited dictum of Confucius concerning the proper conduct for archery: "The *chün tzu* [superior man] competes in nothing." Some dynasties or reigns have been more *wu* than others, among them the T'ang (though it was a great *wen* dynasty as well). In sports this is exemplified by the popularity at the T'ang court of football, tug-of-war, polo, and wrestling. Gradually, however, under later dynasties, the *wu* way of life lost prestige while that of *wen* achieved unquestioned supremacy. Yet, as noted earlier (VI.1c), from pre-imperial times through most of Chinese history, the *wen* side of life usually enjoyed strong support both in and out of government. This fact is important because it was the *wen* scholars, not the *wu* generals, who played the decisive role in China's scientific development.

2. Among our seven chosen sports, only two—archery and polo—are indubitably aristocratic, and of these, polo is a foreign importation and therefore a special case. About the beginnings of wrestling nothing is known, although we may suspect a plebian origin. The other sports all

seem to be of folk origin and in most cases to have roots in popular religion, even though two (football and tug-of-war) came to be played by members of the upper class. But there is no evidence that any of these plebian games reduced the gap between social classes, as football allegedly did in England. No meeting took place, for example, between the officials and courtiers on the one hand who "pulled the river" *(pa-ho)* in the T'ang imperial Pear Garden and, on the other, the ordinary folk who presumably had their own tug-of-war outside.

3. Four of the seven sports involve physical contact, strenuous exertion, danger, or all three. They are the "competitive games" (what little we know of them), the dragon-boat races (noncontact but certainly strenuous and dangerous), and polo and wrestling. The remaining three were noncontact, nondangerous, and probably not too strenuous (although tug-of-war may have involved a fair amount of exertion).

4. Spectator sports, in the sense of competitions systematically organized and performed for the entertainment of spectators, seem in general to have been relatively rare and small-scale in China. The one glaring exception to this statement (as to some others in these conclusions) is wrestling. Another exception, much less certain though possible, is the "competitive games," about which we know so little. However, these latter lasted only a little over a century and a half before suppression and were probably witnessed only by members of the gentry, not the public at large.

5. Most important of all is that all of the seven sports, save again wrestling and possibly the "competitive games" to some extent, involved *group* rather than *individual* competition. We have seen how this principle was made to apply even to archery—the earliest and most intensely aristocratic of all the sports—where one would ordinarily expect individual competition to be the rule. Creel (1954:322) has perceptively pointed out the lasting significance of this situation:

> Thus one-half of the contestants vied with the other half, but no one person could feel the entire onus of defeat. One is tempted to see in this an early example of a Chinese characteristic which is very prominent today, that is, an intense aversion to subjecting any individual to public humiliation—what is commonly called the desire to preserve "face," for others as well as for oneself.[154]

On every one of these points, especially the last, the dominant Chinese attitude toward competitive sports, especially on the upper social

---

154. These words were written more than a decade before public accusations and self-confessions became familiar aspects of life in the People's Republic of China, both before and during the Cultural Revolution.

level, seems for the most part radically different from the prevailing attitude in the West. The difference is most evident at the very beginning, in the carefully muted group competitiveness of the Chinese archery contests as against the intense personal competitiveness of the Greek athletic games.

One important point should be noted here. The Greek spirit of competition was closely associated with birth and therefore not all-inclusive; as we saw above, it did not include the artisans. For the same reason competition was softened in the economic sphere, where "what was lacking was the modern form of the profit motive."[155] This situation continued by and large through the Middle Ages until the vast changes in social thinking and practice that came with the Renaissance. At that time the competitive spirit which had so strongly characterized the earliest stages of Western civilization gained renewed strength as it burst its aristocratic boundaries and poured into the increasingly respectable channels of commerce, industry, and entrepreneurial activity. From this arose the modern society of capitalism, technology, and science.

In China the spirit of direct personal competition had been considerably weaker even during the feudal age, and it remained so through most of Chinese history with only a few exceptions on the *wu* or military side —wrestling was of course one. Yet having said this, we must point paradoxically to another exception, even more conspicuous, on the *wen* or civilian side. This was the civil service examinations.

Nowhere else in the world could competitiveness be better exemplified than in these examinations, taken throughout the country on at least three ascending levels of difficulty by endless generations of literati. Of the tens and hundreds of thousands of aspirants who sat in the examination cells, all but a tiny fraction were doomed to disappointment owing to the very small quota needed to fill the ranks of the civil service. Here was no anonymous competition in which the success or failure of the individual was conveniently blanketed by his group. Here, rather, on the national level, for everyone to see, was a ruthless competition between each individual and all the other individuals, in which the success of one inevitably meant the failure of others. Such success, when it did come, brought fully as much glory to the scholar and his native town as ever did any Olympic victory to the Greek winner and the city-state of which he was a citizen. This sublimation, as one might call it, of the spirit of competition, turning it from physical struggle to competition with the written word, and from success in a private career to success in government bureaucracy, was, as it continued to function

155. See V.4b near end, quoting Burford (1972:119).

over the centuries, to have decisive consequences for the development of Chinese civilization.[156]

5. Summary

Let us end with a very brief recapitulation of some of this chapter's highlights. The chapter began by exploring several ramifications of what may be called the moral view of life: the evaluation of literature, art, and music according to whether or not they convey a high moral message or have been created by persons of high moral stature; the evaluation of history according to whether its events have been dominated by good or bad persons rather than by analyzing the impersonal facts of geography, economics, and institutions; the persistent dominance of human-centeredness in Neo-Confucian thinking, effectively thwarting the creation of a suitable methodology for studying the phenomena of nature. Other aspects of Chinese moralism were the strong antipathy of most literati to warfare and political expansionism; their approval, at least in theory, of simple and frugal living; and their tendency to distrust new manufactures produced by the artisanate as possibly being conducive to extravagance and frivolity.

Turning to sexual mores, it would seem that this all-important aspect of life was long dominated by a male-centered and puritanical ethic that was primarily Confucian but reinforced by Buddhism, and apparently became increasingly repressive during the later dynasties. Although the Taoist view of sex and women was far more enlightened from a modern standpoint, it nevertheless accepted the Confucian premise that the male is always dominant. Unfortunately, despite the importance of sexual morality for each individual within a society, it is apparently very difficult to establish any correlation between that society's general degree of sexual freedom and degree of cultural creativity (including creativity in science and technology).

As to individualism, meaning individualism as commonly thought of in the West, this has always been exceedingly rare in China, where even today it is regarded with suspicion. Chinese writers and artists, likewise, have usually been much less ready to project their egos ("self-expression") than have their Western counterparts, particularly from Renaissance times onward.

Finally, the field of sport reveals several significant differences

156. For a clear, concise, and comprehensive account of the Chinese examination system as it functioned during the Ch'ing dynasty, see Miyazaki (1976). For a much longer, more detailed and technical description, see Zi (1894).

between China and the West. In China physical sports have almost always played a very subordinate role; contact sports have been rare, spectator sports barely existent; and competition between individuals has often been muted by subordination to group competition. The one great exception, completely removed from physical sports, was the intense and public competition between individuals that occurred at the civil service examinations, hurdles to be overcome before entering a government career.

Occasional exceptions can surely be found to the morals and values discussed in this chapter; one is the rise of erotic literature during the very centuries (late Ming and early Ch'ing) when sexual puritanism was probably increasing. Likewise, occasional parallels between China and Europe can also surely be found; one is William Hogarth's brief excursion into the moralistic evaluation of art (see sect. 1a, n. 18).

What is important, however, is not these exceptions or parallels but the overall influence exerted by this network of Chinese morals and values upon Chinese science and technology. In its totality, this influence was surely very unfavorable (even though ideals like pacifism and simplicity of living deserve high praise, judged purely on their own ethical premises). Moreover, the antiscientific influence continued down to the present century in China together with the morals and values that generated it, whereas in Europe, whatever parallels may once have existed had largely disappeared considerably before that time.

VII. MANKIND AND NATURE

1. Seven Approaches to Nature

"Love of nature" and "harmony of man and nature" are clichés invariably brought up in any discussion of Chinese thought. However, they require refinement if they are to carry much meaning. Was there more than one significant Chinese approach to nature? If so, by whom and when, and what were the implications for scientific development? In European philosophy and literature of classical times onward, some hundred or more usages of the word "nature" have been identified and discussed by Lovejoy and Boas (1935), Lovejoy (1948), and Wilson (1941). Here our aim is considerably more modest and less semantic. What interests us is not so much the semantic permutations the nearest Chinese equivalents for "nature" have undergone in the writings of different Chinese intellectuals.[1] Rather it is the psychological reverberations evoked in these intellectuals when they were confronted by, and spoke or wrote about, that nonhuman but also nonmetaphysical realm that in the West is vaguely termed the world of nature. In what follows I have tried to group these varying reverberations along a spectrum of seven descriptive categories, ranging from rejection of the natural world to its total acceptance.

1a. The Antagonistic/Indifferent Approach

"Antagonistic" is included in this first rubric more to define the uttermost negative limit of the spectrum (antagonistic/indifferent) than because it actually represents any readily identifiable Chinese attitude. Perhaps Chinese Buddhism approaches this negative extreme most closely because of its overriding doctrine that the entire world of sensory perception is a mere phantasmagoria, a product of *māyā* or illusion. Yet even this Buddhist attitude is more a matter of simple denial than of positive hostility. And outside of Buddhism, such hostility would be extremely difficult to find. Chinese literature has nothing comparable to

1. Actually, there is no precise Chinese equivalent. However, the word that in appropriate contexts is most often rendered as "nature" is *t'ien,* "heaven," originally the name of an anthropomorphic deity, but already used in early Chou times in other senses, including that of the physical "sky." Another term sometimes rendered as "nature" is, of course, Tao.

the "ascetic distrust" with which, in the words of Babbitt (1919:270), the European "man of the Middle Ages often saw in nature not merely something alien but a positive temptation and peril of the spirit" nor to the revulsion with which this medieval man's neoclassical descendant viewed a nature untouched by man. To quote Babbitt once more (1919: 274): "Wild nature the neo-classicist finds simply repellent. Mountains he looks upon as 'earth's dishonor and encumbering load.' The Alps were regarded as the place where Nature swept up the rubbish of the earth to clear the plains of Lombardy."

Much closer to one strain of Chinese thinking is the "indifference" that makes up the second half of the rubric. In early Confucianism, with its concentration on the world of human beings, such indifference is very apparent. "It is impossible to associate with birds and beasts," declared Confucius (*Lun yü* 18.6), criticizing two hermits whom he had encountered. "To lack a father and lack a ruler: such is to be a bird or beast," reiterated Mencius (*Meng-tzu* 3b.9.9), attacking what, to him, were the extremist doctrines of Mo Tzu and Yang Chu. But perhaps the most vivid token of indifference to nature is the anecdote—still significant even if apocryphal—according to which Tung Chung-shu, the major Confucian of the Han dynasty, was so imbued with the spirit of learning that for three years he did not even look out into his garden.[2] This anecdote is the more significant in that Tung was the chief architect of the dominant sociocosmological worldview of Han times. His philosophical edifice obviously owes far more to the texts in his study and his own cogitations than to any observations of natural phenomena. In post-Han Confucianism, especially that of Sung, a profound change of attitude was to take place. No doubt it was generated in good part by centuries of Confucian-Taoist interaction.

Returning to early Chinese thought, we find in Mo Tzu a passage that typifies that thinker's indifference to and ignorance of animal behavior:

> Birds, beasts and insects use their feathers and hair for clothing and furs, their hoofs and claws for sandals and shoes, water and grass for food and drink. Therefore the male does not cultivate, . . . nor does the female spin, . . . because their supplies of clothing and food are already well provided. Now man is different from these: those who rely on their strength will live, those who do not rely on their strength will not live. (*Mo-tzu* 32; tr. Mei 1929:178)

One would think that the slightest observation of the activities of non-herbivorous animals—the food-gathering of bees and ants, the nest-

2. Fung (1953:16), quoting Tung's biography in *Ch'ien Han shu* 56/1a.

building of birds, the activities of the vast majority of animals in bringing up their young—would have been enough to disprove this observation. Yet apparently it was only during the T'ang period (texts of the eighth and tenth centuries) that Chinese thinkers first noted similarities between human society and the social organizations of bees and ants.[3]

A similar ignorance of animal social behavior is evident in Hsün Tzu's insistence that the basic distinction between human beings and animals is the former's ability to create social organizations (*ch'ün,* lit. "to collect together").[4] Here again, if Hsün Tzu had been more cognizant of the organizing activities of many of the larger animals as well as of bees and ants, he could have focused his attention on the distinctions that really matter, such as a human being's ability as a two-legged animal to make tools or the human capacity for speech.

In this connection there may be significance in the fact that since very early times the Chinese were apparently much more interested in crops and plants than in animals. In particular, they seem to have been decidedly backward in animal husbandry as compared with what was done in other early civilizations. Ho Ping-ti points out:

> Throughout China's long historic periods the agricultural system . . . has always been lopsided in favor of grain production, with animal husbandry playing a subsidiary role. . . . Among relevant [early cultural] traits, the most noticeable was the lack of sufficient knowledge to make and use dairy products. . . . The Chinese had yet another peculiar trait, namely, the unusually late beginnings and persistent underutilization of draft animals for cultivation. (Ho 1975:113–114; see also Bray 1984:3–7)

Later in the same study, Ho (1975:204–207) points to the very tardy development of apiculture and use of honey and beeswax in China as compared with the ancient Near Eastern and classical Western civilizations. In contrast to references to honey found in the Old Testament and other Near Eastern texts, Chinese references seemingly begin only in late Warring States times (perhaps the third century B.C.), and honey was apparently still rare and consumed only by the wealthy when mentioned by Wang Ch'ung in A.D. 82/83.[5] Only in the third century A.D. is

---

3. See two quotations in Needham (1956:447–448): "The sages learnt social order from bees . . . and war from fighting ants" (*Kuan Yin-tzu* 3/7b; viii cent.). "Ants have a prince, and all share in common a palace as big as a fist . . . and store their grains of food in common. . . . One mind interpenetrates them all. . . . This was the unity, too, of ancient [human] civilization" (*Hua shu* 4/5b; tenth century).

4. *Hsün-tzu* 9(5/7b); tr. Dubs (1928:136).

5. For evidence as to the occasional use of honey in late Chou and Han times, see Chang Kwang-chih (1977b:32) and Yü Ying-shih (1977:67).

there mention of the domestication of bees, and only in the early fourth of the use of beeswax for candles. Yet a striking exception to the general thesis here being suggested should be noted. That of course is sericulture, which was definitely practiced in Shang times and probably goes back even to Neolithic China.[6]

Of all early Chinese rejections of nature, the most unequivocal comes once more from Hsün Tzu, in his chapter "On Heaven":

> To understand the distinction between Heaven *(t'ien)* and man: this is to be a great man. . . . Although it [Heaven] is deep, man will not give deep thought to it; although it is great, he will not use his ability [for its investigation]; although mysterious, he will not scrutinize it: this is what is meant by refraining from contesting in one's activities with Heaven. . . . It is precisely the sage who does not seek to know Heaven. (*Hsün-tzu* 17 [11/9b–10a]; tr. Dubs 1928:174–175)

As will be seen directly, there is an inconsistency between this pronouncement and another even more famous one by Hsün Tzu.

1b. The Exploitative/Utilitarian Approach

The most notable utterance on the exploitative/utilitarian approach to nature comes once more from Hsün Tzu. From his anti-Taoist hymn I quote the first four lines:[7]

> You glorify Nature (*t'ien,* Heaven) and meditate on her;
> Why not domesticate her and regulate her?
> You obey Nature *(t'ien)* and sing her praises;
> Why not control her course and use it?

Chan (1963:122) comments on this famous pronouncement:

> Nowhere else in the history of Chinese thought is the ideal of controlling nature so definite and so strong. It is a pity that this did not lead to a development of natural science. One explanation is that although Hsün Tzu enjoyed great prestige in the Han dynasty, his theory of overcoming nature was not strong enough to compete with the prevailing doctrine of harmony of man and nature, which both Confucianism and Taoism promoted.

---

6. See Kuhn (1988:272–284). For another less well known example, see also Huang Hsing-tsung (1986:531–544). This describes how, according to Chinese sources possibly going back as far as ca. A.D. 300, orange growers in Vietnam and South China have protected themselves against insect pests by installing nests of yellow citrus ants in the branches of their orange trees, where the ants devour the pests.

7. *Hsün-tzu* 17(11/13b); tr. Dubs (1928:183). Here I have followed the more literary rendition by Hu Shih (1922:152).

The reference to harmony raises a consideration that, under the topic of cosmic organicism, will be discussed in the final pages of this chapter. With respect to Hsün Tzu himself, one may suspect that the major reason for his failure to stimulate the study of natural science lies in his earlier just-quoted injunction not to seek to understand nature. The conquest of nature, when divorced from its study, can lead only to a crude plundering, accompanied at best by an exploitative technology. It cannot lead to true science. Hsün Tzu's view of nature was part of a broader intellectual rigidity which, in the human sphere, resulted in political authoritarianism, and in the natural sphere left no room for anything but a traditional technology.

The plundering of nature is referred to also in Taoist writings. In *Chuang-tzu,* as one might expect, it meets with strong disapproval.[8] Curiously, however, similar actions are regarded more tolerantly in the later *Lieh-tzu* (1; tr. Graham 1960:30–31). In this a certain Mr. Kuo tells us:

> I have heard that Heaven has its seasons, Earth its benefits. I rob Heaven and Earth of their seasonal benefits, the clouds and rain of their irrigating floods, the mountains and marshes of their products, in order to grow my crops and plant my seed. . . . Birds and animals I steal from the land, fish and turtles from the water. All these I steal, for they are all produced by Heaven.

The *Lieh-tzu* justifies these "robberies" on the grounds that they "are carried out in the spirit of the Tao of common life," in contrast to the robberies of private property committed by ordinary burglars who break into people's houses.

It can be argued that the Chinese alchemists were not really turning against nature when they attempted to concoct their elixirs of immortality.[9] The reasoning is that they regarded nature's time-scales as variable and hence saw nothing "unnatural" in using especially prepared elixirs as "time-controlling substances" to slow down the natural process. Yet even if they in fact accepted this rationalization, they must also have realized that such a slowing down of the time-scale could only come through deliberate human intervention. Ko Hung (ca. A.D. 320), the well-known alchemical theoretician, seems to recognize as much when he writes: "Of those things that have been smelted and fashioned [by

8. *Chuang-tzu* 11; tr. Watson (1968:118–119), where the Yellow Lord's wish to use the products of the earth and the forces of heaven so as to "provide nourishment for my people" is criticized as meaning that he would interfere with the natural descent of rain and the maturing of plants and trees at the right season.

9. Needham (1976:83) and especially, for a detailed exposition, Sivin (1980:242–279).

Heaven], none is more numinous than man, who, therefore, if he but reach the shallower [levels of knowledge], can put all things to his service, and if he gain the deeper [levels], can then enjoy eternal life."[10]

The achievements that Ko Hung promises to successful adepts are certainly anything but "natural," for they include the ability to walk on rivers; see ghosts; write in the dark; keep plague at a distance; cause enemy soldiers to mutiny; cause the waters of a stream to open so that one may cross dry-shod; fast for a hundred days without suffering hunger; be immune to swords or arrows; make oneself invisible; make a tree wither; cause a tiger, wolf, or snake to die by throwing a pill at it; and cause people to imagine that they see mountains and trees moving.[11] It is difficult to see how miracles of this kind could possibly be regarded as compatible with natural processes or how belief in them could help the development of science.

The exploitative/utilitarian view of nature, though it often leads either to the crude plundering of natural resources or sometimes to efforts to gain miraculous powers, can, when channeled more intelligently, also sometimes lead to natural conservation. This represents a significant improvement in the way nature can be approached, even when it is motivated by purely pragmatic considerations. Self-interested ecological concern of this sort is evident precisely among those early Confucians whom we have seen to be most intent on keeping man separated from nature. Mencius, for example, urges that close-meshed nets be banned from waterways and that axes be allowed in the forests only at the proper season. The result, he says, will be that "fish and turtles will be more than can be consumed" and "wood will be more than can be used" (*Meng-tzu* 1a.3; tr. Legge 1960:2.130). Hsün Tzu, too, repeats the injunction, but with greater vehemence and detail (*Hsün-tzu* 9[5/8a–b]; tr. Dubs 1928:138–139).

A still more striking concern for agricultural and natural resources is to be found in the legal texts excavated in 1975 from the grave of a Ch'in provincial official who had died in 217 B.C. The laws contained in these texts very probably form part of the previously long-lost legal code of the state of Ch'in. One of them, despite linguistic uncertainties, seems to say that from after the first spring month until the first month of autumn, the cutting of wood in forests, damming up of waters, taking of birds' eggs, poisoning of fish, and setting out of traps are all forbidden. The only allowed exception is the cutting of wood to make a coffin for a newly deceased person (an interesting concession to traditional family ethics). Other laws specify how much seed is to be used per "Chinese

10. *Pao-p'u-tzu* Inner ch. 3; tr. Ware (1966:52).

11. Ibid., Inner ch. 16; tr. Ware (1966:276–277).

acre" *(mou)* for the planting of the major kinds of grain, pulses, and textiles; or they direct the magistrates of prefectures throughout the state to record and report to the central administration the rainfall received in their area and the effects on crops, as well as droughts, floods, insect pests, and other natural disasters.[12]

Such pragmatic concern about the development and conservation of natural resources is the only indication of any positive interest on the part of the Legalists in the processes of nature. This is a pity because, as we have seen (III.4), the Legalists were much concerned with the use of statistical and quantitative techniques in government. Had they been equally interested in applying such techniques to the classification and analysis of natural phenomena, the results for Chinese science might have been considerable.

The exploitative use of natural resources is very often coupled with a theistic view of the universe, to which we now turn.

1c. The Theistic/Anthropocentric Approach

So familiar to all adherents of the monotheistic religions is this approach that sometimes they find it difficult to conceive of any other. The approach is epitomized in Genesis 1.27:

> So God created man in his own image . . . male and female created he them.
>
> And God blessed them, and God said unto them,
>
> Be fruitful, and multiply and replenish the earth, and subdue it: and have dominion over the fish of the sea, and over the fowl of the air, and over every living thing that moveth upon the earth.

In Chinese thinking the idea that the creatures and things of an anthropocentric universe have been expressly created by a divine providence for the sake of man is rare. Yet very occasionally it occurs among a few theistically minded writers. The earliest is Mo Tzu, who conceived of Heaven *(t'ien)* as a personal deity presiding over the universe, assisted by a multitude of spirits:

> Now Heaven loves the whole world universally. Everything is prepared for the good of man. Even the tip of a hair is the work of Heaven. . . . It sends down snow, frost, rain and dew to grow the five grains, hemp and silk, thereby enabling the people to gain and be benefited by these. . . . It has contributed metals, woods, birds, beasts, . . . the five grains, hemp and silk . . . to be materials for man's clothing and food. (*Mo-tzu* 27; tr. Mei 1929:145)

---

12. For transcriptions, translations, and discussions of these Ch'in legal materials, see esp. Hulsewé (1985:22, 41, 21 [A2, A27, A1]).

Tung Chung-shu, whose "Heaven" is sometimes naturalistic and physical, but more often personalistic or even seemingly anthropomorphic, expresses a very similar idea:

> Man has received life from Heaven in a manner markedly different from the great mass of living creatures. . . . The five grains have been produced to feed him, the mulberry tree and hemp to clothe him, the six domestic animals to nourish him. He tames the ox, rides the horse, traps the panther, pens the tiger. It is thus that he has gained a spirituality greater than that of other creatures.[13]

In the penultimate sentence there is perhaps an echo of Hsün Tzu's glorification of the conquest of nature.[14]

Probably the Han dynasty was the epoch when the theistic/anthropocentric view of the universe became most prevalent. We find it briefly but forcefully expressed in a prayer for rain preserved in one of the apocryphal writings on the classics that probably belongs to the first century B.C.:

> The invocation for rain states: "Glorious Heaven *(hao t'ien)* produces the five grains for the nourishment of man. But now these five grains have been blighted by drought and we fear they will not mature. We respectfully present the clear wine, offer the dried meat, and twice prostrate ourselves as we beg for rain. May the rain greatly come down."[15]

Heaven as a purposeful deity was a concept probably widespread in Chinese folk religion from early until recent times. It was also prominent in the Chinese state religion which, like official religions everywhere, tended to preserve archaic beliefs and practices. Among philosophically minded intellectuals, however, the view that the universe had been created (or at least arranged) by a beneficent Heaven expressly to serve the needs of man was by post-Han times either rejected or simply ignored. Already in the first century A.D., Wang Ch'ung included it

---

13. A statement attributed to Tung in his biography in *Ch'ien Han shu* 56/4b.

14. See previous section. Also *Hsün-tzu* 9(5/7b); tr. Dubs (1928:136), where Hsün Tzu expatiates on man's ability to use the ox and horse for his purposes, despite his lesser speed and strength. The reason, he says, is that man is able to organize socially.

15. *Ch'un-ch'iu wei Han-han-tzu* (Apocryphal Treatise on the Spring and Autumn Annals: Cherished Beginnings of Growth of the Han Dynasty) 56/4b. The same prayer appears in Tung Chung-shu's *Ch'un-ch'iu fan-lu* 74(16/6b); tr. Loewe (1987: 207). Loewe, however, believes that the account of the rite for inducing rain, in which the prayer is quoted, may have been composed after Tung Chung-shu's time. For our purpose, it is immaterial whether the prayer was recorded in the second or first century B.C.

among the doctrines of his age that he explicitly attacked. "Some," he wrote, "say that Heaven has produced the five grains to feed man and silk and hemp to clothe man. But this would mean that Heaven serves man as man's farmer boy or mulberry girl, and would not be in accordance with spontaneity *(tzu jan)*. Therefore this opinion is dubious and unacceptable" (*Lun heng* 54[18/1]; tr. Forke 1962:1.92).

Lieh Tzu makes the same point charmingly with his story of the banquet host who, eyeing the groaning board, exclaimed unctuously: "How generous is Heaven to man! Heaven causes the five kinds of grain to grow and brings forth the finny and the feathered tribes, especially for our benefit." To which the twelve-year-old son of one of the guests boldly replied that man is a creature like any other in the universe; hence if it be accepted that other animals have been produced by Heaven for the sake of man, it should equally be accepted that Heaven has produced man for the sake of tigers and wolves, mosquitoes and gnats (*Lieh-tzu* 8; tr. Graham 1960:178–179). The naturalism of these statements leads us to the fourth approach in our spectrum.

1d. The Naturalistic/Analytical Approach

By the naturalistic/analytical approach is meant a rational and nonsupernatural approach, coupled with an interest in natural phenomena that, though sympathetic, is also detached and questioning. I believe Wang Ch'ung (A.D. 27–ca. 96) to be the supreme example of this approach. Repeatedly he emphasizes that human beings, despite their superior knowledge, are essentially creatures like any others, and very insignificant compared with the universe as a whole:

> Man is a creature *(wu)* among the myriad creatures who possesses knowledge *(chih-hui)*, but in receiving his destiny *(ming)* from Heaven and his vital powers *(ch'i)* from the Originator *(yüan)*, he is no different from any other creature. Birds have their nests and perches, animals have their dens and lairs, fish and scaly creatures their particular habitats, just as men have their homes and mansions. . . . Man lives and dies, and creatures too have their beginnings and ends. . . . In blood, veins, head, feet, ears, eyes, nose and mouth, they are in no way different from man. Only in likes and dislikes is there a difference. . . . They share the same Heaven and same Earth, and equally gaze up at the same sun and moon. (*Lun heng* 72[24/14–15]; tr. Forke 1962:1.528)

> Men live betwixt Heaven and Earth just as lice live on the human body. If a louse, wishing to know a man's intention, were to call to him beside his ear, the man would still not hear it. Why? Because, owing to the inequality in sizes, its utterances would fail to get through. (*Lun heng* 71[24/7]; tr. Forke 1962:1.183)

On the same grounds, Wang ridicules the Taoist belief that it is possible by means of elixirs to become an immortal: "Now man is a creature *(wu).* Although his nobility be that of king or marquis, his nature is no different from that of other creatures. Since there is no creature that does not die, how can a man become an immortal *(hsien)?*" (*Lun heng* 24[7/3]; tr. Forke 1962:1.335).

In contradistinction to Mo Tzu, who believed that Heaven wants all men to love one another, Wang maintains that Heaven's engendering of creatures is without purpose; this is why creatures, instead of loving one another, prey upon and destroy one another (*Lun heng* 14[3/24]; tr. Forke 1962:1.104).

Tirelessly and sometimes almost ad nauseam, Wang strives to refute the popular beliefs of his time, notably the idea that people through their own actions can influence the functioning of the universe. Yet even he fails to escape wholly from the intellectual environment of his time. He believes, for example, that there is a fate for each individual, a larger and more powerful fate for community or state, and that astrology provides signs by means of which these fates can be known. Nonetheless, Wang Ch'ung remains a man of unique independence and intellectual courage. What this uniqueness has meant for Chinese science will be considered in subsection 1h.

1e. The Animistic/Moralistic Approach

So imbued are people past and present with animistic concepts that often they are scarcely aware of their existence. Personifications of inanimate objects, as in the statements "The soil is thirsty" or "The sun is trying to break through the clouds," are implicity animistic unless deliberately intended as metaphors. So is a man's (or more likely a child's) angry kicking of a stone after he has hurt himself on it. It is sometimes hard to decide when such statements are to be taken literally and when only figuratively, perhaps not least by the person uttering them. The difficulty is heightened by the fact that sometimes only a very slight difference separates a statement of fact from a statement with animistic implications. "The sun tries to break through the clouds" is an animistic statement (unless intended only figuratively), whereas "the sun breaks through the clouds" is essentially a statement of fact (not quite so, however, because "breaking through" is itself a figurative term).

In China, what was said in subsection 1c above about a purposeful Heaven applies also to animism. That is to say, both the folk religion and the state cult were overwhelmingly animistic. Since early times, such forces and objects of nature as the sun and moon, stars, thunder, rain, drought, the soil, mountains, rivers, and many more were all deified and could be recipients of offerings. Their cults continued until the present century.

In 219 B.C, for example, when the First Emperor of Ch'in was descending from the "sacred" mountain of T'ai Shan after performing an important sacrifice, he was beset by a severe rainstorm and found shelter under a pine. In gratitude, he conferred on the pine the title of "Fifth Rank Grandee" (one of the recognized honorary ranks of that time).[16] In 1355 the man who was later to found the Ming dynasty is said to have wrapped a red robe of investiture around the trunk of a persimmon tree and to have proclaimed: "I hereby invest you with the title 'Marquis of Ice and Frost.' " His act was in recognition of the fact that in the course of his military campaigns some years earlier he had once been without food for two days, during which time he had happened upon this tree standing in a destroyed garden and had eaten ten of its ripe persimmons.[17] Comparable episodes could be cited down to the end of the empire. The Altar of Heaven in Peking was not dedicated, as often popularly supposed, solely to Heaven (in this context a deity, not "nature"); in actual fact, the other subordinate deities included at the same altar were those of the sun, moon, Great Dipper, the five planets, the twenty-eight constellations, the other stars, the clouds, rain, winds, and thunder (Williams 1913:29).

The same situation obtained among the various nonofficial cults. Within so-called alchemical Taoism, for example, believers in animism included such a notable as Ko Hung.[18] However, for the most extreme examples of animism, one should probably turn to those Taoist adepts who sought prolongation of life through regimens of physiological hygiene. They conceived of the different parts of the human body as being under the jurisdiction of an incredibly large number of divinities. Typical is the following:

---

16. *Shih chi* 6 and 28; tr. Chavannes (1895–1905:2.140 and n. 4 and 3.431).

17. Cited in Tun Li-ch'en, *Yen-ching sui-shih-chi* (Annual Customs and Festivals in Peking); tr. Bodde (1965:74).

18. To cite only a very few of many examples: Ko Hung believed that spirits, which he called "intrinsic powers" *(ching ch'i),* are present in mountains, rivers, plants, trees, wells, water holes, and pools, as well as the human body. See *Pao-p'u-tzu,* Inner ch. 6; tr. Ware (1966:116). He described the spirit of one such mountain as having the shape of a small boy and hopping backward on one foot, that of another mountain as resembling a red drum, and still a third as an anthropoid nine feet high, wearing fur-lined clothing and a straw hat. He also believed that some of the huge trees on mountains were capable of speech. See ibid. 17; tr. Ware (1966:282, 287). Perhaps his idea of the talking trees derives from Chuang Tzu's famous description (in his second chapter) of the varying sounds of the wind heard in a mountain forest.

> The divinity of the hair is called "Deployed," the divinity of the two eyes is called "Abundant Light," the divinity of the top of the head is called "Father King of the East," . . . [and so for other parts of the body] . . . In the body of every man there are three Palaces, six Administrations, 120 Barriers and 36,000 gods.[19]

One of the characteristics of animism is its attribution of human moral qualities to the forces and objects of nature. In the preceding chapter (sect. 1e) we saw how the various aspects of water could be compared to virtue, benevolence, righteousness, courage, discrimination, forebearance, and several other qualities; how bamboo was said to possess firmness, uprightness, integrity; mountains were characterized by propriety, caution, massiveness, and so on; jade could be likened to no less than eleven different qualities and things; and, in Sung times, flowers and trees (the orchid, chrysanthemum, plum blossom, pine tree and others, as well as some animals) were viewed as embodiments of qualities to be looked for in the ideal scholar-gentleman, the *chün tzu*.

Such moralistic personification must have impeded efforts to express ideas in more scientific language. Take, for instance, the statement made by Chu Hsi in the twelfth century: "By observing the fact that water necessarily flows downward, we know that the nature of water *(shui chih hsing)* is to go downward."[20] This is more than just an echo of Mencius's argument by analogy for the goodness of human nature (*Meng-tzu* 6a.2). More importantly, it is also an animistic statement in which water is conceived as a living entity whose desire (*hsing* or "nature") is to move always downward. This attribution of will to water, that is, attribution of a particular vital force to a particular entity, forms part of the larger conception of the Five Elements as five living forces, each having its own long list of distinctive characteristics and powers. Both within and beyond these five basic components of nature there exist a vast number of lesser components, each in turn having its own particular powers and attributes.

Thus the animism of nonphilosophical popular belief could become the basis for a multivitalistic cosmological system. Belief in such a system, because it meant viewing the universe as an aggregate of separate vitalistic forces, inhibited formulation of a worldview in which the universe is subject to certain broad, unvarying principles. Such principles, in the Western monotheistic tradition, came to be known as "laws of nature" and were believed to emanate from a supreme, unitary law-

19. Maspero (1981:346–347), quoting from two Taoist texts of pre-T'ang date, now lost save for scattered quotations.

20. *Chu-tzu ta-ch'üan* (Great Compendium of Chu Hsi's Writings) 67/17a; tr. Chan (1963:598).

giver. Thus, in the case of water, the idea that it is a separate entity that, owing to its own inner impulse, always wishes to go downward, could not but impede formulation of the idea of gravitation as a single universal "law" to which all things whatsoever are subject. More will be said about this all-important topic in a later section.[21]

1f. The Semireceptive Approach

"Semi" is used here to denote an approach that takes a person halfway but no further toward the acceptance of nature. In other words, it denotes an approach that divides the universe into a greater world of nature (the macrocosm) and a lesser world of man (the microcosm). Formulators of this analogy usually focus primary attention on the microcosm which is the human world despite their theoretical recognition that this microcosm is subordinate to the macrocosm. The reason is understandable: being more familiar with the human world than the world of nature, they unconsciously regard the former as a prototype and the latter as merely a larger replica.

Much earlier in this book (II.5), it was pointed out that the view of man and nature as the symbiotic constituents of a larger unity finds utterance already in the earliest Chinese poetry, before the Chinese had begun to express themselves philosophically. What this means is that already in the *Shih ching*, there occurs the literary device of drawing a comparison (or sometimes a contrast) between a vignette of nature and a human situation or action:

> Gorgeous in their beauty
> Are the flowers of the cherry.
> Are they not magnificent in their dignity,
> The carriages of the royal bride?[22]

Not only did such analogical parallelism continue as an important factor in later Chinese prose and poetry. It also marked the start of the

---

21. See sect. 2a. With regard to Chu Hsi's statement, see also Needham (1959:157 and esp. 160), where the same kind of thinking is discussed for Europe. The suggestion is there made that Leonardo da Vinci's backwardness in theorizing (as contrasted with his inventive genius) is connected with his belief that, for example, the moisture in a wet rag has an intrinsic tendency to move toward fire, and that fire in turn possesses a quasi-spiritual power of carrying light things with it in its ascent toward the empyrean. Needham's conclusion for Chinese science and technology is that, expressed in Western terms, it always remained essentially Vincean, not Galilean.

22. *Shih ching* no. 24; tr. Waley (1937:76).

macrocosmic/microcosmic kind of thinking that thereafter was to crop up repeatedly in Chinese intellectual history. Such thinking, sometimes expressed in painfully literal terms, became particularly prominent during Han times. Table 4 shows some of the analogies between "Man" and "Heaven" drawn by Tung Chung-shu in his *Ch'un-ch'iu fan-lu*.[23]

Table 4 **Analogies in the *Ch'un-ch'iu fan-lu***

Man	Heaven
1. 12 large joints	1. 12 months
2. 366 lesser joints	2. 366 [*sic*] days of year
3. 4 limbs	3. 4 seasons
4. 5 viscera	4. 5 elements
5. Eyes and ears	5. Sun and moon
6. Opening and closing of eyes	6. Day and night
7. Orifices and veins	7. Valleys and rivers
8. Breathing of mouth and nostrils	8. Wind
9. Hair on head	9. Stars and constellations
10. Head's roundness	10. Heaven's round countenance
11. Foot's rectangular shape	11. Earth's square shape
12. Body's flesh and bones	12. Earth's thickness
13. Mind's power of thinking	13. Heaven's power of deliberation
14. Alternation of sadness and pleasure	14. Alternation of yin and yang
15. Ruler's likes and beneficence	15. Spring's warmth and germinating
16. Ruler's joy and rewarding	16. Summer's heat and nourishing
17. Ruler's dislikes and punishing	17. Autumn's coolness and destroying
18. Ruler's anger and executing	18. Winter's cold and storing up

Equally prosaic correlations continued to be made long after Tung Chung-shu. A Taoist tract of the twelfth century, for example, equates man's inhalations and exhalations with thunder, his sweat and tears with rain and dew, his four limbs with the four directions, and so through twenty correlations.[24] Some of these and similar correlations

---

23. Chs. 44 and 55–56; tr. Fung (1953:30–31, 47–48). See also Yao Shan-yu (1948:57–58 and esp. 54) for other correlations made by Tung Chung-shu between the human polity and the human body. When Tung speaks of "Heaven," he often uses the word in a broad sense to refer to the entire world of nature, but sometimes also in a narrower sense to refer to the celestial part of the universe in apposition to "Earth," its terrestrial portion.

24. See the *Tao shu* (Pivot of the Tao; completed by 1145) 5/20b–21a; tr. Homann (1976:32).

obviously pass from Heaven to Man rather than vice versa. It is unlikely, for example, that Tung Chung-shu and other similarly minded men would have thought of the human body as having twelve major and 366 lesser joints were they not already familiar with the twelve months and 366 *(sic)* days of the year pertaining to nature. For the most part, however, such writers were more interested in the human world (the microcosm) than in the world of heaven/nature (the macrocosm).[25]

The Han naturalistic thinkers, for example, were much more interested in seeing how human actions might affect the world of nonhuman phenomena than on how "heaven" might influence human affairs. One example out of many is their assertion that the greed of civil officials results in plagues of grain-eating insects having black heads and red bodies, whereas if the insects have black bodies and red heads, it is because the military officials are the greedy ones; punishment of one or the other kind of official will cause the corresponding category of insects to disappear.[26] Thus the prevailing Han approach to nature was more human-centered than nature-centered, which is scarcely surprising in view of the amalgam of social and naturalistic thinking that constituted Han Confucianism.

Wang Ch'ung, as we have seen (sect. 1d), strongly opposed the naturalistic claim that human actions could affect the natural world. Yet even he was very much a creature of his intellectual environment, so that in his eagerness to refute some popular theories of his time, we see him resorting to the very same nature-man analogy on which those theories rest. This confusion appears clearly in a passage in which Wang denies that humans originally came into existence because of any deliberate purpose:

> The literati *(ju)* declare that Heaven and Earth [initially] engendered man on purpose, but this statement is absurd. For when Heaven and Earth united their vital forces *(ch'i),* man was fortuitously and spontaneously engendered, just as when husband and wife unite their vital forces *(ch'i),* a child is then spontaneously engendered. When the husband and wife thus unite their vital forces, it is not because they at that moment wish to engender a child, but they do so because their feelings and desire have been aroused, and then having united, a child is engendered. From this fact that the husband and wife do not engender the child on purpose, it may be known that Heaven and Earth [likewise] did not [initially] engender man on purpose. Thus man was

---

25. It should be stressed once more (see also V.2 near beginning) that "macrocosm" and "microcosm" are purely Western terms, without Chinese equivalents, and are used in this book simply for the sake of convenience.

26. Wang Ch'ung reports this as a doctrine of his time in *Lun heng* 49 (16/10–11); tr. Forke (1962:2.363).

> [initially] engendered by Heaven and Earth just as are fish [today] in the depths or lice on human beings. The [initial] engendering [of man and other creatures] was brought about by the vital forces [of Heaven and Earth, but since that time] the various species have reproduced themselves [in the normal fashion]. This is the single reality for all things existing between Heaven and Earth.[27]

The same sort of analogizing, sometimes from the human world to nature and sometimes the other way, continued through the ages. Ko Hung argues that just as people are unable to know why their bodies may suffer pain or illness, so Heaven is unable to know why within the natural sphere omens or calamities may occur. "If man cannot ensure that his ears and eyes are ever acute and clear-sighted, then likewise Heaven cannot ensure that the sun and moon may not [sometimes] suffer eclipses."[28] In the eleventh century, one of the most famous ethical statements in Neo-Confucianism, Chang Tsai's *Hsi ming* or *Western Inscription,* rests on an analogy between human parents and Heaven and Earth, which are personified as two universal parents. One should serve these two parents (the universe) as one does one's own parents and should regard all people as one's siblings, because they, like us, belong to these universal parents (Fung 1953:493–495). In the seventeenth century, Wang Fu-chih (1619–1693) argued that just as the hands, feet, ears, eyes, and mental activities constitute a person, so the yin and yang and the Five Elements constitute Heaven (Fung 1953:642). And in the late nineteenth century, T'an Ssu-t'ung (1865–1898) maintained that just as the power of the brain unites the five senses and the bodily framework into a single organism, so the power of electricity unites Heaven, Earth, the myriad creatues, and human beings into a single organism (Fung 1953:694).

Analogies of this sort all reflect an anthropocentric point of view and so could contribute little to natural science. Although similar theories arose in Europe and long persisted, they failed to dominate philosophical thinking to the same degree. By 1600, or 1650 at the latest, they had all disappeared from scientific writings or continued only as occasional rhetorical survivals (Needham 1956:294, 298).

1g. The Wholly Receptive Approach

Under the rubric of the wholly receptive approach we at last reach the total union with nature that is the essence of Chinese mysticism and is

27. *Lun heng* 14(3/23); tr. Forke (1962:1.103). The analogy of Heaven and Earth with husband and wife occurs again in 54(18/1); tr. Forke (1962:1.92).

28. *Pao-p'u-tzu* Inner ch. 7; tr. Ware (1966:126).

probably what most people have in mind when they talk about Chinese "love of nature." Its supreme example in philosophy is "Chuang Tzu," meaning the several authors who wrote the mystical passages in the book bearing his name. Its fundamental credo is the presence of the Tao in all things, including even the meanest and most repulsive. In a series of answers to a persistent questioner, Chuang Tzu points out that the Tao is to be found in ants (a humble form of animal life), in weeds (a humble form of plant life), in tiles (a humble form of inanimate matter), and even in excrement (a repulsive form of inanimate matter) (*Chuang-tzu* 22; tr. Watson 1968:240–241).

Taoist mysticism was "the only system of mysticism which the world has ever seen which was not profoundly anti-scientific."[29] It did not practice asceticism nor did it distrust nature. Chuang Tzu's profound identification with nature is expressed in many passages. One of the most magnificent is his description of the diverse sounds of the wind in the mountain forest which constitute a cosmic symphony (*Chuang-tzu* 2; tr. Watson 1968:36–37, Graham 1981:48–49). In other episodes he joyously identifies himself with various forms of life, as in his famous walk with Hui Shih, the dialectician, along the bank of the Hao River, in the course of which he tells Hui Shih how happy are the minnows as they dart through the river. "You're not a fish," says Hui Shih, "so how do you know what fish enjoy?" "You're not I," responds Chuang Tzu, "so how do you know I don't know what fish enjoy?" And after a further interchange, Chuang Tzu insists that he is able to know the feelings of the fish simply by standing there beside the river (*Chuang-tzu* 17; tr. Watson 1968:188–189).

Most famed of these identifications is Chuang Tzu's butterfly dream, from which he awoke not knowing whether he was now Chuang Tzu who had dreamt he had been a butterfly, or a butterfly now dreaming that he was Chuang Tzu (*Chuang-tzu* 2; tr. Watson 1968:49, Graham 1981:61). There are other similar passages as well, as when Confucius is made to say to his disciple Yen Hui: "How do you know that this 'I' we talk about has any 'I' to it? You dream you're a bird and soar up into the sky; you dream you're a fish and dive down in the pool" (*Chuang-tzu* 6; tr. Watson 1968:88, Graham 1981:91). Or there is the story of Mr. T'ai, an ancient worthy who slept peacefully, awoke blank-eyed, and "sometimes thought he was a horse, sometimes a cow."[30]

The same theme continues among the later philosophical Taoists. In the *Lieh-tzu* we are told that between diverse living creatures there are really few great intellectual differences. Animals, like human beings,

---

29. Fung Yu-lan, oral statement, as quoted in Needham (1956:33).

30. *Chuang-tzu* 7; tr. Watson (1968:92). See also 13; tr. Watson (1968:150).

wish to preserve their lives, which is why male and female live in pairs, mother and child keep close together, and the herding animals keep their young within the herd and the adults outside. Very anciently, men lived together with animals and walked side by side with them, and even today, in a certain eastern realm, many people can be found who understand the speech of domestic animals (*Lieh-tzu* 2; tr. Graham 1960: 54–55). In contrast to this Taoist merging of the human and animal worlds, it is illuminating to compare Wang Ch'ung's more human-oriented viewpoint. For even as he stresses the many ways in which a human being is a creature like any other, he at the same time points to the human's cognitive faculty as a vital point of difference (sect. 1d).

Centuries of interaction with Taoism helped post-Han Confucianism to move slowly from its early negative view of nature toward one much more sympathetic. Probably the change was also helped by Buddhism which, despite its philosophical skepticism about the reality of the phenomenal world, nevertheless consistently emphasized—especially on the popular level—compassion for all living creatures as an ethical value. The resulting emotional warmth with which, beginning in the eleventh century, Neo-Confucian thinkers view the physical world around them is one of the profoundest psychological differences between them and the classical Confucians. The changed attitude is beautifully exemplified in the well-known anecdote of Chou Tun-yi, who refused to cut the grass growing outside his window because, as he said, he felt toward the grass as he felt toward himself. Chou was perhaps closer to Taoism than any of the other Neo-Confucians.[31] Yet his just-mentioned attitude was by no means unique. A somewhat bathetic parallel is the alleged remark by Chou's contemporary, Chang Tsai, that when he (Chang) heard the braying of a donkey, he too was affected by it just as Chou had been by the grass.[32]

Variations of the same theme appear among several other Neo-Confucians:

> Man is not the only perfectly intelligent creature in the universe. The human mind is the same as that of plants and trees, birds and animals. (Ch'eng Hao)[33]

---

31. His famous cosmological diagram, the *T'ai-chi t'u* (Diagram of the Supreme Pole), probably had a Taoist origin. See Fung (1953:438–447).

32. This and Chou's refusal to cut the grass are recorded in *Erh Ch'eng ch'üan-shu* (Complete Works of the Two Ch'engs), sect. *Yi-shu* (Surviving Works) 3/2a; tr. with further discussion in Chan (1963:302–303).

33. Ibid., *Yi-shu* 1/3a; tr. Chan (1963:527).

In identifying himself with Heaven and Earth, sun and moon, the four seasons, the yin and the yang, the great man identifies himself with the Tao. (Ch'eng Yi)[34]

The universe [lit. Heaven and Earth] has never separted itself from man. It is man who separates himself from the universe. (Lu Chiu-yüan)[35]

Chu Hsi's attitude toward nonhuman life appears in an interchange with a questioner who asks him: "Man and birds and animals all have consciousness, although with varying degrees of penetration or impediment. Do plants also have consciousness?" Chu replies that they do, and cites how potted flowers will flourish if given water but wither if they are crushed down. However, he continues, the consciousness of plants is inferior to that of animals, which in turn is inferior to that of human beings. Rhubarb, for example, acts as a purgative when eaten, because its consciousness acts in one direction only. Even decayed things have consciousness, as shown by the fact that when burned into ashes and made into a broth and drunk, they will be caustic or bitter.[36] Chu Hsi's assertion here that plants and even inanimate objects possess "consciousness" *(chih-chüeh)* is very similar to his imputation of will or purpose to water when he says elsewhere (see sect. 1e) that it is the "nature" (i.e., the will) of water to flow downward.

The Taoist mystical vision of nature is summed up most graphically in the landscape paintings of the great Sung artists, even though most of them would probably have thought of themselves as belonging primarily to the Confucian tradition.[37] But Taoist naturalism was also an important influence on Ch'an (Japanese:Zen) Buddhism (in many respects a Buddhist version of Taoism), with its emphasis on closeness to nature, love of living things, and simplicity of living.

Whether this total approach to nature was really helpful to the development of natural science is one of several topics to be considered in the next subsection. Before we turn to this, however, I cannot refrain from quoting a relevant remark made by my friend, the late George N. Kates, at a conversation in Chinese during a dinner in a Philadelphia restaurant sometime in the winter of 1946–1947. Kates addressed his remark to the famed Chinese philosopher Fung Yu-lan, then a visiting professor at the University of Pennsylvania. In the West, Kates pointed

---

34. Ibid., *Yi-shu* 1/7b–8a; tr. Chan (1963:571).

35. *Hsiang-shan ch'üan-chi* (Complete Works of Lu Hsiang-shan) 34/5b; tr. Huang (1944:55) and Chan (1963:582).

36. *Chu-tzu ch'üan-shu* 42/31b–32a; tr. Chan (1963:623).

37. The artistic theories of the Sung and later literati painters are described in Bush (1971).

out, major gardens are often beautified by fountains whose jets of water gush high upward. In any Chinese garden, on the other hand, such fountains would be inconceivable because what makes water gush upward is human artifice, and this artifice effectively destroys the garden's harmony with nature.

Of course, Kates was profoundly right. As we have seen, Chu Hsi refers to the tendency of water to flow downward as its "nature"—an idea that eventually goes back to the debate on human nature between Mencius and Kao Tzu (*Meng-tzu* 6a.2). In this debate, Kao Tzu argues that human nature is as indifferent to good and evil as is water to whether it will flow eastward or westward. In reply, Mencius asks whether water is equally indifferent to flowing downward or upward? By striking the water, he continues, a man may cause it to leap above his head, but this is the result of external force and not of the water's own inherent nature. In the same way, a man may be caused to do evil, but such action is then the result of the force applied to him and not of the man's own inherent nature.

1h. Nature, Mankind, and Science

By now it is evident that the seven suggested approaches to nature are by no means mutually exclusive. Overlaps can readily be found, as can instances in which quoted thinkers appear under more than one rubric: Hsün Tzu, for example, under the first and second; Tung Chung-shu under the first, third, and sixth; Wang Ch'ung under the fourth and sixth; Ko Hung under the second, fifth, and sixth. Despite this looseness, I believe the seven rubrics provide a useful framework for indicating general tendencies.

The first approach (antagonistic/indifferent) could, by definition, play only a negative role in Chinese scientific development. It characterized much of pre-Han Confucianism but fortunately gradually gave way thereafter to a much more positive attitude.

The second approach (exploitative/utilitarian) was also, in my opinion, largely negative, even though it could perhaps lead to some development of practical technology for exploitative purposes and in some instances may have provided intellectual encouragement to the researches of the alchemists.[38] Most positively, it could sometimes lead to measures for the conservation of natural resources. On the whole, however, it is probably fortunate that this approach was rarely prominent in Chinese thinking. In the Western world, where it was much

38. See Ko Hung's dictum about "putting all things to one's service" (sect. 1b at n. 10).

more prominent, the disastrous results of the wanton exploitation of natural resources are today only too apparent.

The third approach (theistic/anthropocentric) was dominant in the Western world, where its role, though very complex, may have been crucial to the achievement of modern science.[39] In China, however, this kind of thinking, though not totally unknown in early times, was never common, so that its positive consequences for our subject were insignificant.

The fourth approach (naturalistic/analytical) will be put aside for the moment in favor of more extended consideration shortly.

The fifth approach (animistic/moralistic), though not conspicuous in most Chinese philosophical writings, was basic to much folk religion as well as to the official state cult. The influence of this kind of thinking on scientific development, both in China and the West (where too its effects persisted more strongly than is sometimes realized), must be regarded as strongly inhibitory.

The sixth approach (semireceptive) divided the universe into a macrocosmic world of nature and a microcosmic world of humanity. This worldview had its roots in early prephilosophical Chinese notions. Thereafter it became prominent in Han thinking and remained significant down to fairly recent times. Although seemingly going a long way in the direction of the sympathetic acceptance of nature, this approach was really anthropocentric because of its much greater implicit interest in the world of human beings than in the world of nature. Such analogical thinking also existed in Europe, where its effects on science during the sixteenth and early seventeenth centuries may have been greater than commonly supposed (Needham 1956:298). Whether these effects were really favorable remains a question, but in China, where thinking of this kind was both more pervasive and longer lasting, it is difficult to see how the effects for Chinese science could have been other than harmful.

The seventh or wholly receptive approach differs greatly from the sixth in that it is totally nature-centered. On several counts it may be said to be "scientific" in spirit: its denial of anthropocentrism, recognition of the infinite relativity of all things, acceptance of even seemingly insignificant or repulsive things as worthy of study, and rejection of human ethical judgments when conducting such study.

On the other hand, this total approach is also incompatible with science on several counts: its distrust of reason, exaltation of intuition, dislike of measurement and classification, and readiness to obliterate all distinctions in order to achieve a state of ineffable oneness with nature.

---

39. This is a central theme in the study by Hooykaas (1972).

For the philosophical Taoists, any attempt at rational analysis would destroy the vision of the universal Tao. Total acceptance and absorption, rather than the retention of human standards, were essential if the mystical experience was to take place.

Needham (1956:89–98) has discussed at some length the seeming relationship of mysticism, especially nature-mysticism, to the rise of experimental science in Europe during the sixteenth and seventeenth centuries, as well as its progressive social role in the political struggles of the same period. On the scientific side its influence is evident in men like the Flemish chemist John Baptist van Helmont, Paracelsus, and Francis Bacon. The rise of science was helped during that time by a peculiar combination of circumstances: an alliance between religious mysticism and experimental empiricism as joint agents of change against the, by then, conservative bastions of theological philosophy and Aristotelian rationalism. A significant factor in the alliance was the importance attached by mystics and empiricists alike to manual operations, in contrast to the Aristotelians and Thomists who, like the Confucians, had no sympathy for or interest in such operations.

In China, however, this combination of circumstances was lacking. The nature-mysticism that had originated in Taoism, and from there flowed into Confucianism, was responsible for many of the greatest achievements in Chinese art and literature. It produced a world outlook both beautiful and profoundly "wise" in the best sense of that much abused term. In the absence of other factors, however, such as existed in Europe, it could not be expected by itself to produce a modern science.

This leaves us with the fourth or naturalistic/analytical approach as a sober middle course between the underinvolvement or overinvolvement of the other six approaches. This middle course implied, theoretically speaking, a rejection of supernaturalism and anthropocentrism, but also a refusal to merge humans with nature at the expense of those qualities that affirm their distinctiveness as rational beings. The words "theoretically speaking" are added advisedly because even such a great naturalistic thinker as Wang Ch'ung was so influenced by the ideas of his time that he believed in fate and astrology and, as we have seen, made use of the fallacious macrocosm-microcosm analogy to criticize views of his contemporaries.

Despite these flaws, I believe it was a major tragedy in the history of Chinese science that it had only one Wang Ch'ung. A few skeptics followed him at widely scattered intervals, but no continuing development of his naturalistic rationalism took place.[40] Essays were written from

---

40. What happened after Wang Ch'ung's time is described in Needham (1956: 386–389).

time to time attacking particular "superstitions" and fallacies, but nothing emerged comparable in scope, size, and iconoclastic vigor to Wang's *Lun heng*. And when the great flowering of rationalism took place during the seventeenth, eighteenth, and nineteenth centuries, it was only peripherally interested in nature and turned its major energies to fields of humanistic scholarship like history and textual criticism—this during the very centuries when the European explosion of natural science was taking place.

Wang's methodology was flawed in two important respects. One was his failure to match destructive criticism of the fallacies of others with positive theories of his own. The other—a common defect in Chinese philosophical writings generally—was his probably unconscious reluctance to record and define in sufficiently explicit detail the phenomena to which he addressed his attention. Had he made more deliberate effort in this direction, the effort, enhanced by his generally clear and straightforward style, would have established a model of scientific expository writing of highest value to successors.

This latter deficiency emerges very clearly if we compare Wang Ch'ung's presentation with, say, Aristotle's purely descriptive accounts of animal life as found, for example, in his *History of Animals*. Thus where Wang merely hints that he believes in the spontaneous generation of both lice and fish (see at n. 27), Aristotle writes very explicitly that lice are generated from the flesh of animals and that among fish a very few species are spontaneously generated from the mud and sand in ponds.[41] Or again, it is instructive to see how Aristotle describes the human foot:

> The hinder part of the foot is the "heel"; at the front of it the divided part consists of "toes," five in number; the fleshy part underneath is the "ball"; the upper part or back of the foot is sinewy and has no particular appellation; of the toe, one portion is the "nail" and another the "joint" . . . Men that . . . walk resting on the entire under-surface of their feet, are prone to roguery. The joint common to thigh and shin is the "knee". (Aristotle, *Historia Animalium;* tr. Thompson 1910:494ᵃ).

In spite of the inclusion of folk belief in the penultimate sentence, this passage is an exceptional example of detailed and precise scientific exposition, made all the more unusual because it concerns a portion of the human anatomy so familiar to everyone that most people would probably wonder why it need be described at all. The rarity in Chinese scientific literature of such precision and such acceptance of the commonplace as an object worthy of study must be recognized as a consid-

---

41. *Historia Animalium,* tr. Thompson (1910:556[b], 564[b]–566[b], 569[a]).

erable weakness. Its unfavorable consequences persisted, I believe, despite the special contributions Chinese science also received from a philosophical Taoism whose outlook was in some respects remarkably "scientific" and well ahead of contemporary European thinking. For a modern science to arise in Europe, it was no doubt imperative that the crystallized dogma inherited from Aristotle be swept away by experimental empiricism aided by religious mysticism. But without the spirit of scientific inquiry initiated much earlier by Aristotle and others, there would have been no basis in the first place for a modern science to arise.

Conspicuously missing both in Wang and Aristotle was the experimental method. The Chinese alchemists were willing to work with their hands when conducting their elixir experiments. It is tempting to think, therefore, that alchemical empiricism, combined with Wang Ch'ung's rationalistic tradition, might have contributed much to scientific progress. Yet so often was what the alchemists did imbued with supernaturalism and outright magic that it is difficult to see how such a marriage could have taken place. The gaps and difficulties are admirably indicated in the following statement by Leslie (1954:165):

> Wang Ch'ung's reasoning and metaphysics were eminently favourable to scientific research. A superior natural philosophy kept within bounds his own speculations . . . ; but the crucial factors of systematic observation and experiment were lacking. Experimentation was found among the Taoists, with their alchemical and physiological search for immortality; but unfortunately Wang's restraining logic was not heeded. In the West, the scientific revolt from the fourteenth to the sixteenth centuries, though associated with mysticism and anti-authoritarianism . . . , managed to combine rationalism with its empiricism to produce modern science. In China this combination was never adequately made.

While accepting this judgment, it seems to me that the tradition embodied in Wang Ch'ung, even considered by itself, carried enormous potential for a fruitful development of science. Unfortunately, Wang inspired no immediate followers who might have perpetuated and elaborated the tradition.

2. Organicism and Laws of Nature

2a. Did "Laws of Nature" Exist in China?

At long last we arrive at a topic discussed at great length by Needham in his *Science and Civilisation in China* (1956:518–583).[42] Therein he distin-

42. This subsection is largely though not entirely the same as Bodde (1979), which in turn is an elaboration of Bodde (1957b).

guishes between "natural law" (juridical law as based in the first instance on the accepted immemorial customs of a given society) and "laws of nature" (consisting of defined regularities operative in the world of nature, such as the law of gravity). Needham traces in some detail the stages of development within Western civilization whereby these two kinds of law emerged as differing aspects of an originally undifferentiated body of supposedly universal law.

The earliest clear-cut conception of the governance by law of the entire world—both in its natural and its human aspects—seems to have been Stoic, though Plato uses the term "laws of nature" once. The concept of a universal law applicable to all men and things alike was powerfully reinforced by the Judeo-Christian belief in God as a supreme lawgiver. It was systematized by Thomas Aquinas, who distinguished between a *lex aeterna* governing all nonhuman things for all time (this became the basis for the modern scientific concept of "laws of nature") and a *lex naturalis* valid for all men (this became "natural law"). By the seventeenth century the two kinds of law were completely differentiated and in the process secularized; with Boyle and Newton, the concept of laws of nature that are "obeyed" by chemical substances and planets alike became a commonplace.

From Europe Dr. Needham then turns to China to see if any similar evolution of juridical ideas has taken place. For this purpose some half dozen terms having the meaning of "law," "regulation," "rule," and so on, are put under analysis. However, this effort fails to reveal any clear-cut instances of the term in question being used in the sense of "laws of nature." Needham attributes this phenomenon primarily to the absence in Chinese thought of the idea of a supreme celestial lawgiver. Instead, he finds the dominant cosmological view to be that of a universe uncreated and self-sufficient, functioning by means of its own internal forces—what he has repeatedly referred to as the organismic concept of the universe. It cannot be doubted that these profound intellectual differences between East and West are of fundamental importance when considering the differing roads taken by science in China and Europe.[43]

---

43. The term *tsao wu che* (maker, founder, or originator of things), as well as one or two similar terms, occurs from time to time in Taoist philosophical writings. See, for example, *Chuang-tzu* 33; tr. Watson (1968:373) and Graham (1981:283). Also Kuo Hsiang, commentary on *Chuang-tzu* 2 end; tr. Fung (1953:210). In such cases, however, *tsao wu che* and/or its parallels are usually used in a metaphorical manner solely for literary effect; or, more rarely, they occur in passages written to disprove the idea that any creative deity such as the Maker of Things *(tsao wu che)* is needed to produce the universe. This is the purpose, for example, of the Kuo Hsiang commentary mentioned above. *(continued next page)*

Since Needham's discussion of this topic in 1956, however, nine passages have come to light that in varying ways suggest the existence in early China of ideas not too far removed from those that in Europe led to the developed concept of "laws of nature." These passages therefore require discussion here. Two of them (nos. 4 and 6 below) have already been presented at length in Bodde 1957b, to which the reader is referred for details beyond those given here.

Passage 1. The first of the nine passages comes from the nineteenth chapter of the *Mo-tzu* (late iv or early iii cent. B.C.; tr. Mei 1929:112):

> When it came to [the time of] King Chieh of the Hsia, Heaven had its harsh commands *(ming):* that sun and moon should be untimely *(pu shih),* that cold and heat should arrive irregularly *(tsa chih),* that the five grains should die of drought, that demons *(kuei)* should shout within the capital, and that cranes should cry out for ten days. Heaven then commanded *(ming)* T'ang in the Piao Palace to assume and exercise the Great Mandate *(ta ming)* of Hsia.

Here is Mo Tzu's version of how the Mandate of Heaven was transferred, traditionally in 1766 B.C, from Chieh, the allegedly tyrannical last ruler of the Hsia dynasty, to T'ang, founder of the Shang. That Heaven commands *(ming)* mankind, and especially mankind's appointed rulers, to conduct themselves in various ways, and that it also punishes wrongdoers who violate its commands, is a commonplace in early texts such as the *Shu ching* and *Shih ching.* But it is exceptional for Heaven also, as here, to command the objects and forces of nature to change their usual patterns of operation. In European thinking it was only in the sixteenth and seventeenth centuries, coincident with the rise of modern science, that the idea of God's reign over the world gradually ceased to be a consideration of the exceptions in nature (the comets and monsters that had disturbed medieval equanimity) and came to be iden-

---

Inspired by these Taoist precedents, a few prominent T'ang literati like Han Yü and Po Chü-yi also very occasionally use the term *tsao wu che* or its parallels in their writings. When they do so, their purpose is readily explainable as being equally metaphorical and nonliteral. Schafer (1965), however, in his study of this phenomenon, oddly fails to recognize this interpretation. Instead he concludes that the appearance of *tsao wu che* (or parallel terms) in T'ang literature points to a minority T'ang belief in a Fashioner *(tsao wu che)* whose supernatural skill acts continuously and timelessly in the shaping and molding of matter—this against the dominant Chinese philosophical belief that all things in the universe are self-generated and hence have no need for an external, supernatural agent. Schafer's conclusion is sharply contradicted by the texts adduced by Lamont (1973–1975) in his survey of pre-T'ang, and especially T'ang, opinions about the relationship between Heaven and man.

tified with its unvarying rules (Needham 1956:542). In the above *Mo-tzu* passage, though no laws of nature are actually mentioned, there seems to be a Chinese parallel to the pre-sixteenth century European approach.

Passage 2. The second passage, too, comes from *Mo-tzu* (27; tr. Mei 1929:145):

> Moreover there is a basis for my knowing how generously Heaven loves the people. It is the fact that it [Heaven] has orbitted[44] the sun, moon and stars in order to illumine and lead [the people]. It has instituted *(chih)* the four seasons of spring, autumn, winter and summer in order to guide and untangle them.[45] It sends down snow, frost, rain and dew to grow the five grains, hemp and [mulberry leaves for] silk, thereby enabling the people to gain and be benefited by them. It has laid out the mountains, streams, gullies and valleys, and allocated various activities [for people to carry out in these various terrains], so that it may oversee the people's goodness or lack of goodness [on the basis of how they carry out these activities.][46] It has created kings, dukes, marquises and earls, causing them to reward the worthy and punish the violent. It depletes its metals, woods, birds and animals, and allows the five grains, hemp and silk to be worked upon, so as to provide materials for the people's clothing and food.

This is the same passage already cited in sect. 1c to illustrate the theistic/anthropocentric approach to nature. In it one sees Mo Tzu's personal Heaven acting consciously to place the heavenly bodies in their orbits and to "institute" the four seasons, while at the same time creating human institutions, all for the good of mankind. It is not surprising that Mo Tzu's strong theism should thus lead to statements reminiscent of those which in European thinking led to the formulation of "laws of nature," even though Mo Tzu never uses the term.

Passage 3. A passage in the *Lü-shih ch'un-ch'iu* (22; tr. Wilhelm 1928:56–57) identifies Supreme Oneness *(t'ai yi)* as another name for the Tao and says of it: "Supreme Oneness produced the Dual Forms, and the Dual Forms produced the Yin and Yang." Here the "Dual Forms" *(liang yi)*

---

44. Emending *mo* 磨 to *li* 歷.

45. "Guide and untangle" is an inadequate rendition of *chi kang,* which as a noun means "net" or "nexus" of natural causation, and is one of the technical terms discussed by Needham (1956:554–556) in his search for Chinese equivalents of "laws of nature."

46. This rather obscure sentence seems to mean that Heaven has allocated various occupations, such as farming, hunting, fishing, foresting, and the like, to the people to perform according to the varying kinds of terrain Heaven has provided for them.

are probably Heaven and Earth.[47] The text goes on to describe how from the permutations of the yin and yang the many aspects of nature came into existence. Then, after apostrophizing the Tao in rhymed clauses very similar to descriptions of it in the *Lao-tzu,* the text says again of Supreme Oneness: "Therefore Oneness instituted ordinances which the Dualities follow and obey." In other words, the Tao, under the name of Supreme Oneness, here replaces Mo Tzu's personal Heaven as the originator of the universe and then institutes *(chih)* ordinances or orders *(ling)* for Heaven and Earth (here thought of as physical entities) to obey.

Passage 4. The *Kuan-tzu* contains an interesting definition of *tse,* "rule"—one of the key terms examined by Needham (1956:559–562, 565ff.) in his search for Chinese "laws of nature." The passage reads:

> What are basic to the [yin and yang] vital forces *(ch'i)* of Heaven and Earth, to the harmonious balance between cold and heat, to the properties of water and soil, to the existence of human beings, birds, animals, plants and trees; and which things, despite their extreme abundance, all possess as standards *(chün),* yet which never undergo change [themselves]—such are called "rules" *(tse).* (*Kuan-tzu* 6[2/1b]; tr. Rickett 1985:128)

Here we are told that *tse* are basic to human, animal, and plant life, to climatic phenomena and inorganic matter. All things, despite their multiplicity, possess these *tse* as standards, and yet the *tse* themselves never undergo change. Unfortunately, the text fails to indicate how these *tse* themselves originate. Are they imposed on natural things and phenomena by a transcendent being or power acting as a legislator for created beings? If so, we seem to be in the presence of a concept close to what in the West would be called "laws of nature." Or, conversely, are the *tse* perhaps thought of as internal rules which the things and phenomena enumerated in the text obey simply because the *tse* are proper to the natures of these things and phenomena?

No conclusive decision can be made between these two alternatives. However, it should be pointed out that in the purely human sphere, the word *tse,* as used in other passages, commonly refers to a man-made rule or law; and in the few instances in which it occurs in the term *t'ien tse* ("rule of Heaven"), it seems to signify the rules or laws that Heaven has promulgated for the human (not the natural) world to follow. In either case it is a norm or standard imposed from above upon those sub-

---

47. See Fung (1952:384), which, however, also cites a closely parallel passage in the Great Appendix of the *Yi ching* (tr. Legge [1899b:373] and Wilhelm [1950:1.342]), wherein the Dual Forms are perhaps themselves to be identified as the yin and yang.

jected to it. This seems to suggest that in the present *Kuan-tzu* passage, in which the *tse* are more broadly referred to as basic to the nonhuman as well as human world, the word has a universality and objectivity such as in the human world would be attributed to codes of law. On the other hand, if the passage simply has to do with internal rules proper to the natures of the beings and things specified, one wonders why some other term less legal than *tse* was not used. More appropriate, one would think, would be words like *hsing,* "nature," especially "human nature," or *ch'ing,* "quality."

Passage 5. This is another *Kuan-tzu* passage, remarkable because in it the word *li* ("pattern," "principle of organization"), which is of such central importance in the dominant Chinese organismic view of the cosmos, is unexpectedly correlated with human *fa* ("law") in a typical macrocosmic-microcosmic analogy. Ordinarily, of course, *li* would not be equated with *fa* in this way. The passage (ch. 64) reads:

> Heaven covers over the myriad creatures.[48] It regulates *(chih)* heat and cold. It moves *(hsing)* the sun and moon. It sequentially arranges[49] the stars. Such are Heaven's regularities *(ch'ang).* It governs *(chih)* these matters by means of its principles *(li),* starting them anew when they have reached their end.[50]
>
> The ruler shepherds his myriad people. He governs *(chih)* all-under-Heaven. He supervises his hundred officials. Such are the ruler's regularities *(ch'ang).* He governs *(chih)* these matters by means of his laws *(fa),* starting them anew when they have reached their end.

The text goes on to list a series of paired "regularities" *(ch'ang),* such as those of father and mother, ministers and their inferiors, son and his wife. It then concludes:

> Therefore so long as Heaven does not disregard *(shih)* its regularities *(ch'ang),* cold and heat will have their proper season; sun, moon and stars will have their proper order. . . . Therefore when these regularities are made to function, there is good order; when they are disregarded, there is disorder. Heaven has never yet changed that whereby it governs *(chih).* That is why it is said that Heaven does not change its regularities *(ch'ang).*[51]

---

48. As a bird covers and shelters with its wings. See *Shih ching* no. 245.

49. *Tz'u,* used as a verb (lit. "it sequences . . .").

50. That is, the movements of sun and moon and other heavenly bodies operate cyclically and are started anew by Heaven at the end of each cycle.

51. *Kuan-tzu* 64(20/1b–2a); tr. Rickett (1965:122–123), where the significance of this passage with regard to "laws of nature" has already been pointed out. The final sentence is a repetition of a sentence in *Kuan-tzu* 2, on which this whole passage in 64 is a commentary.

In later cosmological thinking, especially of the Neo-Confucians, *li,* "principles," are usually thought of as inherent patterns of organization for the objects, beings, and forces to which they pertain. In the present anonymous passage, however, probably belonging to the first century B.C. (Rickett 1965:121), they function as instruments used by Heaven to maintain its regularities *(ch'ang)* in such matters as the alternation of cold and heat, movements of sun and moon, and orderly sequence of the stars. By analogy, the ruler is pictured as similarly using his *fa,* "laws," to maintain his regularities in his governing of the human world. Probably the reason why *fa* is used here only to refer to human government and not to Heaven's administrative activities is the traditional prejudice (especially powerful among Confucians) against the whole idea of *fa* or positive law. This prejudice was intensified by the harsh way in which the Legalists had used *fa* to establish a powerful centralized state (the Ch'in empire) in 221 B.C. Yet despite this reluctance to apply the word for human law to the world of nature, it would seem that the author of this *Kuan-tzu* passage comes rather close to what in European thought would be called "laws of nature."

Passage 6. Our sixth example, a rather lengthy passage from the *Huai-nan-tzu* (5/18bff.; ca. 140 B.C.), has already been translated in Bodde (1957b:714–720) and then retranslated with considerable changes in Needham (1962:15–17). Most of the chapter to which it forms the conclusion is identical in wording, or nearly identical, with the calendrical text known as the *Yüeh ling* (Monthly Ordinances), originally a part of the *Lü-shih ch'un-ch'iu* (240 B.C.) and then reproduced in the *Li chi* (Record of Ceremonial). The *Huai-nan-tzu*'s fifth chapter, following the wording of the *Yüeh ling* (a product of Five Elements thinking), describes month by month the natural phenomena and human behavior proper to each month and the dire consequences (snow in summer and the like) if the behavior prescribed for one month is practiced instead in another.

The chapter bears the suggestive title *Shih tse hsün,* "Teachings on the Rules for the Seasons," about which Kao Yu (fl. 205–221) comments: "*Tse* (rules) are *fa* (laws). They are fixed laws *(ch'ang fa)* for the four seasons, for cold and heat, and for the twelve months. Hence they are spoken of as 'rules for the seasons,' [which phrase] is accordingly used as the title for the chapter."[52]

The conclusion of the passage, which is what concerns us and differs from the *Yüeh ling,* deals with the regulating *(chih)* and measuring *(tu)* of the yin and yang, that is, of meteorological phenomena, as carried out

52. It should be noted that *ling,* "ordinance," in the corresponding *Yüeh ling* title, likewise has a legal connotation.

by Heaven, Earth, and the four seasons. These six powers are respectively symbolized by the carpenter's marking line (Heaven), the water-level (Earth), the drawing compass (spring), the balance (summer), the carpenter's square (autumn), and the steelyard (winter).[53] Acting in the manner of these six kinds of measuring instruments, Heaven, Earth, and the four seasons are each said, with much poetic imagery, to "regulate" and "measure" the yin and yang, and to align, even, and otherwise fit all things of the world into their proper relationships. In this way, says the text, the seasonal movements of natural phenomena will be made to occur smoothly and without hitch.

Two sentences in particular are of crucial importance but are unfortunately ambiguous. At the end of the several sentences describing the work of the heavenly marking line, the text reads: "This is why *shang ti* uses [or used] it [the marking line] as the progenitor *(tsung)* of things."[54] Again at the end of the sentences describing the operations of Earth's water-level there is a parallel sentence: "This is why *shang ti* uses [or used] it [the water-level] as the equalizer of things."

The question is whether *shang ti* is to be translated here as the name of a unitary deity, "Lord on High," which is the meaning the term regularly had in early Chou texts, or should be pluralized and secularized to mean "the rulers of old" (taking *shang,* "above or on high," in the secondary sense of "ancient"), which is a meaning very occasionally attested for it in Han texts. The former interpretation suggests that the whole passage describes the activities of a personal divine engineer, the "Lord on High" of ancient Chinese belief, who uses Heaven and Earth (and by analogy the four seasons, though this is not explicitly stated) as his instruments for operating the universe. This interpretation brings us close to the idea of "laws of nature."

The second interpretation destroys this idea, because it means that these functions were performed by the "rulers of old," that is, by human beings, even though ones of more than ordinary human qualities. Semantically, both interpretations are possible, in that *ti,* originally the name of a divinity, had by late Chou times also come to be a title applied to human rulers of exceptional powers ("emperors").

Shang ti, in the sense of "rulers of old," occurs thrice in the Han medical text *Huang-ti nei-ching su-wen,* where the *shang ti* or "rulers of old" are said to have been interested in taking the pulse and performing other

---

53. See also III.4 at n. 69, where this same passage was briefly cited as an example of quantification in classical Chinese thought.

54. There seems to be an echo here of *Lao-tzu* 4, which says of the Tao that "it is as it were the progenitor *(tsung)* of the myriad creatures *(wan wu)*." Cf. tr. Waley (1934: 146): "the very progenitor of all things in the world."

medical tasks.[55] In early Chou texts, on the other hand, *shang ti* was indubitably the name of a unitary deity. The question is whether this original meaning was still current as late as the second century B.C., when the *Huai-nan-tzu* was compiled. To reach an answer, it will be helpful to study the history of another longer title in which the words *shang ti* appear.

Sometimes in early Chou texts *shang ti* was coupled with *t'ien,* Heaven, to form the compound name of what was still regarded as a single deity, the well-known *huang t'ien shang ti* or "August Heaven Lord on High." By the end of the Chou this lengthy title had no doubt become somewhat archaic. Yet it continued to be used occasionally both then and in Han texts, and when this happens there can be no doubt that *shang ti* means only "Lord on High," never "rulers of old."

Huang t'ien shang ti occurs twice in this way in the *Yüeh ling,*[56] and therefore also in the corresponding places in the *Huai-nan-tzu*'s fifth chapter, preceding the passage in which we are interested. The same title also appears in a proclamation issued by Wang Mang in A.D. 7 in response to the rebellion of Chai Yi; the *Ch'ien Han shu* (History of the Former Han Dynasty) explicitly says of this proclamation that it was archaistically modeled on one of the chapters in the *Shu ching.* Later in the same proclamation the term *shang ti ming,* "mandate of *shang ti,*" likewise occurs and is there glossed by the T'ang commentator Yen Shih-ku (581–645) as equivalent to *t'ien ming,* "mandate of Heaven." There can be no doubt, therefore, that when Wang Mang refers in his proclamation either to *shang ti* or *huang t'ien shang ti,* he has in mind only the "Lord on High" of antiquity, not the secular "rulers of old."[57]

Although an absolute judgment is impossible, the weight of the above evidence favors the interpretation of *shang ti* in our passage as "Lord on High" rather than the much less often mentioned "rulers of old." And this in turn confirms the interpretation of the passage as having to do with a supreme deity who guides and controls natural phenomena by means of measuring instruments. Admittedly this is not quite the same as a deity who actually legislates such phenomena. Nevertheless, a similarity undoubtedly exists, just as the words "law" and "measure" are semantically linked by the fact that every law has a certain quantitative

---

55. *Huang-ti nei-ching su-wen* 9(3/25a) and twice at 13(4/31a); cited in Needham (1962:15). Under the first reference, the commentator Wang Ping (eighth century) states very explicitly: "*Shang ti* [here] means the lord-rulers *(ti chün)* of high antiquity."

56. Under the last months of summer and winter; tr. Legge (1885:1.278, 309) and Couvreur (1913:1.367, 408).

57. For the proclamation, see *Ch'ien Han shu* 84/4a–b.

aspect (Needham 1956:553). Such a view is further strengthened by the title (mentioned earlier) of the *Huai-nan-tzu*'s fifth chapter, "Teachings on the Rules *(tse)* for the Seasons" and also by Kao Yu's gloss of *tse* as equivalent to *ch'ang fa,* "fixed laws." Both he and the author of the chapter (or at least the editor who devised its title) obviously recognized a close connection between using measuring instruments to control natural phenomena and using rules or laws to do the same.

Passage 7. This passage is gratifyingly brief. In his *Ch'un-ch'iu fan-lu,* Tung Chung-shu applies to the yin and yang an administrative word, *chih,* which in the sense of "regulating" or "instituting" has already been encountered several times in the passages under discussion (2, 3, 5, and 6). The sentence in question reads: "That the yang is noble and the yin mean: this is Heaven's regulation *(chih).*"[58] Here again it is conceivable that Tung, as a Confucian, preferred the word *chih* to *fa,* "law," because of the Confucian prejudice against the latter word. Actually, however, the two terms are close in meaning in this particular context. It is not surprising that Tung, whose view of the universe was strongly colored by the macrocosmic-microcosmic analogy (see sect. 1f), should cite the alleged cosmic inferiority of the yin to the yang to justify the actual inferiority in Chinese society of woman to man.[59] Because of his prevailingly personalistic view of Heaven (see sect. 1c), it was easy for him to imagine the cosmic yin-yang hierarchy as having been institutionalized by Heaven.

Passage 8. In A.D. 6, after Wang Mang established himself as Acting Emperor, a memorial was submitted to him that included the words: "You have established the Pi Yung and set up the Ming T'ang to propagate the laws of Heaven *(t'ien fa)* and to spread the influence of the sages."[60] The Pi Yung or Hall of the Circular Moat was closely associated in Han times with the Imperial Academy, and consisted of an open-air structure where, on occasion, the emperor himself might deliver a lecture on scholarly matters. The Ming T'ang or Cosmic Hall was a building consisting of rooms corresponding to the months of the year and oriented around a central axis so as to face the compass points that corresponded to the months. At monthly intervals, within each appropriate room, the emperor, following the prescriptions laid down in the *Yüeh ling* and clad in colors appropriate to the particular season,

58. *Ch'un-ch'iu fan-lu* 46(11/15a). For *chih* (regulation) some editions read *hsing* (punishment), which yields no sense.

59. See Fung (1953:42–43), citing especially *Ch'un-ch'iu fan-lu* 53.

60. *Ch'ien Han shu* 99A; tr. Dubs (1938–1955:3.227).

allegedly performed the ceremonies designed to accord with the cosmic conditions of that month.

It is impossible to know for certain whether the "laws of Heaven" for whose propagation the Hall of the Circular Moat and the Cosmic Hall are said in the memorial to have been established were thought of only as Heaven-given norms for human beings to follow or also included celestial regulations governing natural phenomena. Mention of the Cosmic Hall, however, suggests that possibly the latter as well as the former were intended. The only other textual reference to *t'ien fa,* "laws of Heaven," known to me is less ambiguous. It occurs in the *Tso chuan* history under the year 516 B.C., where there is little doubt that it relates to human affairs only and therefore is comparable to "natural law" rather than to "laws of nature" in the scientific sense.[61]

Passage 9. The final example occurs in Ko Hung's *Pao-p'u-tzu,* where we read (Inner ch. 10; tr. Ware 1966:168):

> The Tao serves internally to control the body and externally to conduct the state. It is able to order *(ling)* the Seven Agents [*ch'i cheng,* i.e., sun, moon, and five planets] to hold to their degrees of measurement *(tu),* the two vital forces [*ch'i,* i.e., the yin and yang] to be in harmonious accord, the four sea-

---

61. *Tso chuan* Chao 26; tr. Legge (1960:5.718) and Couvreur (1914:3.415). Cited in Needham (1956:547). Since the above was written, two further references to *t'ien fa,* "laws of Heaven," have come to my attention:

(1) The Han statesman K'uang Heng, in a memorial submitted to Emperor Ch'eng in 32 B.C., quotes a sentence (now no longer extant) from the *T'ai shih* (Grand Declaration) section of the *Shu ching.* See *Ch'ien Han shu* 25B/3b. The sentence reads: "Correctly observe antiquity, establish achievements, and establish affairs; thereby may Heaven's great statute *(lü)* be received for untold years." The commentator Yen Shih-ku (581–645) glosses *lü,* "statute," as *fa,* "law," and therefore explains the latter part of the sentence as meaning: "Thereby may Heaven's great law *(fa)* be received." The fact that human beings (and not the creatures and things of the nonhuman universe) are to receive this "statute" or "law" is enough to indicate that it has nothing to do with "laws of nature." The same passage and gloss are cited in the T'ang Code of 654; tr. Johnson (1979:51), where by typographical error the words *lü* and *fa* are interchanged.

(2) The *T'ai-p'ing ching* (Canon of Grand Peace), a text of religious Taoism possibly dating from the second century A.D., though many scholars regard it as considerably later (see III.3 n. 50), contains the following passage in its thirty-sixth chapter: "They [the sages] did not dare to deviate however slightly from the laws of Heaven *(t'ien fa).*" See Kaltenmark (1979:22) (where, however, *t'ien fa* is rendered as "celestial norm," not "laws of Heaven"). Here, as in the preceding instance, the fact that it is human beings (the sages) who are said to conform makes it once more evident that *t'ien fa* does not, in this sentence, signify "laws of nature."

sons not to lose [their proper times], cold and warmth to keep their limitations,[62] and wind and rain not to act with violence.

Ko Hung's approach to nature was diverse and unusual. He believed strongly in animism (see sect. 1e, n. 18) and was confident that the vitalistic forces of nature could be controlled or changed by human intervention. But he also began his book with a rhapsodic invocation of the Tao as "the Mystery,[63] the first ancestor of Spontaneity *(tzu jan),* the great progenitor[64] of the myriad different things" (Inner ch. 1; tr. Ware 1966:28). In view of his wide range of thinking, it should not surprise us to find him likewise affirming the power of the Tao to order the yin and yang and various other astral and meteorological forces all to keep to their courses and otherwise function harmoniously. This seems a rather close approach to the idea of "laws of nature" even though the term itself is lacking.

The foregoing nine passages come from seven different sources covering a total time span of perhaps seven centuries (roughly fourth century B.C.–ca. A.D. 320). In only one of them (the eighth, memorial to Wang Mang of A.D. 6) does *t'ien fa,* the literal equivalent of "laws of nature," occur, and its exact significance there is uncertain. In other passages, however, several supreme agencies—Heaven *(t'ien),* the Lord on High *(shang ti),*[65] or the Tao (also referred to as Supreme Oneness)—are said by means of their commands or orders, institutions or regulations, or simply through their own direct action, to cause various celestial bodies (the sun, moon, and stars) to follow their proper orbits, the yin and yang to operate harmoniously, the four seasons to come in due succession, heat and cold and other meteorological phenomena to be equable, and the five grains to be available for human needs. In one of the passages (the sixth, from *Huai-nan-tzu*), the Lord on High is said to use Heaven, Earth, and the four seasons (respectively symbolized by the carpenter's marking line, the water-level, drawing compass, and other measuring tools) as the agencies for carrying out various cosmic

---

62. Curiously, there is no verb in this clause, which literally reads: "the limitations of cold and warmth." However, the meaning is surely that cold and warmth should not transgress the levels of intensity expected of them during the normal course of the year.

63. *Hsüan che.* In *Lao-tzu* 1, the Tao is referred to in an indirect way as "the mystery of mysteries."

64. *Tsung,* the same word that appeared in the sixth passage above (from *Huai-nan-tzu*).

65. Accepting the interpretation of *shang ti* in passage 6 as "Lord on High" rather than "rulers of old."

functions. In another passage (the fifth, from *Kuan-tzu*) no supreme operator is mentioned, but *tse,* "rules," are described as basic standards for the functioning of the yin and yang, cold and heat, and animal and plant life, as well as humans.

Here a significant difference between Chinese and Western thinking should be noted. In Europe, prior to the secularization of the concept of "laws of nature" in the seventeenth century, the idea of a supreme deity who legislates cosmic phenomena seems to be never far away from that of a supreme deity who creates the universe. In the nine cited Chinese passages, on the contrary, the idea of creating is conspicuously lacking. The sole exception is the third (from the *Lü-shih ch'un-ch'iu*), in which the Tao, there renamed Supreme Oneness, is said to have "produced" (*ch'u,* lit. "put forth") the Dual Forms, namely Heaven and Earth. The relative weakness of the idea of creation in Chinese thinking, and thus the relative weakness of the idea of a truly all-powerful deity, is (as stated earlier) a probable major reason why the concept of "laws of nature" developed no further in China than it did.

Does the new evidence invalidate Needham's earlier conclusion that the concept of "laws of nature" was alien to Chinese philosophical thinking? I believe not, as far as the overwhelming bulk of Chinese philosophical writing is concerned. What it does oblige us now to recognize, however, is that in addition to the dominant viewpoint hitherto argued for by Dr. Needham, a minority viewpoint also exists, expressed by a very few early Chinese thinkers, which was a good deal more congenial to the ideas underlying the "laws of nature" than would initially be expected. On second thought it would, indeed, be strange if within an intellectual tradition as rich and varied as that of China, no trace whatever should ever have appeared of a concept that in the European environment was to prove so persistent and significant.

Not surprisingly, the embryonic beginning of "laws of nature" are particularly apparent among those relatively early thinkers—Mo Tzu, the Han Confucian Tung Chung-shu, alchemists such as Ko Hung—who thought in strongly theistic or animistic terms. On the popular level it is probable that such ideas long remained widespread. Remarkably, however, they failed to gain more than a temporary and minority position in the mainstream of Chinese philosophical speculation. Although traces of "laws of nature" may occur in philosophical writings after the time of Ko Hung, I have failed to come across them. In any event, one may doubt that they could long have survived the rise of Neo-Confucianism from the eleventh century onward. This new movement was methodologically unfavorable to natural science in that despite its sympathetic attitude toward nature, its major concern was overwhelmingly human. Nevertheless, Neo-Confucianism gave wide currency and respectability to a view of the universe that had originated among the early philosophical Taoists and had long been China's real *philosophia perennis,* but had not been clearly recognized as such by many thinkers. This *philosophia perennis* was the organismic view of the universe to which the final subsection will now bring us.

2b. Cosmic Organicism and Science

In the successive volumes of his *Science and Civilisation in China,* Joseph Needham has repeatedly used the word "organicism" to describe what he believes to have been the dominant Chinese view of the universe.[66] Before embarking on our own discussion of this subject, it is well to remind ourselves just what the term signifies. Here are two of Needham's descriptions:

> Chinese coordinative thinking . . . [conceived of] . . . an extremely and precisely ordered universe, in which things "fitted," "so exactly that you could not insert a hair between them." . . . But it was a universe in which

---

66. The two scholars who have read this book on behalf of the University of Hawaii Press before publication have commented on Needham's use of the word "organicism" to describe certain aspects of Chinese thinking, a usage that has been followed in the present book.

Professor David L. Hall (Department of Philosophy, University of Texas at El Paso) defines the word "organic" as "a whole with parts that functionally interrelate to achieve an aim or purpose." He then comments: "Certainly this rather standard Aristotelian, biologically grounded, concept of 'organism' doesn't fit the Chinese." To me, however, it very much calls to mind the biological metaphor used by the Neo-Taoist Kuo Hsiang (see V.2) to describe his ideal society. Such a society, for Kuo, consists of individuals ("ruler and subject, superior and inferior" and "servants") who spontaneously "assist each other like the hand and foot, the ear and eye, the four limbs and the hundred other parts of the body, each having his own particular duty and at the same time acting on behalf of others." Is this not "a whole with parts that functionally interrelate to achieve an aim or purpose"?

Professor Roger T. Ames (Department of Philosophy, University of Hawaii) observes that in Western philosophy, the word "organicism" has been "so dominated by Aristotle and his teleology" that it "evokes potentiality and actuality." In Chinese cosmology, on the other hand, he notes, such a distinction between potentiality and actuality is "problematic" and "the absence of cosmogonic beginning in the Chinese tradition" means that "the power of creativity and the responsibility for creative product resides more broadly in the phenomena themselves." The distinction is undoubtedly true, but I do not see why the word "organicism" should therefore have to be confined to the ambit set by Aristotelian teleology, particularly in a book concerned with Chinese rather than Western thought. As I hope this section will make clear, in China, not to speak of Europe, more than one approach to "organicism" is possible.

> this organisation came about, not because of fiats issued by a supreme creator-lawgiver, which things must obey subject to sanctions imposable by angels attendant; nor because of the physical clash of innumerable billiard-balls in which the motion of the one was the physical cause of the impulsion of the other. It was an ordered harmony of wills without an ordainer; it was like the spontaneous yet ordered, in the sense of patterned, movements of dancers in a country dance of figures, none of whom are bound by law to do what they do, nor yet pushed by others coming behind, but cooperate in a voluntary harmony of wills. (Needham 1956:286–287)

> The Chinese world-view depended upon a totally different line of thought [from that which in the West came to formulate the concept of "laws of nature"]. The harmonious cooperation of all beings arose, not from the orders of a superior authority external to themselves, but from the fact that they were all parts in a hierarchy of wholes forming a cosmic pattern, and what they obeyed were the internal dictates of their own natures. Modern science and the philosophy of organism, with its integrative levels, have come back to this wisdom, fortified by new understanding of cosmic, biological and social evolution. (Needham 1956:582)

Diametrically opposed to the outlook of these statements are the texts that in the preceding subsection were quoted and described as representing what might be called a minority Chinese cosmological position. It is not enough, however, merely to say that the organismic viewpoint triumphed and the concept of a supreme divine legislator disappeared. For if we look closer, we can see that within what can be loosely labeled "Chinese cosmological organicism," variations and even some curious contradictions occur.

First should be noted the considerable distance separating the organicism of the philosophical Taoists (notably Chuang Tzu) from the organicism of the yin-yang and Five Elements cosmologists (notably that of their major Han formulator, Tung Chung-shu). Chuang Tzu (or at least the anonymous writers who contributed to the book bearing this name) is the classical enunciator of Chinese cosmological organicism in its "purest" form.[67] He views the universe in natural, not supernatural, terms, and his acceptance of it is total and unconditional (see sect. 1g).

Tung Chung-shu may also be regarded as an organicist, largely on the strength of his doctrine of the significance of correlates in the structure of the cosmos. Between each member of a correlative pair, he maintains, there exists a mysterious communion, attraction, or resonance that functions spontaneously and from a distance, without any

---

67. See Needham (1956:52, 288, 302). It should be stressed that we are *not* here speaking of what may be called "social organicism," for which early Confucian and even pre-Confucian roots exist and which was discussed earlier (V.2).

mechanical impulsion or causation. In the words of the title of one of his chapters, "things of the same genus energize each other." As examples he cites the way in which water, when poured on level ground, moves toward the part of the ground that is already wet; the preference shown by fire for a dry as against a damp piece of wood; how a given note on a stringed instrument will resound when the same note is plucked on another; or the answering whinny of one horse when another horse whinnies.[68]

Yet this same Tung Chung-shu, as has been abundantly shown (e.g. in passage 7 a few pages back), frequently thinks of Heaven as a personal presiding deity which, for example, has instituted the superiority of the yang over the yin as a cosmological model for the relationship between man and woman. Moreover, Tung is a firm believer in the macrocosm-microcosm analogy, which he depicts with painful crudity and literalness (VII.1f). In so doing, he noticeably places more emphasis on the human than the natural half of his cosmos because of his greater interest in how human conduct can influence the movements of nature than the other way around. Thus his cosmos is human-centered and at the same time dominated by a supernaturalism that operates far from spontaneously. In short, this kind of organicism is far removed from that of the philosophical Taoists. Moreover, Tung is far from consistent in his thinking, for at one moment he seems teleological and theistic, at another naturalistic. Such ambivalence often seems to have characterized the thinking of the yin-yang and Five Elements theorists generally.[69]

Again, one finds a wide difference between the naturalistic organicism of the early Taoists (Chuang Tzu) and the much more vitalistic and purposive organicism of the Taoists interested in alchemy (Ko Hung), who believed that by means of "natural," though human-induced, techniques they could modify the regularities of the universe. And still another kind of organicism is that of the Taoist-influenced Neo-Confucians, who avoided the crude anthropomorphism of Han Confucianism, yet remained eminently Confucian in that they firmly kept mankind at the center of the universe.[70]

---

68. Fung (1953:56) and Needham (1956:281–282), where portions of the chapter in question, *Ch'un-ch'iu fan-lu* 57, are translated. See also Fung (1953:23–24) and Needham (1956:288), translating portions of *Ch'un-ch'iu fan-lu* 51, in which Tung describes how the yin and yang, as complementary forces, alternately replace each other in the performance of their activities according to a spontaneously recurring pattern.

69. See Fung (1952:163; 1953:58).

70. Which is why, as indicated earlier (VI.1f), the Neo-Confucian doctrine of "the investigation of things" could not lead to natural science.

Just as the spectrum of Chinese organicist thinking embraces a considerable variation, fading away at one end into a theological outlook where even "laws of nature" become a possibility, so in European thinking, despite its usual exclusive dichotomy between "Democritean mechanical materialism and Platonic theological spiritualism" (Needham 1956:302), room can nevertheless be found for something resembling cosmological organicism.

By far the most important such example, I believe, is that within the worldview of the Stoics. Of these same Stoics, Needham has written (1956:534): "The conception of the governance of the whole world by law seems to be peculiarly Stoic." My own differing appraisal is based on several considerations. One is the considerable difficulty of accurately ascertaining Stoic doctrine, owing to the loss of most of the original Stoic writings and the consequent necessity of working with a variety of fragmentary accounts emanating from other classical writers not infrequently hostile to the Stoic philosophy. This situation, I believe, has in the past facilitated incomplete or one-sided evaluations of that philosophy. Another is the fact that historians of classical philosophy have traditionally tended to be more interested in Stoic ethics than in Stoic cosmology. And the third and decisive factor is the appearance, subsequent to what Needham has stated (as quoted above), of the special study of Stoic physics by Sambursky (1959). This monograph (about which, it must be noted, negative comments have been made)[71] uses statements on Stoic thinking collected from fragmentary sources that in the past have often been ignored. On this basis it suggests the need for a reappraisal of Stoic physical doctrine.

Without attempting any detailed exposition, let us try to catch a glimpse of the Stoic outlook by quoting a very few of these same statements. The legal metaphor that makes the Stoics important in the history of "natural law" and "laws of nature" appears in the following from Philo Judaeus (b. 15/10 B.C.):

> This world is a great city, has one constitution and one law, and this is the word or reason of nature, commanding what should be done and forbidding what should not be done.[72]

---

71. The major one is that Sambursky is overly ready to read ideas from modern physics and mathematics into the Stoics. Thus his book uses terms like continuum, field of force, tensional motion, determinism, and functional analysis. See reviews by Diamopoulos (1961) and Stannard (1961). See also Sivin (1968:6 n. 5), who describes the book as "an exceptionally stimulating example of the historiography of hindsight."

72. *De Iosepho* 29; tr. Colson and Whitaker (1929–1953:6.157) and Saunders (1966:124).

As against this legal statement, the two that follow express the Stoic doctrine that the world is periodically destroyed by fire but then recreated by God; this recreation does not take place ex nihilo but proceeds from God's own substance, which thus, by implication, itself constitutes the world:

> God . . . is indestructible and ungenerated, being the creator of this orderly arrangement, who at stated periods of time absorbs into himself the whole of substance and again creates it from himself.[73]

> The Stoics speak of one world only; God is the cause of its creation but not of its destruction. This is due to the force of the ever-active fire which pervades all existing things, dissolving, in the long periods of time, everything in itself; while from it again a regeneration of the world takes place through the providence of the creator.[74]

The universal substance that unites the universe is its *pneuma* or ether, the Stoic conception of which is strangely reminiscent of the Chinese *ch'i* or vital power:

> There is one *pneuma* which pervades, like a soul, the whole universe, and which also makes us one with them [i.e., with other men, gods, and also animals].[75]

The following passage, too, with its likening of the universe to a living organism held together by an all-pervasive "sympathy" *(sympatheia),* is remarkably reminiscent of Chinese theories of "attraction from a distance," as developed, for example, by Tung Chung-shu:

> The Universe . . . is a body, . . . but it is neither a conjoined nor of separate parts [i.e., it is a single continuum], as we prove from the "sympathies" it exhibits. For in accordance with the waxings and wanings of the moon many sea and land animals wane and wax.[76] . . . In the same way, too, in accordance with certain risings and settings of the stars alterations in the surrounding atmosphere take place. . . . From these facts it is obvious that the Universe is a unified [i.e., continuous] body. For in the case of bodies formed from conjoined or separate elements the parts do not "sympathize" with one

73. Diogenes Laertius (third century A.D.), *Lives of Eminent Philosophers* VII.137; tr. Hicks (1925:2.241) and Saunders (1966:90).

74. Philo Judaeus, *De Aeternitate Mundi* 89; tr. Colson and Whitaker (1929–1953:9.191) and Saunders (1966:93).

75. Sextus Empiricus (early third century A.D.), *Against the Physicists* I.127; tr. Bury (1933–1949:3.69).

76. For similar Chinese theories, see Needham (1956:272) and his Section 39 of *Science and Civilisation in China.* Forthcoming.

> another, . . . but in the case of unified bodies there exists a certain "sympathy," since, when the finger is cut, the whole hand shares in its condition.[77]

How does one reconcile the seemingly conflicting ideas of (1) a universe created and subject to the law of a supreme deity and (2) an apparently self-sufficient universe wherein everything functions harmoniously in the manner of the parts of the human body, owing to "sympathies" between each and every thing? Sambursky (1959:36) has supplied what seems to me, despite its modern terminology, a plausible explanation:

> It was their conception of a continuous field of force interpenetrating matter and spread through space, and thus being the cause of physical phenomena, which formed the central idea of pneuma. . . . The idea of the existence of forces continuous in space and time merged in Stoic doctrine with the conception of the ever-present and all-permeating Deity. Pneuma became a concept synonymous with God, and either notion was defined by the other. On the one hand, natural force (i.e., pneuma) was seen as endowed with divine reason. . . . On the other hand, God was identified with the all-penetrating pneuma.

Another historian of science, Giorgio de Santillana, does not hesitate to use the word "organismic" to describe the Stoic vision:

> The universe creates itself forever out of its own inner forces, which also hold it together. There is no distinction between above and below. It is a completely organismic view, with no outside mind required to rule the whole. . . . The Stoic world is a "block universe," complete in itself. The vitalistic intuition carries with it an almost mystical sense of union with the travail of the forces of nature as we feel them reflected in us. It expands into a sense of vast cosmic solidarity which in turn defines the solidarity among men. (Santillana 1961:294–295)

"No outside mind required to rule the whole" is not, I believe, explicitly confirmed in so many words in any surviving text. Yet the statement seems a reasonable interpretation of a Stoic level of philosophical abstraction in which God and his law become mere metaphors to describe the functioning of the all-pervasive pneuma.[78]

---

77. Sextus Empiricus, *Against the Physicists* I.78–80; tr. Bury (1933–1949:3.45–47).

78. See Needham (1956:564 and esp. n. e) for an example of how the Taoists similarly sometimes personified the Tao as a Maker (lit. Founder) of Things *(tsao wu che)*. The difference, of course, was that the Taoists used this language deliberately for literary effect (see sect. 2a, n. 43), whereas among the Stoics such personification probably had no such purpose but was simply a survival from an earlier time when it had been literally accepted. See also Needham (1956:500), where it is pointed out that the term *shang ti* (Lord on High), anciently the name of a major deity, later lost

If this interpretation can be accepted, it would seem that the worldview of the Stoics was as "organismic" as anything found in China. Possibly there is significance in the fact that such Stoic organicism, in which a place was left for a supreme lawgiver, and Tung Chung-shu's kind of organicism, in which a place was likewise left for Heaven as a consciously surpervising power, should both have arisen during ages of emerging empires, when centralized government had become the rule.[79]

In making such comparisons, it is well to remember that what in European thought was exceptional (granted that the Stoic worldview was in fact organic) corresponded to what on the Chinese side was the dominant view. Can this view, whether found in West or East, be said to have been either favorable or unfavorable to the development of natural science? Sambursky refers to what he calls some basic Stoic assumptions, especially their "dynamic notion of the concept of continuity," as "anticipating in many respects the approach to continuity which dominated the scientific ideas of Descartes, Huygens, Faraday and Maxwell."[80] Yet even he, with his avowed sympathy for the Stoics, speaks only of similarities, not of actual influences.

On the Chinese side, Needham has tried to demonstrate not only that similarities in outlook exist between Chinese organic naturalism and the organicism of "modern" science. He has also suggested (1956:esp. 339ff. and 497ff.) that positive Chinese influence is perhaps visible in a line of European thinkers beginning with Leibniz (1646–1716) and continuing

---

its anthropomorphic character and eventually came to be used by the Neo-Confucians as a metaphor for their *li* (pattern of organization). On *shang ti* see also the discussion in sect. 2a beginning at n. 54.

79. On the Greek side, see Zilsel (1942:251), who ignores the question of Stoic organicism but nevertheless states of the Stoics that "living in a period of rising monarchies they viewed the universe as a great empire, ruled by the divine Logos." On the Chinese side, Needham (1956:337–338) has suggested that the attempts of the correlative thinkers to fit the variegated data of human and extrahuman existence into an ordered whole were inspired by the formal beauty of the bureaucratic state system. "Perhaps the entire system of correlative organismic thinking was in one sense the mirror image of Chinese bureaucratic society." This would of course be especially true during Han times, when the bureaucratic system was being elaborated and consolidated and when correlative thinking was at its height.

80. Sambursky (1959:vii). See also 1959:35ff., where he compares the Stoic concept of the pneuma as a continuum with the ether theories of Descartes, Boyle, Newton, and others. Other comparisons appear on pp. 48, 58–60, and 105–106, as for example: "There is a remarkable similarity of the Stoic doctrine to the ideas on time of some modern philosophers, and especially noticeable in this respect is the theory of Whitehead" (1959:105).

through Hegel and Engels to Whitehead. Still more specifically (Needham 1956:458), he has pointed to what he terms a seemingly irreconcilable dichotomy in European thinking at the end of the seventeenth century, one in which the sole choice seemed at the time to be between theological vitalist idealism and mechanical materialism. Leibniz, he believes, with his doctrine of pre-established harmony, was the first major thinker to attempt to surmount this dichotomy, and although the doctrine per se did not persist, it started a train of organicist thinking that came to provide an alternative to the body-mind dilemma.

Leibniz's life-long interest in China, beginning in his early twenties, is well known, and his knowledge of Chinese thought, and especially of Chu Hsi's Neo-Confucianism, was quite exceptional for the age in which he lived. By specific examples, Needham has tried to show that echoes of Chinese thought are detectable in Leibniz's philosophy. In particular, he has offered the hypothesis that Leibniz's theory of the monads, hierarchically manifested in their "pre-established harmony," resembles, and may well have been influenced by, the Neo-Confucian concept of the innumerable individual manifestations of *li* in every pattern and organism (Needham 1956:496–505).

Without going into undue detail, it should be noted that since this hypothesis appeared in *Science and Civilisation in China* in 1956, doubts about it have been voiced by several scholars. Mungello (1977:15), for example, suggests that "the Chinese influence on Leibniz was more corroborative than germinal." In support of this conclusion he argues that although Leibniz, in 1686, had already completed his *Discours de métaphysique* or *Discourse on Metaphysics* (accepted by Mungello as a mature presentation of Leibniz's philosophy), his knowledge of China as late as that year still remained superficial (Mungello 1977:99).

Similarly, but in a more detailed manner, Cook and Rosemont (1981) have chronologically examined the sources and growth of Leibniz's knowledge of Chinese philosophy during the last several decades of his life. Their conclusion is negative:

> The conclusion seems compelling that . . . scholars who have urged a Neo-Confucian influence on Leibniz's philosophy will not be borne out in their interpretations. . . . Materials thus far analyzed provide little warrant for such interpretations; on the contrary, they go some way toward establishing the originality of Leibniz's metaphysics of monads and pre-established harmony. (Cook and Rosemont 1981:264)

With equal certitude, Olivier Roy's study in French questions the validity of assuming any basic similarity between Leibniz's ideas and Chinese thinking:

> As the monads of Leibniz have neither doors nor windows, it is necessary for them to be regulated in advance by someone so that they may accord with one another. Here we touch on the fundamental point of divergence between Leibniz and the Chinese, namely, the "atomistic" aspect of Leibniz's thought. . . . Since the activity of one monad possesses no causality vis-à-vis the others, God is necessary to ensure this [causality]. . . . Leibniz believed he was completing the system of the Chinese by bringing pre-established harmony to them, whereas the Chinese have no need for this harmony nor for the Christian God; causality is immanent in their system, and the world is made up of connections and relations without substance, and of a harmony which is immanent and self-regulating.[81]

If one turns to Leibniz himself, his most mature and extended exposition of Chinese philosophy is to be found in his *Discours sur la théologie naturelle des chinois* (Discourse on the Natural Theology of the Chinese), written in 1716, a few months before his death. However, the work reveals serious errors of understanding of Chinese thought, some deriving from Leibniz's own Christian outlook (with which he wished to reconcile the Chinese position), others from the biases and errors contained in the writings of his two major informants on the subject, the China missionaries Nichola Longobardi (1565–1655), a Jesuit, and Antoine de Sainte-Marie (1602–1669), a Franciscan. Here, for example, are a few of Leibniz's remarks on the *li* or "principle of organization" of Neo-Confucianism:

> The creation of all things is its [*li*'s] proper science. . . . *Li* . . . sees all, knows all and can do all.[82]

> *Li* is the natural law of Heaven and by its operation all things are governed.[83]

> *Li* is eternal and endowed with all possible perfections; in a word one can take it for our God.[84]

Other equally misleading remarks could readily be cited. In short, it is evident that the real extent and nature of Chinese influence on Leibniz's thinking cannot be decided without further study.

Irrespective of such considerations, however, a final question now has

---

81. Roy (1972:116–117). The entire chapter (pp. 107–119) from which this passage is here translated is worth reading.

82. *Discourse* sect. 11; English tr. by Rosemont and Cook (1977:70–71). I am much indebted to Professor Rosemont for sending me copies of both Mungello (1977) and Rosemont and Cook (1977). The latter is the first translation into English of Leibniz's *Discours sur la théologie naturelle des chinois*.

83. Sect. 17; tr. Rosemont and Cook (1977:83). See also sect. 21; tr. ibid. 87.

84. Sect. 28; tr. Rosemont and Cook (1977:101). See also sect. 22; tr. ibid. 89.

to be posed: In Europe, could the growing river of post-seventeenth-century organically oriented science have sprung directly from an organicist world of thinking, had such a world previously existed there, as it did in China? Or, for it to appear, did there first have to be a Cartesian-Newtonian stage of mechanical materialism? The answer suggested by comparison with organically oriented China would seem to lean toward the latter alternative.

In the first place, as suggested not far back (sect. 1h), scientific development in China would probably have been facilitated had it attracted more men able to examine natural phenomena not merely with interest and sympathy, but also with detachment and critical questioning. Probably Wang Ch'ung most nearly represents such a man, but unfortunately there was only one Wang Ch'ung in Chinese history.

In Europe, as suggested earlier (IV.2), there seems to have been a significant congruency, and possibly even a cause-and-effect relationship, between the doctrines, commitments, and organizational solidarity induced there by the monotheistic religions, and the rise of a commercial-industrial society centered upon a new experimental science.

In China the situation was sharply different. There the Mohists were the only sizable and long-lived group among the classical schools of Chinese philosophy to show serious analytical interest in logic, natural science, and technology. At the same time, they were the only pre-Buddhist Chinese group to create a tightly knit socioreligious organization consisting of men fanatically ready to practice ascetic denial in order to achieve idealistic goals. My belief, expressed earlier, is that had this school survived as a sort of ordinary man's counterpart to the Confucianism of the elite minority, China's entire subsequent sociopolitical development might have been profoundly different. Scientifically speaking, it is conceivable that a mechanistic (as opposed to an organismic) kind of science might have developed, perhaps in the end not too unlike what eventually arose in Western Europe. Surely it is no accident that this same Mohist school provided one of the best textual exemplifications of the theistic/anthropomorphic approach to nature discussed not too many pages above, as well as two of the nine passages adduced to support the thesis that anticipations of "laws of nature" are occasionally to be found in early Chinese thinking.

Organicism provided the Chinese with deep insights into the nature of the universe. These insights are particularly appealing to us today because of their seeming congruence with some of the findings of the modern physical and biological sciences. Organicism also contributed immeasurably to the poetry, beauty, and richness of Chinese cultural achievements. But, to repeat our earlier question, could organicism provide a suitable basis, in China or elsewhere, for the rise of what we today term "modern science?" Or did it lack certain features essential for such a science, such as the concept of "laws of nature"? In short, if science were to become what it is today, was it necessary for it initially to experience a stage of mechanical materialism, only after which it could and may still conceivably move onward to an organic level?

The imponderables are too many and complex to permit any absolute answer. However, we can point to the differing roads taken by science in China and the West and speculate, on their basis, that the predominance of organicist thinking in the one and its rarity in the other were probably among the causes—of course there were many others—responsible for these differing roads.

VIII. CONCLUSIONS

We have now completed our exploration of the many intellectual and social factors that, conceivably, may have encouraged or discouraged scientific thinking and creativity in pre-modern China before roughly 1600. They include various features of the Chinese written language known as Literary Chinese; Chinese concepts of time, space, and the arranging of things in terms of symmetry and numerical categories; the role and influence of religion in Chinese society; the Chinese world of human activities, beginning with orthodoxy, authoritarianism, and the organicist view of man; Chinese class attitudes, particularly with reference to the intellectual, the merchant, and the artisan; moralistic attitudes with respect to the arts, history, militarism, technology, nature, and science; sexual mores, individualism, and competition in the world of sports; and finally, the relationship of man to nature in Chinese thinking, including considerations of Chinese differing attitudes toward nature, the existence or nonexistence of "laws of nature" in Chinese philosophical thinking, and the significance for science of Chinese concepts of cosmic organicism.

Of the foregoing ideas and social factors, some, in some ways, were no doubt favorable to scientific development. The Chinese written language, for example, was a powerful conserver and propagator of Chinese cultural and social values, both chronologically and geographically. A few Chinese thinkers and historians conceived of time in linear or continuous rather than compartmentalized terms. The dominant Chinese view of the universe as an organismic unity led to important scientific advances, notably in magnetism and acoustics. And the Chinese ideal of government as a strongly centralized organ, resting on an elaborate bureaucracy, stimulated the development of those particular sciences, notably mathematics and astronomy, that seemed important to the government and hence received government support. Such support also facilitated the construction of roads, canals, bridges, walls, and other massive forms of public technology such as could never have been achieved by private or local initiative alone.

As against these and other favorable factors, it is my contention that most of the factors we have discussed seem on balance either to have been unfavorable (sometimes strongly so) or at least unhelpful and neutral. For example, the role of religion in China (meaning especially Buddhism) differed greatly from that of religion in Europe (meaning especially Christianity), and its influence on science was generally less favorable.[1]

Indeed, even among the four factors mentioned a few lines above—those of language, concepts of time, organismic thinking, and ideas of government—each seems to have had its negative as well as positive aspects. Granted, for example, that the written language was a powerful instrument for cultural cohesion. This fact nevertheless hardly compensated, in my opinion, for the many difficulties associated with Literary Chinese, as discussed in chapter two. Granted, too, that Literary Chinese, as suggested in chapter one, may have been theoretically suitable for the development of logic. This fact did not prevent logic, conceived as a separate discipline, from being ignored by all but a tiny handful of early Chinese thinkers. Likewise, as against a few individuals who thought of time in linear terms, the great majority apparently took for granted that it functioned compartmentally or in recurring cycles. And Chinese organismic thinking, though regarded by some today as remarkably compatible with modern science, did not, within pre-modern China—and perhaps could not—lead by itself to modern science. Finally, although Chinese centralized bureaucracy helped astronomy and certain other sciences in some ways, it also eventually stultified astronomy by binding it in a tight government monopoly which legally prohibited its serious study by all but a small group of specialists (see V.4a). The question of Chinese bureaucracy is an important one to which we shall return later.

These remarks unfortunately seem to lead us to the unhappy position of the historian who, prevented by the nature of his discipline from proving his hypothesis by means of controlled repetitive experiments, is obliged instead to formulate theories to explain events whose outcome is already known. Faced by this situation, the temptation becomes strong to "prove" the "inevitability" of everything that has happened simply because it has happened. The danger of such juggling between past and present is vividly pointed out by Sivin for science when he writes:

> A fixation on finding knowledge in the past which merits recognition in the light of the present, leads to a seriously distorted understanding of the past. Once one forms the habit of dividing the thought of ancient scientists into a "scientific" part and an "unscientific" part, there is no reason to put it together again. (Sivin 1974:xi)

No doubt the warning is justified. Yet what solution does it leave to the historian of science (as contrasted with the practicing scientist), anx-

1. This remains so despite the remarks about Buddhism and astronomy some pages below.

ious to interpret as fairly and meaningfully as he or she can the known scientific facts of the past? Perhaps Professor Sivin will agree that the answer does not lie in minimizing the differences between the old science and the new, but rather in clearly understanding those differences and their causes. In this way one can avoid the twin dangers of either attributing to the past what can only belong to the present or condemning the past because it does not live up to the present, as well as the similar dangers that often arise when one civilization is compared with another. The definitions of science that follow have been chosen deliberately because they are Western and modern, as well as authoritative. They begin with one made in 1951 by a Cambridge professor describing the methodology advocated by Francis Bacon (1561–1626):

> Let them [the scientists] concentrate . . . on discovering by suitably designed experiments and appropriate reasoning the fundamental laws and structure of nature. (C. D. Broad 1951:31, describing Bacon's concept of science but not actually quoting him)

> In scientific inquiry it becomes a matter of duty to expose a supposed law to every possible kind of verification, and to take care, moreover, that this is done intentionally, and not left to a mere accident. (Thomas Henry Huxley 1964:5, from an essay originally written in 1863)

> Science brings . . . the conception of definite, positive, and fixed law. For the vague conception of a law as a predominating, though variable, association, . . . science substitutes . . . the association which is absolutely invariable, unchanging, universal. (Norman Campbell 1953:179–180, orginally published in 1921)

> [Science is] a branch of study which is concerned either with a connected body of demonstrated truths or with observed facts systematically classified and more or less colligated by being brought under general laws. (*Oxford English Dictionary,* 1933)

> [Science is] a branch of study which is concerned with observation and classification of facts and especially with the establishment . . . of verifiable general laws, chiefly by induction and hypothesis. (*Webster's New International Dictionary,* 3d ed., 1963)

Noteworthy is the association made in every one of these statements between science and "law" or "laws," which are further defined as "fundamental," "general," "universal," "invariable," "unchanging," and "verifiable." Underlying all the statements is an awareness and acceptance of the idea of "laws of nature." Only in an environment both modern and Western (or at least Western-oriented) could this consciousness have happened. No comparable definitions of science are to be found in pre-modern Chinese literature or, for that matter, in that of pre-modern Europe.[2]

From science in general let us turn to science in China and more particularly to the chronology and categories of that science—a topic already discussed to some extent earlier (I.1). Chronologically, the two most important periods were probably the Han and Sung dynasties. This is certainly so for mathematics (Needham 1959:153) and probably also to a considerable extent for most of the other sciences. The Han, as the first long-lived Chinese centralized empire, provided centuries of relative internal peace and security. During this time much literature from the past could be recovered and edited following the Ch'in "burning of the books" of 213 B.C., and much new literature could be written. The Sung, by contrast, was politically far less stable. Intellectually and socially, nonetheless, it was a vigorous age, characterized by a growing literacy helped by the spread of printing, an increasing urbanization of culture, and a remarkable development of trade, industry, and exploitation of natural resources. These and other features made it in some ways a kind of Chinese "Renaissance," though one that failed to evolve much further. Thus Chinese interest and creativity in several important scientific fields tended to decline or at least crystallize after the Sung, in sharp contrast to developments in Europe from Renaissance times onward.

Comparable in importance to chronology is the categorizing of what we call Chinese science. The classification scheme outlined in the first chapter is that commonly applied to sciences generally (mathematics, astronomy, physics, civil engineering, etc.). In its place, for present purposes, I wish to suggest a smaller number of categories, formulated from a layman's point of view and with Chinese science particularly in mind. A four-part grouping seems possible:

1. Those procedures pertaining to what people today often refer to rather contemptuously as "pseudo-sciences," such as geomancy and alchemy
2. Those procedures more respectfully referred to as "pure" sciences, such as mathematics and astronomy, as well as a wide range of other more "mixed" sciences and techniques, such as natural history, medicine and agriculture
3. The government-sponsored activities of observing and recording

2. With regard to the initial statement concerning (but not by) Bacon, it should be noted that although Bacon himself used the term "Summary Law of Nature" in his *Advancement of Learning* (1605), his concept of "laws of nature" seems to have been ambiguous. See Needham (1956:540).

unusual natural phenomena, such as eclipses, floods, droughts, earthquakes, monstrous births, and so on—all reported and studied in China more assiduously and continuously than in any other long-lived ancient civilization

4. The many kinds of technology, such as construction, invention, or operation of roads, canals, irrigation works, walls, ships, instruments for boring deep salt wells or measuring natural phenomena (e.g., the rain gauge), paper and printing, and much more, extending to the tools and techniques used in daily life by countless humble artisans

All four of these categories, on close examination, appear to have been susceptible to the unfavorable effects of one or more adverse factors. The "pseudo-sciences," for example, though sometimes responsible for by-products of great scientific and social importance (e.g., the mariner's compass in the case of geomancy and gunpowder in the case of alchemy), were throughout their histories unfavorably influenced by their unquestioning acceptance of the dominant Chinese theories of numerology and of the yin-yang and Five Elements cosmologies. To varying degrees, the other three enumerated categories, including even the so-called pure sciences, were also adversely influenced by the same theories and cosmologies, which continued to enjoy wide currency down to the present century, despite occasional piecemeal criticisms made by certain scholars from late Ming times onward. In addition, the third scientific category, that of observing and recording natural phenomena, was sometimes further distorted—though just how far is difficult to say—by political considerations (VI.1b). Likewise, the fourth category, that of the technologies, suffered significantly from the traditional gap between the scholar and the artisan (V.4c).

The effects of this gap are clearly visible in such fields as engineering. Few original texts, for example, are today extant on pre-modern Chinese mechanical engineering because this subject lay in the hands of artisans, who rarely wrote about what they did. Machines were invented and massive technological feats accomplished despite, rather than because of, the lack of a developed Euclidean deductive geometry in Chinese mathematics. Many of the best Chinese shipwrights were illiterate and used no templates or blueprints for their work, largely depending for knowledge on personally transmitted know-how rather than written manuals.[3]

On the other hand, one suspects that in some ways Chinese technol-

3. See, respectively, Needham (1965a:1, 66; 1971:378, 403, 413). Also chap. V above, near end.

ogy may have gained more than it lost from the gap between the scholar and the artisan. The reason, I would suggest, is that such a gap provided better isolation from the adverse effects of Chinese numerology and correlative theorizing than that enjoyed by the other several categories. After all, to cite only a few notable examples, early Chinese invention of a superior kind of breast strap in horse harness, or of the wheelbarrow, or of paper and printing, could all be more easily carried out, and probably was in fact carried out, without resort to numerology or the Five Elements. When to these major inventions one adds the Great Wall, the world's longest transport canal, the world's largest wrought-iron suspension bridge, Cheng Ho's fifteenth-century ships, which were so much larger than their Portuguese counterparts at the end of the same century—when one puts all these things together and a great deal else besides—it then becomes evident that it was the country's pre-modern technologies, in particular, that repeatedly enabled China to anticipate similar developments in other parts of the world. In many instances, in fact, they probably or certainly provided direct stimuli for these developments (see end of I.1).

Significantly, many such technological achievements could be successfully carried out without any recognition of the concept of "laws of nature," with little or no performance of controlled experiments, and with few of the combinations of practice with theory that marked the rise of the "new science" in Europe from the seventeenth century onward. Also to be remembered is the massive support some of the technologies received from the powerful central goverment. All this may help explain pre-modern China's frequent superiority in technology.

The second chapter's discussion of language several times stresses (II.9, 10) a serious methodological weakness characterizing much traditional Chinese scholarship, namely, its all-too-common preoccupation with discrete events and details at the expense of overall synthesis, generalization, and analysis. The same preoccupation often appears in Chinese scientific writing. So too does the companion tendency of concentrating on pragmatic application at the expense of theory. One might expect this for the technologies, but such practicality is to be found also in the other categories enumerated above, including even that of the "pure" sciences. Han mathematicians, for example, unlike their Greek and Hellenistic opposites, showed little interest in explaining their techniques (Needham 1926:59 n. a). Yet despite (or because of?) this inattention to theory, they and their successors succeeded very well in handling practical problems concerned with land mensuration, the dimensions of granaries, rates of taxation, and other matters closely connected with bureaucratic administration. In similar fashion, Chinese astronomy, "like all Chinese science," was "fundamentally empirical and observational" (Needham 1959:460).

Perhaps this disinterest in theory helps explain why the Chinese failed to follow up several inventions or discoveries in which they had been pioneers. One is Chang Heng's seismograph of A.D. 132, which was probably deemed more ominous than "practical" and therefore had no real successors (Needham 1959:627–628). Another is Chu Tsai-yü's discovery in 1584 of equal temperament in music, of which little subsequent use was made in China, whereas quite the opposite is true of the same discovery made by Simon Steven in Europe a couple of decades later; Steven's discovery was very possibly inspired by that in China (Needham 1962:228, with Kenneth Robinson). Something similar took place in the case of magnetism, of which the Chinese, with their organismic view of the universe, had long been aware. Perhaps as early as 1080, and possibly considerably earlier, they had applied the magnetic compass to navigation—a whole century or more before its earliest mention in Europe. Yet thereafter the Chinese made very little further progress in this field, in contrast to Gilbert's book on magnetism of 1600, and all the great consequences this entailed (Needham 1962:230, 234). Another much earlier product of Chinese organismic thinking was the "south-pointing chariot," whose invention in the early third century A.D. made it the first example of a homoeostatic or "constancy-maintaining" machine in any civilization (Needham 1965a:303). Yet this event brought with it no significant further consequences. Finally, there is Su Sung's great astronomical clock of 1090, long predating anything comparable in the Western world. Despite this priority, Chinese clock making, unlike its European counterpart of the fifteenth and sixteenth centuries, never became a mass industry before the coming of the Jesuits to China (Needham 1965:546).

Geographical isolation, although a topic beyond this book, deserves passing mention as another possible inhibitor. With it should be mentioned the difficulties and distortions caused by the Chinese characters when used as a medium for transmitting ideas and values into China—in contrast to their effectiveness as propagators of Chinese ideas and values to the outside world (II.3). Even Buddhism, China's most powerful alien influence before modern times, despite the new dimension it brought to Chinese religious life and moral values, apparently made remarkably little impact in the scientific sphere. Between roughly 600 and 800, for example, some notable Indian astronomers came to China as the result of Buddhism, among them one who in 729 produced in Chinese the remarkable astronomical work known as the *K'ai-yüan chan-ching* (K'ai-yüan Treatise on Astrology). This compilation is valuable, among other reasons, because it preserves a large number of ancient and medieval Chinese astronomical fragmentary texts. Yet despite the fact that it also contains the Indian symbol for zero and other innovations, its appearance seems to have left Chinese astronomy remarkably unchanged: the Chinese equatorial mansions remained as before, the Chinese circle continued to consist of 365¼ degrees, Indian trigonometry was not taken up, and the zero continued to slumber for another four centuries (Needham 1959:202–203; Yabuuti 1954:595). Chinese mathematics, too, remained handicapped then and later by its failure to adopt mathematical symbols, in place of which its formulas were written with ordinary Chinese characters (Needham 1959:152). This, possibly, is one of the reasons why Chinese navigation, after anticipating Europe by some two or three centuries in its entry into the second of three navigational stages (that of quantitative techniques), never, before recent times, reached the third phase, that of mathematical navigation (Needham 1971:560).

Many purely physical factors undoubtedly also contributed to the shaping of Chinese civilization and therefore of Chinese science. They include topography (China's mountains, plains, deserts, great rivers, and coast line), climate (China's north-south climatic division), agriculture (emphasis on grain production, often helped by irrigation, with relatively little attention to animal husbandry), flora and fauna (the ubiquitous use of bamboo, for example), and others. These being physical rather than intellectual or social factors, we can only name but not discuss them here.

However, there was another physical factor whose combination with an ideological one produced a situation probably unfavorable to Chinese science. The physical factor was China's continental geography, the ideological one that of China's age-old ideal of political and intellectual unity. The interaction of the two powerfully reinforced the drive toward single empire that was such a conspicuous theme throughout Chinese history.[4]

The suggestion that this emphasis on centralism may have been harmful to science perhaps seems strange in view of the assertion several pages back that science flourished both during the highly unified Han dynasty and also during the much weaker and often invaded dynasty of Sung. Important to both epochs, however, is the common

4. See also Balazs (1964:31): "What ultimately prevented the blossoming of capitalistic buds in China seems to me to be the fact that China was undifferentiated geographically, and lacked a system of separate nation-states. . . . Lack of the necessary geopolitical preconditions resulted in the birth, growth, and perpetuation of a centralized state—bureaucratic, absolute, and omnipotent, and embodied in a ruling class of scholar-officials."

fact that despite their contrasting political circumstances, each saw the rise of govenment-favored intellectual orthodoxies: Tung Chung-shu's Confucianism during the Han and the Ch'eng-Chu school of Neo-Confucianism during the Sung. The Sung state, moreover, during all its lengthy struggles with the Mongols and their predecessors, never gave up the centralized bureaucratic form of government which was China's unique creation or the concept of universal empire which was its persistent ideal.

This long-lasting emphasis on oneness contrasts sharply with the geographical, political, and ethnic diversity of Europe's many realms and peoples and all the social, religious, and other cultural differences that accompanied this diversity. In China, the drive toward intellectual and political centrality was persistent, intense, and always eventually successful. In Europe after Rome, attempts in the same direction were always prevented by a variety of factors from gaining more than partial or temporary success.

Had China permanently remained a collection of some half dozen competing principalities after 221 B.C. instead of becoming a single empire, the resulting conditions of life for the ordinary person would no doubt have been far more difficult and less secure. On the other hand, had such political disunity continued, and with it the diversity of ideas that had previously been its byproduct, I venture to suggest that the resulting intellectual environment might well have been more conducive to scientific development than was that afforded by the orthodox state Confucianism of imperial China.

Institutionally, too, China's political centralism is open to criticism. On the one hand, Chinese emphasis on oneness, together with the bureaucratic form of government that maintained this oneness, enabled China to persist as a large empire for more than two millennia, despite many invasions and changes of dynasty. Perhaps even more important is that this oneness facilitated the wide diffusion of China's moral and cultural values to realms far beyond China's political frontiers.

On the other hand, the price paid for these achievements was considerable. It lay in the fact that Chinese political centralism and bureaucratism, coupled with the tendency to organize government schematically rather than realistically (III.4 at n. 82), led to a growing rigidity that eventually precluded the carrying out of various administrative changes necessary for China to become a modern state. Such is the thesis repeatedly stressed by Ray Huang (1986:61, 69, 72, 103; 1988:112–115, 127–134, 169–174, 192–199, etc.).

From these general remarks on China's governmental system let us consider more specifically its possible long-term effects on science. This brings up the question of whether the failure of a modern science to appear in China should be attributed primarily to institutional and related factors as they became fully operative only in late imperial times or whether it goes back to such factors as they functioned at a much earlier date.

The answer, it seems to me, is that both sets of factors—the early and the late—were extremely important, but that the earlier ones were probably more so. True indeed, post-Sung China witnessed the growth of political authoritarianism and with it of a government-fostered intellectual uniformity resting on the Chu Hsi form of Neo-Confucianism. But also true, the tenet of political absolutism, and with it of intellectual conformity as embodied in the theory of the *tao t'ung* or "sequence of the Tao," goes back to early imperial and pre-imperial times.

Contradictions are easy to find. Astronomy, for example, as already noted, was both encouraged and yet ultimately weakened by the monopolistic support provided by the government. In technology, the government supported the construction of great transport canals primarily to ensure the ready transport of grain to the capital (Needham 1971:374 n. g). On the other hand, as stated earlier, Su Sung's astronomical clock, which was likewise supported by the government, failed to lead to a mass industry of clock making, because the clock was built only for official astronomers and the emperor, not the general public.

The Chinese system of government, and especially its Confucian-based civil service examinations for the recruitment of government personnel, compare very favorably with the government systems known to pre-modern Europe. For science, however, the intellectual influence of the civil service system, with its emphasis on literary rather than scientific abilities, was quite unfavorable. As long ago as 1735, this fact was clearly noted by the Jesuit J. B. du Halde in his *Description géographique, historique . . . de la Chine.* The aptness of what he says entitles it to be our final quotation:

> When we cast our Eyes on the great number of Libraries in China magnificently built, finely adorn'd, and enrich'd with a prodigious Collection of Books; when we consider the vast number of their Doctors and Colleges established in all the Cities of the Empire, their Observatories, and their constant Application to watch the course of the Stars, and when we farther reflect that by Study alone the highest Dignities are attained, and that Men are generally prefer'd in proportion to their Abilities; that according to the Laws of the Empire the Learned only have, for above four thousand years [*sic*], been Governors of Cities and Provinces, and have enjoy'd all the Offices about the Court, one would be tempted to believe, that of all the Nations in the world China must be the most knowing and most learned.
>
> However a small acquaintance with them will soon undeceive one; for tho' it must be acknowledg'd that the Chinese have a great deal of Wit, yet it is

> not an inventive, searching, penetrating Wit, nor have they brought to perfection any of the speculative Sciences which require Subtilty and Penetration.
>
> Yet I am not willing to find fault with their Capacity, since it is very plain that they succeed in other things which require as great a Genius and as deep a Penetration as the speculative Sciences; but there are two principal Obstacles which hinder their Progress in these kinds of sciences: 1. There is nothing within or without the Empire to stir up their Emulation; 2. Those who are able to distinguish themselves therein have no Reward to expect for their Labour.
>
> The chief and only way that leads to Riches, Honours, and Offices, is the Study of the Canonical Books, History, the Laws and Morality; it is to learn to write in a polite manner, in Terms suitable to the Subject treated upon; by this means the Degree of Doctor is obtained, and when that is over they are possessed of such Honour and Credit that the Conveniences of Life follow soon after, because then they are sure to have a Government [Post] in a short time; even those who wait for this Post, when they return into their Provinces, are greatly respected by the Mandarin of the Place, their Family is protected from vexatious Molestations, and they there enjoy a great many Privileges.
>
> But as there is nothing like this to hope for by those who apply themselves to the speculative Sciences, and as the Study of them is not the Road to Affluence and Honours, it is no wonder that these sort of abstracted Sciences should be neglected by the Chinese.[5]

With this informative statement—one that throws light on mid-eighteenth-century Europe as well as China—the time has come to bring our own discussion to a close. I regret that it cannot end with clear-cut answers to all the many problems that have appeared in these pages. Instead, the generalizations that follow must suffice.

1. Chinese science and technology have been successively dominated by two major phases or currents. The first is that of prolonged brilliance, originality, and creativity, beginning early and frequently resulting in major contributions to the outside world. The second is that of the slowing and eventual stagnation of this current during the last several centuries of imperial China. This latter phenomenon, which overlapped in part the European transition from pre-modern to modern science, by no means signaled the end of Chinese creativity as a whole.

---

5. Du Halde (1741 [Engl. ed.]:3.63–64). At our V.4a beginning, it was pointed out that starting in the seventh century during the T'ang dynasty, and continuing with increasing emphasis during the Sung, the great majority of civil service examination candidates preferred to take the generalized kind of literary examination leading to the degree of *chin shih* or Doctor in Letters, while only a small minority sat for the more specialized examinations in mathematics, law, and other particular subjects.

Rather, it was accompanied by an intensified interest in China's humanistic past, especially noticeable in nongovernmental scholarship. This resulted in the development of new and much more sophisticated techniques for textual criticism, epigraphy, linguistic study, historical investigation, and so on, which, however, were only occasionally directed toward the physical sciences.

2. This book has on the whole paid greater attention to factors believed to have been unfavorable to Chinese science than to their favorable opposites. This is because the shift from the first to the second phase is obviously of great importance both for understanding the evolution of Chinese civilization and for contrasting it with what happened in Europe. Among the factors influencing Chinese science, those deserving particular mention on the unfavorable side include the growing authoritarianism and intellectual orthodoxy of the Chinese bureaucratic state and the failure of mercantile enterprise to gain a truly respected and independent status. Other factors could also be readily listed, but it is better not to single some out at the expense of others, because all relevant factors were probably important in different ways and also because there were so many of them and their connections with one another and with science were so complex and difficult to measure.

3. Among the unfavorable factors, some became prominent only in late (i.e., post-Sung) China, but others were already apparent much earlier. Hence it seems unwise to attribute China's scientific decline in later times solely or even primarily to the later factors. Even without them, it seems quite possible, in view of the importance and persistence of the earlier factors, that the course of Chinese science, downward as well as upward, might not have differed greatly from the one it actually followed.

4. The differences between the sciences and technologies of China are important but often overlooked. The sciences were primarily pursued by learned men whose education was in the classics and whose views of the world stemmed largely from books. The technologies were primarily undertaken by artisans and master craftsmen who were often illiterate or semiliterate and whose achievements depended more on practical experience than on abstract theory. This fact raises the considerable possibility that the intellectual limitations imposed by the yin-yang and Five Elements theories may have been less harmful to the technologies than to the sciences. The same may also have been true of some of the other scientifically unfavorable factors, such as the widespread Chinese tendency to concentrate on discrete details rather than on broad syntheses. In short, we may have a partial explanation here for the fact that so many of China's "firsts" have been in technology and not in science. Some may even argue that China's technological

achievements, broadly speaking, have had greater significance for China itself and the outside world than have its achievements in science. Yet it would be unwise to press this point too far. Chinese magnetical science, for example, is not only a notable product of the Chinese organismic view of the cosmos, but also constitutes, according to Needham (1962:334), one of the three basic pillars from which rose the worldwide science of today (the other two, Euclidean geometry and Ptolemaic planetary astronomy, being of Greek origin).

5. Acceptance of the third proposition, namely that factors unfavorable to Chinese science were already strongly operative at an early time, leads to the further conclusion that without the impact of powerful outside stimuli (as in fact happened from the early nineteenth century onward), Chinese pre-modern science not only did not, but could not, evolve by itself into anything comparable to what is today called modern science. Supportive of this thesis on the socioreligious side is the failure of the Mohist school to survive after the second century B.C., and the major consequences this failure very possibly entailed both for Chinese science and for Chinese society generally. Or again, on the philosophical side, we saw at the end of the preceding chapter that Chinese organicism—a worldview long very rare in the Western world but now seemingly increasingly congenial to science—was nevertheless, within China itself, perhaps one of the factors that hampered the rise of an indigenous modern science.

And so we reach the end of our inquiry, rewarded more by broad generalities than by precise certainties. No doubt this must be so because of the nature of the subject, and probably it will long remain so. What has been attempted here is a reconnaissance upon a largely uncharted ocean; it cannot be definitive. If, nevertheless, it induces others to embark on further voyages, I shall be very happy. It is the great privilege of scholars to have the opportunity of receiving the knowledge put together for them by their predecessors, building it a little higher, giving it if possible a little more clarity and meaning, and transmitting the result to those who follow them.

APPENDIX: THE FOUR SOCIAL CLASSES

Section 3 of Chapter V discusses the Chinese concept of social classes and particularly the traditional descending scale of the four major categories called *shih* (scholar-gentry), *nung* (farmers), *kung* (artisans), and *shang* (merchants). The technical details of how this categorizing started are presented in this Appendix as a basis for the less technical presentation in Chapter V. Basic to both is the table below, consisting of twenty-nine text passages of Chou or Han dynasty date, in each of which the four social classes (or sometimes variations of these classes) are mentioned. The table also indicates the sequence of classes in each passage.

Among the twenty-nine passages, the six from the Han dynasty are datable with fair exactness. Of the remaining twenty-three that are loosely labeled "Chou," however, only a few (no. 21 and probably nos. 15–19) can be dated with fair assurance to the final decades of the Chou. The chronology of the others is very uncertain, with some of them, although Chou in origin, assuming their final form only in Han times. This makes it possible that the sequence of categories found in them may in certain instances reflect Han thinking.[1]

In seven of the Chou passages (nos. 1, 5, 19, 21, 25–27) the *shih* are omitted, because the authors happened to be concerned only with the three remaining categories of "small men." There is no doubt, however, that had the *shih* in fact been included, they would have been placed in the initial position to which I have therefore assigned them. In other passages (nos. 3–4), the four categories are embedded within a wider spectrum of enumerated groups, ranging from gradations of the nobility on the one hand to lowly persons like shepherds and grooms on the other. In still other passages (nos. 2–4, 8, 19–21), terms other than the standard social designations are used. The most common is *shu jen,* "the multitudinous persons," replacing *nung,* "farmers"—an understandable substitution when one remembers that the great bulk of the population consisted of farmers. All these variations confirm the tentativeness with which the four-part classification was worked out.

The "standard" sequence of *shih, nung, kung,* and *shang* appears in each of the six Han texts but in only eight of the twenty-three that are

1. Eleven of the twenty-nine passages have already been assembled and briefly discussed in Ch'en Teng-yüan 1958–1962:228–229 (essay no. 100). Ch'en's presentation has in turn been repeated by Vandermeersch (1965:134–135).

Table A1 **The Four Social Classes in Chou and Han Texts**

	Texts	Sequence of Social Classes			
	Chou				
1.	*Tso chuan*	(missing)	*shang*	*nung*	*kung*
2.	*Tso chuan*	*shih*	*nung*	*shang*	*kung*
3.	*Tso chuan*	*shih*	*nung*	*kung*	*shang*
4.	*Tso chuan*	*shih*	*nung*	*kung*	*shang*
5.	*Kuo yü*	(missing)	*nung*	*kung*	*shang*
6.	*Kuo yü*	*shih*	*nung*	*kung*	*shang*
7.	*Kuo yü*	*shih*	*kung*	*shang*	*nung*
8.	*Kuo yü*	*shih*	*nung*	*kung*	*shang*
9.	*Kuan-tzu*	*shih*	*nung*	*kung*	*shang*
10.	*Kuan-tzu*	*nung*	*shih*	*shang*	*kung*
11.	*Kuan-tzu*	*shang*	*kung*	*nung*	*shih*
12.	*Kuan-tzu*	*shih*	*nung*	*kung*	*shang*
13.	*Kuan-tzu*	*shih*	*nung*	*shang*	*kung*
14.	*Kuan-tzu*	*nung*	*shih*	*shang*	*kung*
15.	*Hsün-tzu*	*shih*	*nung*	*shang*	*kung*
16.	*Hsün-tzu*	*shih*	*kung*	*nung*	*shang*
17.	*Hsün-tzu*	*nung*	*shih*	*kung*	*shang*
18.	*Hsün-tzu*	*shih*	*shang*	*kung*	*nung*
19.	*Hsün-tzu*	(missing)	*nung*	*shang*	*kung*
20.	*Shang chün shu*	*nung*	*shih*	*shang*	*kung*
21.	*Lü-shih ch'un-ch'iu*	(missing)	*nung*	*kung*	*shang*
22.	*Ku-liang chuan*	*shih*	*shang*	*nung*	*kung*
23.	*Chou li*	*shih*	*kung*	*shang*	*nung*
	Han				
24.	*Huai-nan-tzu*	*shih*	*nung*	*kung*	*shang*
25.	*Shih chi*	(missing)	*nung*	*kung*	*shang*
26.	*Shih chi*	(missing)	*nung*	*kung*	*shang*
27.	*Yen t'ieh lun*	(missing)	*nung*	*kung*	*shang*
28.	*Shuo yüan*	*shih*	*nung*	*kung*	*shang*
29.	*Ch'ien Han shu*	*shih*	*nung*	*kung*	*shang*

SOURCES:

1. Hsüan 12 (596 B.C.); tr. Couvreur (1914:1.614).
2. Hsiang 9 (564 B.C.); tr. Couvreur (1914:2.238–239). In place of *nung* in the second column, the original text reads *shu jen;* however, its further statement that "the *shu jen* devote their energies to husbandry" makes it evident that the term serves as an equivalent of *nung*.
3. Hsiang 14 (557 B.C.); tr. Couvreur (1914:2.309). Again in the second column the original text reads *shu jen* instead of *nung*. It also prefixes the four categories

with several ranks of feudal nobility, beginning with the Son of Heaven, and follows them with several still lowlier groups, such as shepherds and grooms.

4. Ai 2 (493 B.C.); tr. Couvreur (1914:3.607). The original text once more reads *shu jen* in the second column instead of *nung.* As in the preceding entry, some noble and some mean groupings are respectively listed before and after the four main categories.
5. Chou 1.1 (attributed to time of King Hsiang of Chou, r. 651–620).
6. Ch'i 1.1, in inquiry addressed by Duke Huan of Ch'i (r. 685–643) to his adviser, Kuan Chung.
7. Ch'i 1.1, in Kuan Chung's reply to Duke Huan a few words later.
8. Chin 4.13 (635 B.C.). Again the original text reads *shu jen* in the second column, which (as in texts 2–4) is to be understood as equivalent to *nung.*
9. 5(1/22a), title of 6th subsect.; tr. Rickett (1985:118).
10. 5(1/21a–b), in the text itself, tr. Rickett (1985:122–123). In the third column, the original text reads *ku,* a fairly common alternative for *shang.*
11. 5(1/21b); tr. Rickett (1985:123); again *ku* for *shang.*
12. 20(8/6b); tr. Rickett (1985:325).
13. 30(10/13a); tr. Rickett (1985:403).
14. 48(15/15a); tr. T'an and Wen (1954:94); not in Rickett (1985).
15. 8(4/3b); tr. Dubs (1928:96–97). In place of *shih* in the first column, the original text reads *chün tzu* (princely man).
16. 8(4/13a); tr. Dubs (1928:118).
17. 9(5/7b); tr. Dubs (1928:136).
18. 11(7/14b); not in Dubs (1928).
19. 21(15/6b); tr. Dubs (1928:270). In the third column, the original text reads *ku* for *shang* (as in nos. 10–11 above).
20. 3; tr. Duyvendak (1928:193). In the first column, the original text reads *min,* "the people," instead of *nung,* with the remark that they "busy themselves morning and night with agriculture *(nung).*" The original text also reads *chi yi,* "skilled arts," in the fourth column, which is equivalent to *kung.*
21. 156; tr. Wilhelm (1928:454). For *shang* in the fourth column, the original text uses the alternative word *ku.*
22. 13/1b: Ch'eng 1 (590 B.C.).
23. 40; tr. Biot (1841:2.458).
24. 11/12b.
25. 129; tr. Swann (1950:422).
26. 129; tr. Swann (1950:449, 450 [two mentions of the identical sequence within a few paragraphs]).
27. 36 end; not tr. in Gale (1967). The text, besides deliberately omitting *shih,* fails in its present form to mention *shang.* Syntax and meaning alike, however, make it evident that the omission has resulted from faulty transmission, and that the words *shang jen,* "merchant men," should be restored. See Wang Li-ch'i (1958:256 n. 26).
28. 7 end; by Liu Hsiang (79–8 B.C.).
29. 24A; tr. Swann (1950:115).

labeled Chou (nos. 3–6, 8–9, 12, 21). The other fifteen Chou entries show remarkable instability. In no. 6 *(Kuo yü),* for example, Duke Huan of Ch'i is represented as addressing a question to his adviser Kuan Chung, in the course of which he enumerates the "standard" sequence. To this Kuan Chung immediately replies in no. 7 *(Kuo yü)* with quite a different sequence. But in no. 12 *(Kuan-tzu),* where essentially the same interchange takes place, it is Kuan Chung who utters the standard sequence; the duke, because his question is differently worded, does not mention the four categories at all. Particularly puzzling are nos. 9–11 (all *Kuan-tzu*): the section title constituting no. 9 provides the standard sequence, but within the text itself there occur two other different sequences (nos. 10 and 11). Equally varied are the five *Hsün-tzu* passages (nos. 15–19, of which the first two belong to the same chapter), in which no less than four different sequences occur. The conclusion seems inescapable that only following the Chou dynasty, that is, beginning with the second century B.C., did the standard sequence of *shih, nung, kung,* and *shang* become firmly established. Thereafter it was so taken for granted that though cited from time to time, it was almost never discussed.[2]

The variations in the twenty-three Chou texts may be quantified by adding up all occurrences of each social category under each of the four columns and multiplying the resulting totals according to a scale ranging from four for each first-column occurrence down to a single point for each in column 4. The results of this process are shown in table A2.

Table A2 **The Four Social Classes in Chou Texts Quantified**

Categories	Column 1: 4 points	Column 2: 3 points	Column 3: 2 points	Column 4: 1 point	Total
Shih	56	12	0	1	69
Nung	16	36	8	3	63
Kung	0	12	20	9	41
Shang	4	9	18	10	41

It can be seen that when the twenty-three texts are thus quantified, they present a profile in the "total" column that conforms in a general way to the standard sequence. The unevenness of this conformity, however, is evident in the wide gap that separates the "totals" of the *shih/*

2. See, for example, the two casual citations in the writings of the Neo-Confucian Ch'eng brothers (eleventh century), as quoted in Chu Hsi's compendium, the *Chin ssu lu* (A Record for Reflection) (1175); tr. Chan (1967:206, 220).

nung cluster above (69 and 63) from those of the *kung/shang* cluster below (41 and 41). More importantly, the latter figures indicate that between *kung* and *shang* themselves there is no difference. In other words, the upper and lower scores indicate that late Chou writers already regarded both artisans and merchants as sharply inferior to both the office-holding *shih* and the food-providing farmers but that they saw little to choose between the artisans and merchants themselves.

We must conclude that the sharp decline of merchant prestige as compared with that of the artisans began only during the opening century of the Han dynasty (second century B.C.), when the standard four-part sequence was crystallizing. It has been suggested that the change in attitude may have been a reaction to the very considerable growth of private trade and industry that, despite a good deal of effort at government regulation, took place during the Former Han.[3] This could well have been a factor, but another one probably equally important was the growing Confucianization of official thinking during the Han.

We come now to the question of where the four-part classification originated. Despite its five occurrences in the *Hsün-tzu,* a Confucian background seems unlikely. Confucius and Mencius themselves know only the traditional two-layer division; the formulation of this division by Mencius has already been quoted (V.3).

Nor does the four-part classification appear either in Mohism or Taoism. Its real roots, I believe, lie in the thinking of the political scientists loosely referred to as Legalists. These men wanted to reduce the vagaries of personal government by creating an impersonal machinery of institutions. Thus they laid the foundations for what became the bureaucratic way of government in imperial times. They also wanted to build up the state's wealth and power through a planned economy that would involve the total population and utilize quantitative techniques (see III.4). Thus it is natural that they, more than other thinkers, should be interested in classifying the population along socioeconomic lines.

It should not be supposed, however, that all Legalists were equally sophisticated on such matters. Worth noting, for example, is that only one of the twenty-three Chou instances of the four-part classification occurs in the *Shang chün shu* (Book of Lord Shang). Moreover, even this instance is atypical in that it praises only the farmers, criticizing the other three. Thus the sequence of categories in this text is not really one of graduated preference at all.

More striking yet is the failure of the *Han Fei-tzu,* despite its reputation as the epitome of Legalist theory, even to mention the four social

---

3. See Ch'en Teng-yüan (1958–1962:229) and, following him, Vandermeersch (1965:135).

categories. Han Fei Tzu, in fact, appears to have very little idea of what a social class really means. Like other Legalists, he urges a reduction in the number of merchants and artisans and a redirection of their activities toward the "fundamental occupation," by which he means agriculture. Yet in the same breath, and on the same level, he speaks disparagingly of what he calls "itinerant food-seekers"—apparently riffraff or beggars, who scarcely can be said to constitute a social class (*Han Fei-tzu* 49; tr. Liao [1959:296]). Or again, the "five termites" against whom he directs his famous diatribe consist of a potpourri of obnoxious people (probably mostly hangers-on at court), to which he adds two genuine social classes, lumped into one. The obnoxious people include men of learning whose eloquence throws the laws into doubt, talkers who promulgate false statements, wearers of swords who assemble their own adherents, and courtiers who use bribery to advance their own interests. By contrast, the lumped-together social classes are merchants and artisans who fabricate useless objects, accumulate wealth, and exploit the farmers (*Han Fei-tzu* 49; tr. Liao [1959:297]). Finally, despite the apparent importance of Shen Pu-hai (d. 337 B.C.) as an originator of bureaucratic techniques, there is no mention of anything resembling a social class in the fragments painstakingly recovered from his lost writings.[4]

It would seem, therefore, that the four-part social classification derives from that particular variety of Legalism represented by the economically oriented chapters in the *Kuan-tzu.* The Legalism of these chapters is much softer and more humane than the stern dictates found in the *Shang chün shu* and *Han Fei-tzu.* It is a Legalism less interested in harsh punishments, totalitarian thought-control, and secret surveillance, and more interested in state-handled economic controls and monopolies designed to increase the output of agriculture and natural resources, reduce the gap between rich and poor,[5] and generate a stable and ongoing economy. In short, the major aim of these chapters seems to be the creation of what would today be regarded as an authoritarian but relatively humane welfare state.

The differing orientations of the *Kuan-tzu* and the *Han Fei-tzu* are clearly revealed by the concordances available for these two works. Thus the *Han Fei-tzu* concordance lists only twelve entries under *nung,* "agriculture," whereas the same word appears more than a hundred

---

4. See Creel (1974b), whose concordance to Shen's surviving literary fragments does not list the words *nung, kung,* and *shang* at all.

5. See for example the statement in *Kuan-tzu* 8(23/7a); tr. T'an and Wen (1954:154): "If you can take away from the rich and give to the poor, then it will be possible to handle all-under-Heaven *(t'ien hsia).*"

times in the *Kuan-tzu* concordance. Conversely, the binome *shang fa*, "rewards and punishments," appears twice as often in the *Han Fei-tzu* concordance (53 entries) as in that for the *Kuan-tzu* (26 entries).[6]

It is not surprising, then, that six of our twenty-three Chou references to the four classes should come from the *Kuan-tzu* and that two others (nos. 6–7, both *Kuo yü*) should be occasioned by the same apocryphal conversation between Duke Huan and Kuan Chung reported in no. 12 *(Kuan-tzu)*. Despite the diversity of schools of thought represented in the *Kuan-tzu*, the theory that has won widest acceptance is that the work as a whole goes back to the scholars attached to the Chi-hsia Academy founded by King Hsüan around 302 B.C. in the capital of the Ch'i state, where it continued to flourish for some three-quarters of a century. To the original *Kuan-tzu* corpus composed there during the first half of the third century B.C., some accretions were probably later added before the text was finally stabilized around 26 B.C. through the editorial activity of Liu Hsiang (79–8 B.C.).[7] Among the many scholars who attended the Chi-hsia Academy, one of the most distinguished, although then still a young student, was Hsün Tzu (Knoblock 1982–1983:33–34). Thus, if our thesis is correct, it would have been from his associates there, rather than from his Confucian forebears, that Hsün Tzu derived the notion of the four-part division that occurs five times (with four different sequences) in the book bearing his name.

For further discussion of the four-part division in the *Kuan-tzu* and later, the reader should turn to the latter pages of V.3.

---

6. See Johnson (1970:1022–1023 and 1975:815), both *sub nung;* Johnson (1970:1100 and 1975:877), both *sub shang fa.* A systematic comparison of other key terms would further help delineate the differing foci of the two texts.

7. See Rickett (1985:15). The fact that the historical Kuan Chung (d. 645 B.C.), the apocryphal author or hero of so many chapters in the *Kuan-tzu,* was a native of the same state of Ch'i in which the Chi-hsia Academy was situated, strengthens the theory that associates the *Kuan-tzu* with the Chi-hsia scholars.

CHINESE GLOSSARY

For the sake of typographical simplicity, no italics are used for the names and terms that follow, although many, when cited in the main text, appear there in italics. In this Glossary, the names of persons are commonly followed by dates.

An Lu-shan 安祿山 (eighth century)
ch'a 察 (discrimination)
Ch'ai 柴 (type of ceramic)
Chai Yi 翟義 (d. A.D. 8)
Ch'an (Japanese Zen) 禪 (Buddhist sect)
ch'an 讖 (prognostication texts)
Chang 章 (type of ceramic)
ch'ang 常 (regularity, fixed)
Ch'ang-an 長安 (city)
Chang Ang 張卬 (first century A.D.)
ch'ang fa 常法 (fixed laws)
Chang Heng 張衡 (78–139)
Chang Tsai 張載 (1020–1077)
Chang Tzu-mu 張自牧 (nineteenth century)
che pei 折臂 (broken arms)
chen 貞 (integrity)
ch'en 臣 (servant, subject, minister)
chen jen 真人 (True Man)
cheng 正 (rectitude)
cheng fa 征伐 (attacks and campaigns)
Ch'eng Hao 程顥 (1032–1085)
Cheng Hsüan 鄭玄 (127–200)
cheng sheng 正聲 (correct sounds)
cheng t'ung 正統 (Orthodox Sequence)
Ch'eng Yi 程頤 (1033–1108)
chi 技 (art, dexterity)
chi 機/幾 (germs)
chi 機 (mechanical device)
Ch'i 齊 (pre-imperial state)
ch'i 氣 (energy, vital power)
ch'i 奇 (strange, extraordinary)
ch'i 器 (utensil, implement)
ch'i cheng 七政 (Seven Agents)
Chi-hsia 稷下 (ancient scholarly academy)
chi kang 紀綱 (net, nexus; to guide and untangle)
Chi K'ang Tzu 季康子 (contemporary of Confucius)
ch'i p'an wu 七槃舞 (dance of the seven plates)

chi yi 技藝 (skilled arts)
chia-tzu 甲子 (first combination in the sexagenary cycle)
ch'iao 巧 (skill, cleverness)
chiao 教 (a teaching, religion)
chiao ti 角抵 / 角觝 (competitive games, horn-butting, wrestling)
Chieh 桀 (ancient tyrant)
chieh 節 (joint, node, limit, restraint, moral integrity)
chieh po 劫波 (kalpa)
chien 間 (crevice, internal, etc.)
chien ai 兼愛 (all-embracing love)
ch'ien lan 淺藍 (light blue)
ch'ien pi 淺碧 (light jade-green)
chien sheng 姦聲 (depraved sounds)
chih 治 (to govern)
chih 制 (to institute, regulate; regulation)
chih 知 (knowledge)
chih 直 (upright)
chih 志 (will, determination, purpose)
chih 智 (wisdom)
Chih, Robber. *See* Robber Chih
chih-chüeh 知覺 (consciousness)
chih hsing ho yi 知行合一 (unity of knowledge and conduct)
chih-hui 知慧 (knowledge)
Ch'ih-yu 蚩尤 (a mythological divinity)
ch'in 琴 (lute)
Chin Sheng-t'an 金聖嘆 (1610?–1661)
chin shih 進士 (advanced scholar, Doctor in Letters)
ching 精 (essence, semen)
ch'ing 青 (green, blue, etc.)
ching hsi tzu chih 敬惜字紙 (respect and care for written paper)
ching shih 靜室 (houses of retreat)
Ching-te-chen 景德鎮 (porcelain manufacturing center)
ching-t'ien 井田 (well-field land system)
ch'ing t'ien 青天 (blue sky)
ch'ing ts'ao 青草 (green grass)
chiu 久 (duration)
chiu chou 九州 (nine provinces or continents)
chiu ch'ung 九重 (nine levels)
chiu t'ien 九天 (nine heavens)
chou 宙 (beams; time). *See also* yü
Chou (or Chou Hsin) 紂辛 (ancient tyrant)
Chou Ch'ang 周昌 (Han statesman)
Chou Tun-yi 周敦頤 (1017–1073)
chü. *See* kuei-chü
ch'ü 曲 (type of song)
ch'u 出 (be produced, put forth)
Chu Hsi 朱熹 (1130–1200)

chü jen 舉人 (raised man)
chu k'o 諸科 (several [other] fields)
Chu Tsai-yü 朱載堉 (1536–after 1595)
Chü-tzu 鉅子 (Grand Master)
Chu Yi-hsin 朱一新 (nineteenth century)
chüan 卷 (part or division of a book)
chün 均 (standards)
ch'ün 羣 (to collect together)
ch'un hua 春畫 (spring pictures)
chün tzu 君子 (lord's son, true gentleman, superior man, princely man, etc.)
chung 中 (middle, centrality)
chung min 衆民 (the numerous people)
chung shu 衆庶 (common herd)
erh 爾 (thou, you)
fa 法 (pattern, model, law)
fan li 凡例 (general rules)
fang 方 (direction, region)
Fei Yi 肥義 (fourth century B.C.)
fen 分 (distinctions, divisions, social distinctions)
feng huang 鳳凰 (phoenix)
fu (clothing). *See* yi-fu
fu 父 (father)
Fu Hsi 伏羲 (Subduer of Animals, a legendary culture hero)
fu mu kuan 父母官 (father-and-mother officials)
hao t'ien 昊天 (glorious heaven)
Ho Hsiu 何休 (129–182)
Ho Hsü 赫胥 (ancient worthy)
Ho-shen 和珅 (1750–1799)
Ho t'u 河圖 (River Chart)
Hsi K'ang 嵇康 (223–262)
Hsi Shih 西施 (legendary beauty)
hsia 下 (inferiors)
Hsiang 湘 (river)
hsiang p'u 相撲 (wrestling)
hsiao jen 小人 (small or petty man)
hsiao k'ang 小康 (small tranquility)
hsieh 俠 (knights-errant)
Hsieh Chi-shih 謝濟世 (1689–1756)
hsien 縣 (district, country)
hsien 仙 (immortal)
hsien 賢 (man of moral worth)
hsien sheng 先生 (gentleman, Mr.)
hsin 信 (good faith)
hsing 性 (human nature)
hsing 行 (to move)
hsing 刑 (punishment)
hsiung 凶 (malignant)

Hsü Ch'i 徐祺 (eighteenth century)
Hsü Hsing 許行 (ancient agricultural thinker)
hsüan che 玄者 (the Mystery)
Hsün Tzu 荀子 (ca. 298–ca. 238 B.C.)
hua kung 畫工 (artisan painter)
Huang Kan 黄幹 (1152–1221)
Huang-Lao 黄老 (the Yellow Lord and Lao Tzu)
Huang Ti 黄帝 (the Yellow Lord)
Huang T'ien Shang Ti 皇天上帝 (August Heaven Lord on High)
Hui Shih 惠施 (third century B.C. dialectician)
Huo Ch'ü-ping 霍去病 (general, 146/145–117)
i. *See* yi
jang 讓 (yieldingness)
jen 仁 (benevolence, humanity, love, perfect virtue, etc.)
jen 人 (person)
Jen An 任安 (friend of Ssu-ma Ch'ien)
jen lun 人倫 (human relationships)
jen tao chih cheng 人道之正 (correct Way of Man)
Ju 汝 (type of ceramic)
ju 汝 (thou, you)
Ju chia 儒家 (School of the Scholars or Literati)
kai 垓 (realms)
kai t'ien 蓋天 (Heavenly Cover)
kalpa. *See* chieh po
k'ao cheng hsüeh 考證學 (School of Evidential Research)
k'an 凵 (open container)
Kao Tzu 告子 (opponent of Mencius)
K'ao yi 考異 (Examination of Differences)
Kao Yu 高誘 (fl. 205–212)
Ko 哥 (type of ceramic)
ko-jen chu-yi 個人主義 (single-man-ism, individualism)
ko wu 格物 (investigation of things)
ku 固 (firm)
ku 賈 (merchant, shopkeeper)
Kuan 官 (type of ceramic)
kuan hua 官話 (Mandarin, official speech)
K'uang Heng 匡衡 (Han statesman)
kuei 鬼 (demons)
kuei-chü 規矩 (compass and square; normative, proper)
kuei-hsieh 庋擷 ("pigeon-holing" filing system)
k'ung 空 (empty)
Kung-shu Pan 公輸般 (ancient artisan)
Kuo Jo-hsü 郭若虛 (eleventh century)
kuo-yü 國語 (national language)
Lao-Chuang 老莊 (Lao Tzu and Chuang Tzu)
lei 類 (category)
Lei Fa-ta 雷發達 (1619–1693)

li 理 (principle, pattern)
li 利 (profit, gain)
li 禮 (rites, ceremonials, rules of civilized behavior, traditional mores)
Li Chih 李贄 (1527–1602)
Li Hung 李弘 (Han magistrate)
Li Kung 李塨 (1659–1746)
Li Ling 李陵 (general, d. 74 B.C.)
li min 黎民 (the black-haired people)
Li Ssu 李斯 (d. 208 B.C.)
Liang Ch'i-ch'ao 梁啟超 (1873–1929)
liang chiang 良匠 (able artisan)
liang yi 兩儀 (Dual Forms)
ling 令 (ordinances, orders, to order)
Liu Hsin 劉歆 (ca. 46 B.C.–A.D. 23)
Liu-hsia Hui 柳下惠 (ancient paragon of virtue)
Liu Hsü 劉胥 (128–54 B.C.)
Liu Kung-ch'üan 劉公權 (eighth century)
liu po 六博 (Han gaming board)
liu yi 六藝 (six accomplishments)
lo 樂 (delight in). *See also* yüeh (music)
Lo shu 洛書 (Lo Writing)
Lo-yang 洛陽 (city)
Lü Wen 呂溫 (ca. 774–ca. 813)
luan 亂 (disorder, confusion; to bring into order)
luan ch'en 亂臣 (rebellious ministers; order-bringing ministers)
Lung-ch'üan 龍泉 (type of ceramic)
ma 馬 (horse)
Ma Chün 馬鈞 (fl. A.D. 260)
Ma-wang-tui 馬王堆 (a tomb site)
men 門 (gate, sect)
Meng-ch'ang, Lord of 孟嘗君 (fourth–third centuries B.C.)
Mi 密 hsien (place)
min 民 (the people)
min chih fu mu 民之父母 (people's father and mother)
ming 命 (destiny, fate, decree)
ming 名 (name, reputation)
ming chu 明主 (enlightened ruler)
Ming T'ang 明堂 (Sacred or Cosmic Hall)
mo 末 (secondary occupations)
mou 畝 (Chinese land measure or acre)
Nei 內 (type of ceramic)
nei tan 內丹 (internal elixir, physiological hygiene)
ni ch'i 逆氣 (vital force in its unruly aspect)
nien p'u 年譜 (yearly register)
pa-ho 拔河 (pulling the river, tug-of-war)
pa-ku wen 八股文 (eight-legged essay)
pai hsing 百姓 (the hundred surnames)

pai hua 白話 (colloquial Chinese)
pao meng 包蒙 (forebearance)
pao pien 褒貶 (praise and blame)
pei shu 背書 (turn the back on the text)
pen 本 (primary occupation)
pi jen 敝人 (humble person)
Pi Yung 辟雍 (Hall of the Circular Moat)
p'iao 縹 (pale blue)
pieh 别 (differentiation)
pien 辯 (dispute, argument, debate)
p'ien 篇 (chapter)
pien-che 辯者 (arguers)
pinyin 拼音 ("linked sounds," a romanization system)
Po Chü-yi 白居易 (772–846)
Po-yi 伯夷 (ancient paragon of virtue)
p'u 樸 (Unwrought Simplicity). *See also* su p'u
pu shih 不時 (untimely)
p'u t'ung hua 普通話 (ordinary or standard speech)
Robber Chih 盗跖 (ancient exemplar of violence)
san ts'ai 三材 (the three powers)
se 色 (color, appearance, beauty, sex)
shan 善 (merit)
shan 山 (mountain)
shan hua 善化 (transforming power of goodness)
shan shu 善書 (good writings, moral tracts)
shang 上 (superiors)
shang fa 賞罰 (rewards and punishments)
shang ti 上帝 (Lord on High; rulers of old)
Shao Yung 邵雍 (1011–1077)
Shen Nung 神農 (Divine Husbandman, a legendary culture hero)
Shen Pu-hai 申不害 (d. 337 B.C.)
sheng p'ing 升平 (Approaching Peace)
shih 矢 (disregard)
shih 士 (gentleman, scholar)
shih 式 (Han divination board)
shih 時 (period, season, timeliness)
shih 實 (reality, substance)
shih 事 (things, affairs, occurrences)
shih chih kung jen 世之工人 (artisans of the world)
shih nung kung shang 士農工商 (scholars, farmers, artisans, merchants)
Shih tse hsün 時則訓 (Teachings on the Rules for the Seasons)
shu 數 (numbers, statistics, statistical methods)
shu min 庶人 (the multitudinous persons)
shu min 庶民 (the multitudinous people)
shuai luan 摔亂 (Disorder)
shuang sheng 雙聲 (doubled sounds)
shui chih hsing 水之性 (the nature of water)

Shun 舜 (legendary sage ruler)
shun ch'i 順氣 (vital force in its compliant aspect)
Shun-yü Yi 淳于意 (Han physician)
so 所 (place)
ssu 嗣 / 司 (to regulate)
ssu 私 (private, personal, selfish, etc.)
ssu-chiao hao-ma 四角號碼 (four-corner numbering filing system)
ssu min 四民 (the four peoples)
ssu t'i 四體 (four limbs)
ssu yüeh 四嶽 (four peaks)
Su Ch'in 蘇秦 (third century B.C. diplomat)
su p'u 素樸 (Unwrought Simplicity). *See also* p'u
Su Sung 蘇頌 (eleventh century)
Su Tung-p'o 蘇東坡 (1037–1101)
sui shou 碎首 (battered heads)
Ta Hsiang-kuo Monastery 大相國寺
Ta-hu-t'ing 打虎亭 (place)
ta jen 大人 (great man)
ta ming 大命 (Great Mandate)
ta t'ung 大同 (Great Unity)
Tai Chen 戴震 (1723–1777)
T'ai Ho Tien 太和殿 (Hall of Grand Harmony)
t'ai p'ing 太平 (Grand Peace)
t'ai yi 太一 (Supreme Oneness)
T'an Ssu-t'ung 譚嗣同 (1865–1898)
tao 盜 (brigand)
Tao 道 (Way, road)
Tao chia 道家 (Taoist school)
T'ao Ch'ien 陶潛 (365–427)
Tao hsüeh chia 道學家 (School of the Study of the Tao)
tao t'ung 道統 (sequence or transmission of the Tao)
te 德 (virtue, power)
Te-hua 德化 (place)
ti 地 (earth)
ti chün 帝君 (lord-rulers)
tieh yün 疊韻 (piled-up rhymes)
t'ien 天 (Heaven, sky)
t'ien chen 天眞 (Genuineness of Heaven)
t'ien fa 天法 (laws of Heaven)
t'ien hsia 天下 (all-under-Heaven)
t'ien ming 天命 (Mandate of Heaven)
t'ien tse 天則 (rule of Heaven)
Ting 定 hsien (place)
t'ing liao 庭燎 (courtyard torches)
tou chi 鬭雞 (cock fighting)
tou mu 鬭鶩 (drake fighting)
tsa chih 雜至 (arrive irregularly)

Ts'ai Chih 蔡質 (d. prob. A.D. 178)
Ts'ao Chih 曹植 (192–232)
tsao wu che 造物者 (maker, founder, or originator of things)
tse 則 (rule)
tsei hai 賊害 (do violence)
Tseng Tzu 曾子 (disciple of Confucius)
tso 作 (to make)
Tso Kuang-tou 左光斗 (1575–1625)
tsou kou/ch'uan 走狗/犬 (dogs or hounds racing)
tsou ma 走馬 (horse racing)
Tsou Yen 騶衍 (fl. ca. 305–ca. 240 B.C.)
ts'u chü 蹴鞠 (kick-ball, football)
tsui 翠 (kingfish color)
tsui 最 (perfection)
tsui 罪 (sin)
tsung 宗 (clan)
tsung chiao 宗教 (religion)
tu 度 (measuredness, measuring)
tui-tzu 對子 (juxtaposed things)
Tun-huang 敦煌 (place)
Tung 東 (type of ceramic)
Tung Ch'i-ch'ang 董其昌 (1555–1639)
Tung Hsien 董賢 (d. 1 B.C.)
tung liang 棟梁 (roof-beams)
tzu 紫 (purple)
tzu 子 (son, Master)
Tz'u-chou 磁州 (place)
Tzu-hsia 子夏 (disciple of Confucius)
tzu hsü 自序 (self-narration)
tzu jan 自然 (spontaneity, so of itself)
Tzu-kung 子貢 (disciple of Confucius)
Tzu-ssu 子思 (grandson of Confucius)
Wan-ch'üan 萬全 hsien (place)
wan tuan kuei chü 萬端規矩 (a phrase)
wan wu 萬物 (myriad creatures)
Wang Fu-chih 王夫之 (1619–1693)
Wang Hsü 王旭 (eighth century)
Wang Ken 王艮 (1483–1540)
Wang Liang 王良 (ancient charioteer)
Wang Mang 王莽 (ruled A.D. 9–23)
Wang Ping 王冰 (eighth century)
wang tao 王道 (Kingly Way)
Wang Yang-ming 王陽明 (1472–1529)
wei 緯 (apocrypha)
Wei chi 未濟 (Unfinished; name of sixty-fourth hexagram)
wei ch'i 圍棋 (Chinese chess)
Wei Hung 衛宏 (fl. A.D. 25–57)

wei wo 為我 (acting for self, for the "I")
wen 文 (culture, writing, civil tradition). *See also* wu
wen-jen hua 文人畫 (literati painting)
wen yen 文言 (Literary Chinese)
wo 我 (I, my)
wu 吾 (I, my)
wu 武 (military tradition). *See also* wen
wu 無 (nonexistence, nonbeing, "what-is-not"). *See also* yu
wu 物 (things, creatures, affairs)
wu ch'ang 五常 (five constants)
Wu Han 吳漢 (first century A.D.)
wu hsing 五性 (five aspects of human nature)
wu hsing 五行 (Five Elements)
Wu Liang Tz'u 武梁祠 (Wu Liang Shrine)
wu p'in 五品 (five categories)
Wu Tao-tzu 吳道子 (d. 792)
wu te 五德 (Five Powers)
wu ts'ai 五材 (Five Substances)
wu yen 屋簷 (eaves)
wu yüeh 五嶽 (five peaks)
yang 陽 (positive cosmic principle). *See also* yin
Yang Chu 楊朱 (ancient thinker)
Yang Hsiung 揚雄 (53 B.C.–A.D. 18)
Yang Kuang-hsien 楊光先 (1597–1669)
Yao 堯 (legendary sage ruler)
yeh 野 (expanses)
Yen Fu 嚴復 (1853–1921)
Yen Hui/Yen Tzu 顏回/子 (disciple of Confucius)
Yen Shih-ku 顏師古 (581–645)
Yen Su 燕肅 (d. 1040)
Yen Yüan 顏元 (1635–1704)
yi (i) 易 (change, easy, etc.)
yi (clothing). *See* yi-fu
yi 義 (duty, righteousness, right conduct, justice, what is right)
yi 意 (thought)
yi-ch'ou 乙丑 (second combination in the sexagenary cycle)
yi-fu 衣服 (clothing)
Yi-hsing 宜興 (place)
Yi Ti chih jen 夷狄之人 (man of the Yi and Ti tribes)
yi tuan 異端 (strange shoots)
yin 陰 (negative cosmic principle). *See also* yang
yin 淫 (obscene, licentious)
ying chien 營建 (construction)
ying tsao 營造 (architecture)
Yo 岳 (type of ceramic)
yu 有 (existence, being, "what is"). *See also* wu
Yü 禹 (legendary founder of Hsia dynasty)

yü 宇 (eaves; space). *See also* chou
yü 猶 (to be like)
yüan 元 (originator)
Yüan Ang 袁盎 (first century B.C.)
Yüeh 越 (type of ceramic)
yüeh 樂 (music). *See also* lo
Yüeh Fei 岳飛 (a patriot, 1103–1141)
Yün-meng 雲夢 (place)
yung 勇 (courage)
Zen. *See* Ch'an

BIBLIOGRAPHY

General comments concerning Sections A and B below will be found in I.3d.

A. Chinese Original Sources

Texts are here listed by title when this is better known than their author, but by author when the author is better known. Many early texts are anonymous or, even when named after given thinkers, have, to varying degrees, been compiled by the followers of those thinkers rather than the thinkers themselves. Many texts, too, although originating during the Chou dynasty, have undergone subsequent accretions or other changes before reaching their present form. Particularly important has been the editing and commenting done by Han scholars. These are some of the reasons why it is so often difficult to determine the dating or authorship of early Chinese texts with any precision.

Collectanea Abbreviations

KHCPTS — *Kuo-hsüeh chi-pen ts'ung-shu* 國學基本叢書 ed. Taipei reprint: Commercial Press, 1968.

Pal. ed. (1923) — 1923 photographical reprint by Chung-hua (China) Book Co., Shanghai, of the 1739 Palace ed. (Tien pen 殿本) of the Twenty-four Dynastic Histories. The reprint carries forty-two characters to a column and twenty columns to a page, making it exactly twice the size of the original Palace ed., with which it is otherwise identical.

Ma Kuo-han — Ma Kuo-han 馬國翰 (1794–1857), comp., *Yü-han shan-fang chi-yi-shu* 玉函山房輯佚書 (Lost Writings Restored in the Jade-Receptacle Studio). 120 vols. Changsha ed. of 1883.

SPPY — *Ssu-pu pei-yao* 四部備要 ed. Shanghai: Chung-hua (China) Book Co., 1927–1935.

SPTK — *Ssu-pu ts'ung-k'an* 四部叢刊 ed. Shanghai: Commercial Press, 1929–1936.

Taoist Canon — *Tao tsang* 道藏. Taoist collection of 1,467 works completed in 1626. Reprinted Shanghai: Commercial Press, 1923–1926.

TSCC — *Ts'ung-shu chi-ch'eng* 叢書集成 ed. Shanghai: Commercial Press, 1936–1939.

Cited Works

Analects. See *Lun yü*

Chan-kuo ts'e 戰國策 (Stratagems of the Warring States). Late third century B.C. Anon. 10 *chüan*. Tr. Crump (1970).

Chang Heng 張衡 (78–139). *Tung-ching fu* 東京賦 (Rhapsody on the Eastern Metropolis). In *Wen hsüan* 文選 (Selections of Refined Literature), *chüan* 3, ca. A.D. 530. Comp. by Hsiao T'ung 蕭統 (501–533). Tr. Knechtges (1982: 243–310).

Chang Po-hsing 張伯行 (1652–1725). *Tao t'ung lu* 道統錄 (Record of the Sequence of the Tao). 1708. 2 *chüan*. *TSCC* ed., vol. 3338.

Chang Tsai 張載 (1020–1077). *Hsi ming* 西銘 (Western Inscription).

Changes or *Changes Classic*. See *Yi ching*

Ch'i-min yao-shu 齊民要書 (Essentials for the People's Welfare). Ca. A.D. 535. By Chia Ssu-hsieh 賈思勰.

Chieh-chou hsüeh-hua p'ien 芥舟學畫篇 (A Tiny Treatise on the Study of Painting). 1781. By Shen Tsung-ch'ien 沈宗騫. Peking: Fine Arts Publishing Co., 1959. Tr. Lin Yutang (1967).

Ch'ien Han shu 前漢書 (History of the Former Han Dynasty). Also known as *Han shu* (Han History). By Pan Ku 班固 (A.D. 32–92) et al. 120 chs. Covers period 209 B.C.–A.D. 25. Pal. ed. (1923). Tr. Dubs (1938–1955).

Ch'ien-tzu wen 千字文 (Thousand Characters Classic). Attributed to Chou Hsing-ssu 周興嗣 (d. A.D. 521). Tr. Julien (1864).

Chin P'ing Mei 金瓶梅. Ca. 1619. Erotic Ming novel.

Chin shu 晉書 (Chin History). By Fang Hsüan-ling 房玄齡 (578–648) et al. 130 chs. Covers period 265–419. Pal. ed. (1923).

Chin ssu lu 近思錄 (A Record for Reflection). 1175. By Chu Hsi 朱熹 and Lü Tsu-ch'ien 呂祖謙. Tr. Chan (1967).

Ching Ch'u sui-shih-chi 荊楚歲時記 (Record of the Annual Seasons in Central China). Short text by Tsung Lin 宗懍 (ca. 498/503–ca. 561/566), with much longer commentary by Tu Kung-shan 杜公瞻 (fl. ca. 600). Tr. Turban (1971).

Ching-te-chen t'ao-lu 景德鎮陶錄 (Report on the Potteries of Ching-te-chen). 1815. By Lan P'u 藍浦 with major editions and editing by Cheng T'ing-kuei 鄭廷桂. 10 *chüan*. Taipei: Wen Hai (Sea of Literature) Publishing Co., 1969 (reprint of 1891 ed. based on a text then in Kyoto). Tr. Julien (1856), Bushell (1910), Sayer (1951).

Chiu T'ang shu 舊唐書 (Old T'ang History). A.D. 945. By Liu Hsü 劉昫 (887–946) et al. 200 chs. Covers period 618–906. Pal. ed. (1923).

Chou li 周禮 (Institutes of Chou). Late Chou/early Han. Anon. 42 *chüan*. *SPPY* ed. Tr. Biot (1841).

Chou Tun-yi 周敦頤 (1017–1073) *T'ai-chi t'u* 太極圖 (Diagram of the Supreme Pole).

Chu Hsi 朱熹 (1130–1200). *Chu-tzu ta-ch'üan* 朱子大全 (Great Compendium of Chu Hsi's Writings). 1714. 66 *chüan*. *SPPY* ed.

———. *Chu-tzu yü-lei* 朱子語類 (Classified Conversations of Chu Hsi). 1270. Comp. by Chu Hsi's disciples. 140 *chüan* in 8 vols. with continuous pagination (reprint of 1473 ed.). Taipei: Cheng-chung (Correct and Central) Book Co., 1962. See also *Chin ssu lu*

Ch'u tz'u. See Ch'ü Yüan; *T'ien wen*

Ch'ü Yüan 屈原 (ca. 300 B.C.). Putative author of *Li sao* 離騷 poem in *Ch'u tz'u* 楚辭 (Songs of Ch'u). Tr. Hawkes (1959).

Chuang-tzu 莊子 (Master Chuang). By Chuang Chou 周 (ca. 369–ca. 286) and other anon. Taoists. 33 chs. Chekiang Book Co. ed. of 1876. Tr. Watson (1968), Graham (1981).

Ch'un ch'iu 春秋 (Spring and Autumn Annals). Concise chronicle of the state of Lu, 722–481 B.C. Incorrectly attributed to Confucius. Tr. Legge (1960:5), Couvreur (1914).

Ch'un-ch'iu fan-lu. See Tung Chung-shu

Ch'un-ch'iu-wei Han-han-tzu 春秋緯漢含孳 (Apocryphal Treatise on the Spring and Autumn Annals: Cherished Beginnings of Growth of the Han Dynasty). Among the apocrypha of first century B.C. Fragments re-collected in Ma Kuo-han, vol. 65.

Chung Hung. See *Shih p'in*

Chung yung 中庸 (Doctrine of the Mean). Late Chou or early Han. One of the Confucian "Four Books." Tr. Legge (1960:1.382–434). Also constitutes *Li chi* ch. 28; tr. Legge (1885:28.300–329), Couvreur (1913:2.427–479).

Confucius. See *Lun yü*

Doctrine of the Mean. See *Chung yung*

Documents or *Documents Classic*. See *Shu ching*

Erh Ch'eng ch'üan-shu 二程全書 (Complete Works of the Two Ch'engs), sect. *Yi-shu* 遺書 (Surviving Works). By Ch'eng Hao 程顥 (1032–1085) and Ch'eng Yi 頤 (1033–1107). 164 *chüan*. *SPPY* ed.

Erh ya 爾雅 (Literary Expositor). Perhaps third–second centuries B.C. Anon. China's first dictionary (words listed under topics).

Fan Sheng-chih shu 氾勝之書 (Book of Fan Sheng-chih). Ca. 10 B.C. By Fan Sheng-chih. Agricultural treatise, today known only in re-collected fragments in Mao Kuo-han, vol. 81. Tr. Shih Sheng-han (1959).

Fang Yi-chih 方以智 (1611–1671). *Wu li hsiao shih* 物理小識 (Small Encyclopedia of the Principles of Things). 1644.

Feng-su t'ung-yi 風俗通義 (Comprehensive Meaning of Customs). By Ying Shao 應劭 (ca. 140–ca. 206). 10 *chüan* plus 6 *chüan* of fragments re-collected from other sources. Refs. to *Le Fong Sou T'ong Yi*. Peking: Centre Franco-Chinois d'Etudes Sinologiques, 1943.

Great Learning. See *Ta hsüeh*

Han Fei-tzu 韓非子 (Master Han Fei). By Han Fei (d. 233 B.C.) and other anon. Legalist thinkers. 55 chs. Refs. to Wang Hsien-shen 王先愼 (1859–1922), ed. *Han Fei-tzu chi-chieh* 集解 (Collected Commentaries on the *Han Fei-tzu*). Taipei: Shih-chieh (World) Book Co., 1973. Tr. Liao (1939, 1959).

Han shu. See *Ch'ien Han shu*

Han Yü 韓愈 (768–824). *Han Ch'ang-li ch'üan-chi* 韓昌黎全集 (Complete Writings of Han Ch'ang-li [Han Yü]). 40 *chüan*. *SPPY* ed. Cited items: *Shih shuo* 師說 (On the Teacher), 12/1b–3a; *Yen Fo ku piao* 言佛骨表 (Memorial on a Bone of the Buddha), 39/3a–5b; *Yüan Tao* 原道 (Origin of the Tao), 11/1a–5a.

Hou Han shu 後漢書 (History of the Later Han Dynasty). By Fan Yeh 范曄 (398–445) with treatises by Ssu-ma Piao 司馬彪 (d. 305). 90 chs. Covers period A.D. 25–220. Pal. ed. (1923).

Hsin hsü (A New Narration). *See* Liu Hsiang

Hsin Wu-tai shih 新五代史 (New History of the Five Epochs). By Ou-yang Hsiu 歐陽修 (1007–1072). 74 chs. Covers period 907–960. Pal. ed. (1923).

Hsün-tzu 荀子 (Master Hsün). By Hsün Ch'ing 卿 (ca. 298–ca. 238). 32 chs. *SPPY* ed. Tr. Dubs (1928), Watson (1963).

Hsün Yüeh 荀悅 (148–209). *Shen Chien* 申鑒. 5 *chüan*. Tr. Ch'en (1980).

Hua shu 化書 (Book of Transformations). Attributed to T'an Ch'iao 譚峭 (tenth century). *Taoist Canon*, vol. 724, no. 1038.

Huai-nan-tzu 淮南子 (The Huai-nan Master). Anthology of anon. essays comp. ca. 140 B.C. under auspices of Liu An 劉安, King of Huai-nan (d. 122 B.C.). 21 chs. Refs to Liu Wen-tien 劉文典, ed. *Huai-nan hung-lieh chi-chieh* 鴻烈集解 (Illustrious Essays of Huai-nan with Collected Commentaries). Shanghai: Commercial Press, 1933. Tr. Morgan (1934); also several individual chs. by other scholars.

Huang Po-ssu 黄伯思 (1079–1118). *Lun Chang ch'ang-shih shu* 論張長史書 (On the Calligraphy of Senior Secretary Chang). In Huang's *Tung-kuan yü-lun* 東觀餘論 (Further Discussions in the Eastern Lodge), A/53b–54b. Refs. to *Hsüeh-ching t'ao-yüan ts'ung-shu* 學津討原叢書 ed. (1965 Taipei reprint).

Huang-ti nei-ching su-wen 黄帝内經素問 (Plain Questions of the Yellow Lord: The Canon of Internal Medicine). Han dynasty. Anon. *SPTK* ed.

Jih-chih lu 日知錄 (Daily Additions to Knowledge). First version, 1670; definitive version, 1695. By Ku Yen-wu 顧炎武 (1613–1682). 32 *chüan*. *KHCPTS* ed.

K'ai-yüan chan-ching 開元占經 (K'ai-yüan Treatise on Astrology). A.D. 729. By Ch'ü-t'an Hsi-ta 瞿曇悉達 (Gautama Siddhārtha). 120 *chüan*.

K'ang-hsi tzu-tien 康熙字典 (K'ang-hsi Dictionary). 1716.

K'ang Yu-wei 康有為 (1858–1927). *Ta t'ung shu* 大同書 (Book on the Great Unity). Completed ca. 1901–1902. Tr./paraphrase Thompson (1958).

K'ao kung chi 考工記 (Artificers' Record). In *Chou li* (Institutes of Chou) *chüan* 40–44; tr. Biot (1841:2. 456–611).

Ko Hung. See *Pao-p'u-tzu*

Ku-liang chuan 穀梁傳 (Ku-liang Commentary). One of three major commentaries on the *Ch'un ch'iu* (Spring and Autumn Annals), dubiously attributed to a disciple of a disciple of Confucius. *SPPY* ed.

Ku Yen-wu. See *Jih-chih lu*

Kuan-tzu 管子 (Master Kuan). Attributed to Kuan 仲 (d. 645 B.C.) but actually an anon. eclectic work of 86 chs. dating from late Chou into early Han. *SPPY* ed. Tr. Rickett (1965, 1985); T'an Po-fu and Wen Kung-wen (1954).

Kuan Yin-tzu 關尹子 (Master Kuan Yin). Eighth century A.D. Probably by T'ien T'ung-hsiu 田同秀. *SPPY* ed.

Kuang yün 廣韻 (Expanded Rhyming Dictionary). 1011. Revision and enlargement of earlier rhyming dictionaries.

Kung-yang chuan 公羊傳 (Kung-yang Commentary). Late third–early second century B.C. One of the three major commentaries on the *Ch'un ch'iu* (Spring and Autumn Annals).

Kuo Hsiang 郭象 (d. A.D. 312). Commentary on *Chuang-tzu*. Excerpts tr. in Fung (1953:207–216).

Kuo yü 國語 (Discourses of the States). fourth–third century B.C. Refs. to divisions and subdivisions under names of states.

Lao-tzu 老子 (The Old Master). Also known as *Tao te ching* 道德經 (Canon of the Way and Its Power). Piously attributed to an older contemporary of Confucius, but probably third century B.C. 81 short chs. Tr. Waley (1934) and many others.

Li chi 禮記 (Record of Ceremonial). Chou/early Han. 46 chs. Tr. Legge (1885), Couvreur (1913).

Li Shih-chen. See *Pen-ts'ao kang-mu*

Li-tai ming-hua chi 歷代名畫記 (Compendium of Famous Paintings of Successive Dynasties). A.D. 847. By Chang Yen-yüan 張彥遠. 10 *chüan*. Tr. Acker (1954, 1974).

Li yi chih 禮儀志 (Treatise on Ritual). Name of certain chs. included in many of the dynastic histories.

Li Yüan-kang 李元綱 (twelfth century). *Sheng-men shih-yeh t'u* 聖門事業圖 (Diagrams of the Works of the School of Sages). 1173. 1 *chüan*. Contained in *Pai-ch'uan hsüeh-hai* 百川學海 2.999–1029. Taipei reprint in 7 vols. Hsin-hsing (New and Flourishing) Book Co., 1969.

Li yün 禮運 (Revolutions of Ceremonial). In *Li chi* ch. 7. Tr. Legge (1885: 27.364–393), Couvreur (1913: 1.496–537).

Lieh-tzu 列子 (Master Lieh). Allegedly by a pre-Han Taoist but probably a third century A.D. compilation based in part on earlier materials. *SPPY* ed. Tr. Graham (1960).

Liu Hsiang 劉向 (79–8 B.C.). *Hsin hsü* 新序 (A New Narration) and *Shuo yüan* 說苑 (A Garden of Sayings).

Lo-yang ch'ieh-lan chi 洛陽伽藍記 (Record of Buddhist Monasteries in Lo-yang). A.D. 547. By Yang Hsüan-chih 楊衒之. 5 *chüan*. Tr. Yi-t'ung Wang (1984) and Jenner (1981).

Lu Chiu-yüan 陸九淵 (1139–1193). *Hsiang-shan ch'üan-chi* (Complete Works of Hsiang-shan [Lu Chiu-yüan]). 36 *chüan*. *SPPY* ed.

Lü-shih ch'un-ch'iu 呂氏春秋 (Mr. Lü's Springs and Autumns). 240 B.C. Comp. by anon. scholars under auspices of Lü Pu-wei 呂不韋. 26 *chüan*. Chekiang Book Co. ed. of 1875. Tr. Wilhelm (1928).

Lü Tsu-chien. See *Chin ssu lu*

Lun heng 論衡 (Doctrines Evaluated). A.D. 82/83. By Wang Ch'ung 王充 (A.D. 27–ca. 96). 84 chs. Refs. to Huang Hui 黃暉, ed. *Lun-heng chiao-shih* 校釋 (*Lun heng* Collected and Explained). 2 vols. with continuous pagination. Changsha: Commercial Press, 1938. Tr. Forke (1962).

Lun yü 論語 (Analects). Sayings of Confucius (551–479 B.C.) comp. by his disciples and their disciples. 20 chs. with many subsects. Tr. Legge (1960: vol. 1) and Waley (1938).

Mao Ch'i-ling 毛奇齡 (1623–1716). *Ssu-shu kai-ts'o* 四書改錯 (Correction of Errors Concerning the Four Books). 22 *chüan*.

Mencius. See *Meng-tzu*

Meng-ch'i pi-t'an 夢溪筆談 (Dream Torrent Jottings). 1086. By Shen Kua 沈括 (1030–1094). 26 *chüan*.

Meng liang lu 夢粱錄 (A Dream during the Cooking of Porridge). 1275. By Wu Tzu-mu 吳自牧. *Chih pu-tsu chai ts'ung-shu* 知不足宅叢書 ed. (1921 reprint).

Meng-tzu 孟子 (Master Meng). Collected conversations of Mencius (fl. 371–279). 7 chs., each divided into 2 pts. with many subsects. Tr. Legge (1960: vol. 2).

Ming shih 明史 (Ming History). 1736. By Chang T'ing-yü 張廷玉 (1672–1755) et al. 332 chs. Covers period 1368–1644. Pal. ed. (1923).

Mo-tzu 墨子 (Master Mo). By disciples of Mo Ti 翟 (ca. 479–ca. 381) and later followers of his school. 71 chs. (18 no longer extant). Refs. to Sun Yi-jang 孫詒讓 (1848–1908), ed. *Mo-tzu chien-ku* 閒詁 (A Penetrating Exposition of the *Mo-tzu*), 1907. Tr. Mei (1929) and Graham (1978).

Mu ching 木經 (Timberwork Manual). By Yü Hao 喻皓 (fl. 965–995). Now lost.

Nan Ch'i shu 南齊書 (Southern Ch'i History). By Hsiao Tzu-hsien 蕭子顯 (489–537). 59 chs. Covers period 497–502. Pal. ed. (1923).

Nan shih 南史 (Southern Histories). Ca. 629. By Li Yen-shou 李延壽. 80 chs. Covers period 420–589. Pal. ed. (1923).

Nung-cheng ch'üan-shu 農政全書 (Complete Treatise on Agriculture). 1639. By Hsü Kuang-ch'i 徐光啟 (1562–1633). 60 *chüan*.

Pai wen p'ien 百問篇 (The Hundred Questions). In *Tao shu* 道樞 (Pivot of the Tao) *chüan* 5. Before 1145. *Tao shu* is by Tseng Ts'ao 曾慥. *Taoist Canon*, vols. 641–648, no. 1011. Tr. Homann (1976).

Pao-p'u-tzu 抱朴子 (The Master Who Embraces Simplicity). Ca. A.D. 320. By Ko Hung 葛洪 (283–343). 20 inner chs., 52 outer chs. Po Yün T'ang 栢筠堂 ed., undated. Tr. Ware (1966) of inner chs.

Pei shih 北史 (History of the Northern Dynasties). Ca. 629. By Li Yen-shou 李延壽. 100 chs. Covers period 386–617. Pal. ed. (1923).

P'ei-wen yün fu 佩文韻府 (Rhyming Treasury for Honoring Literature). 1711. 106 *chüan* plus 106 supplementary *chüan*. Repertory of mostly two-character terms. Shanghai reprint in 6 vols. Commercial Press, 1937.

Pen-ts'ao kang-mu 本草綱目 (Grand Pharmacopoeia). 1596. By Li Shih-chen 李時珍 (1518–1593). 52 *chüan*.

Po Chü-yi 白居易 (772–846). *Yang chu chi* 養竹記 (On Cultivating Bamboo). Essay in *Ch'üan T'ang wen* 全唐文 (Complete T'ang Prose), comp. Tung Kao 董誥 (1740–1818) et al. Palace ed. of 1818.

Po-hu t'ung-yi 白虎通義 (Comprehensive Discussions in the White Tiger Hall); abbr. *Po-hu t'ung* (White Tiger Discussions). Ca. A.D. 80. By Pan Ku 班固 (A.D. 32–92). Digest of a conference on the classics held in A.D. 79. 4 *chüan*. *TSCC* ed., vols. 238–239 with continuous pagination. Tr. Tjan (1949–1952).

Shang chün shu 商君書 (The Book of Lord Shang). Attributed to Shang Yang 商鞅 (d. 338 B.C.) but actually by anon. later Legalists. 29 chs. (4 lost today). Refs. to Yen Fu 嚴復 (1853–1921), ed. *Shang chün shu hsin chiao* 新校 (A New Collation of the *Shang chün shu*). Chekiang Book Co. ed. of 1876. Tr. Duyvendak (1928).

Shen Kua. See *Meng-ch'i pi-t'an*

Shen Te-ch'ien. See *T'ang shih pieh-ts'ai*

Shen Tsung-ch'ien. See *Chieh-chou hsüeh-hua p'ien*

Shih chi 史記 (Historical Records). Ca. 86 B.C. Begun by Ssu-ma T'an 司馬談 (d. 112 B.C.) but mostly written by his son, Ssu-ma Ch'ien 遷 (ca. 145–ca. 86 B.C.). 130 chs. Covers period from the Yellow Lord (trad. 2698–2599) down to ca. 100 B.C. Pal. ed. (1923). Tr. Chavannes (1895–1905, 1969) and Watson (1961).

Shih or *Shih ching* 詩經 (Songs Classic). Early Chou before ca. 500 B.C. Collection of 305 songs/poems. Tr. Waley (1937) and Karlgren (1950a).

Shih p'in 詩品 (Classification of Poets). By Chung Hung 鍾嶸 (early sixth century A.D.). 3 *chüan*.

Shih shuo. *See* Han Yü

Shih-shuo hsin-yü 世說新語 (A New Account of Tales of the World). By Liu Yi-ch'ing 劉義慶 (403–444). 3 *chüan*. Tr. Mather (1976).

Shih-t'ao 石濤, a Buddhist monk also known as Tao-chi 道濟 (ca. 1642–ca. 1717). *Shih-t'ao hua yü-lu* 畫語錄 (Shih-t'ao's Talks on Painting). Ca. 1710. Refs. to Huang Lan-po 黃蘭波, ed. *Shih-t'ao hua yü-lu yi-chieh* 譯解 (Shih-t'ao's Talks on Painting Interpreted and Explained). Peking, 1963.

Shu or *Shu ching* 書經 (Documents Classic). Anthology of political pronouncements and other official documents (58 sects.). These ostensibly date from sages Yao and Shun through early Chou, but actually some date from beginning of Chou to ca. 600 B.C.; others are much later, including forgeries of third century A.D. Tr. Legge (1960: vol. 3) and Karlgren (1950b).

Shui hu chuan 水滸傳 (Water Margin). Perhaps fifteenth century. A historical novel.

Shuo-wen chieh-tzu 說文解字 (Script Explained and Characters Elucidated); abbr. *Shuo-wen*. Ca. A.D. 100. By Hsü Shen 許慎. Most important early Chinese dictionary, defining over 9,000 characters. 30 *chüan*. *SPTK* ed.

Shuo yüan. *See* Liu Hsiang

Songs or *Songs Classic*. See *Shih ching*

Spring and Autumn Annals. See *Ch'un ch'iu*

Ssu min yüeh ling 四民月令 (Monthly Ordinances for the Four Categories of People). By Ts'ui Shih 崔寔 (d. ca. A.D. 170). A now fragmentary almanac.

Sui shu 隋書 (Sui History). A.D. 636. By Wei Cheng 魏徵 (580–643) et al. 85 chs. Covers period 581–617. Pal. ed. (1923).

Sun-tzu ping-fa 孫子兵法 (Master Sun's Art of War). Fourth century B.C. Attributed to Sun Wu 孫武 (ca. 500 B.C.), but much of it probably later. Tr. Giles (1910), Griffith (1963).

Sung 頌, a division of the *Shih ching*.

Sung shu 宋書 ([Liu] Sung History). By Shen Yüeh 沈約 (441–513). 100 chs. Covers period 420–479. Pal. ed. (1923).

Ta hsü 大序 (Great Preface to the *Shih ching*). Attributed to Tzu-hsia 子夏, a disciple of Confucius, but probably by Wei Hung 衛宏 (fl. A.D. 25–57). Tr. Legge (1960: vol. 4, Prolegomena).

Ta hsüeh 大學 (Great Learning). One of the Confucian "Four Books." Trad. attributed to a disciple of Confucius or to Confucius's grandson. Tr. Legge (1960: 1.137–381). Also constitutes *Li chi* ch. 39; tr. Legge (1885: 28.411–424) and Couvreur (1913: 2.614–635).

Ta Tai Li chi 大戴禮記 (Record of Ceremonial of the Elder Tai). Trad. by Tai Te 戴德 (first century B.C.) in 85 chs. and then reduced by his nephew, Tai Sheng 聖, to the work in 46 chs. now known as *Li chi*. Tr. Wilhelm (1930).

T'ai-p'ing ching 太平經 (Canon of Grand Peace). Taoist treatise supposedly dating from the Yellow Turban rebellion of A.D. 184 but many portions probably later.

T'ai-p'ing yü-lan 太平御覽 (Imperial Encyclopedia of the T'ai-p'ing Period). 983.

By Li Fang 李昉 et al. 1,000 *chüan*. Taipei reprint in 12 vols. Hsin-hsing (New and Flourishing) Book Co., 1959.

T'ai shih 太誓 (Grand Declaration). Sect. 1 in Chou portion of *Shu ching*. Tr. Legge (1960:3.281–297).

T'ang liu tien 唐六典 (Six Groups of Statutes of the T'ang Dynasty). A.D. 739. Taipei reprint. Wen Hai (Sea of Literature) Book Co., 1962.

T'ang lü shu-yi 唐律疏議 (T'ang Code with Commentary). A.D. 653. By Chang-sun Wu-chi 長孫無忌 et al. 502 articles. *KHCPTS* ed. Partial tr. Johnson (1979).

T'ang shih pieh-ts'ai 唐詩別裁 (Selections from T'ang Poetry). Prefaces dated 1717 and 1763. By Shen Te-ch'ien 沈德潛. Peking: Chung-hua (China) Book Co., 1954.

Tao-chi. *See* Shih-t'ao

T'ao shuo 陶說 (Description of Pottery). 1774. By Chu Yen 朱琰. 6 *chüan*.

Tao te ching. See *Lao-tzu*

T'ien-kung k'ai-wu 天工開物 (Exploitation of the Works of Nature). 1628. By Sung Ying-hsing 宋應星 (ca. 1600–?). 18 sects. Tr. Sun and Sun (1966).

T'ien wen 天問 (Questions about Heaven). Ca. 300 B.C. Fourth item in *Ch'u tz'u* 楚辭 (Songs of the South). Tr. Hawkes (1959).

Ts'ao Chih 曹植 (192–232). *Ts'ao Tzu-chien chi* 曹子建集 (Collected Works of Ts'ao Tzu-chien [Ts'ao Chih]). *SPPY* ed.

Tso chuan 左傳 (Tso Chronicle). Fourth or third centuries B.C. A history attached to the *Ch'un ch'iu* (Spring and Autumn Annals) and covering the years 722–481. Tr. Legge (1960: vol. 5) and Couvreur (1914).

Ts'ui Shih. See *Ssu min yüeh ling*

T'ung-chien kang-mu 通鑑綱目 (Abridged View of the Comprehensive Mirror). 1172. By Chu Hsi 朱熹 (1130–1200) et al. 59 chs. A chronicle history (403 B.C.–A.D. 959) compiled as an abridgement of the *Tzu-chih t'ung-chien*.

Tung Chung-chu 董仲舒 (ca. 179–ca 104). *Ch'un-ch'iu fan-lu* 春秋繁露 (Luxuriant Spring and Autumn Dew). 82 chs. Refs. to Su Yü 蘇輿 (d. 1914) ed. *Ch'un-ch'iu fan-lu yi-cheng* 義証 (Verification of the Meaning of the *Ch'un-ch'iu fan-lu*), 1910.

Tzu-chih t'ung-chien 資治通鑑 (Comprehensive Mirror for Aid in Government). 1084. By Ssu-ma Kuang 司馬光 (1019–1086). 294 chs. Covers period 403 B.C.–A.D. 959. Partial tr. Fang (1952, 1965).

Wang Ch'ung. See *Lun heng*

Wang Fu 王符 (fl. ca. 76–ca. 157). *Ch'ien fu lun* 潛夫論. *SPPY* ed.

Wang Tang 王讜. *T'ang yü lin* 唐語林 (A Forest of Talks about the T'ang). Ca. 1105. Shanghai: Ku-tien Wen-hsüeh (Ancient Repertories of Literature) Publishing Co., 1957.

Wei shu 魏書 (Wei History). A.D. 554. By Wei Shou 魏收 (506–572). 114 chs. Covers period 386–550. Pal. ed. (1923).

Wen-hsien t'ung-k'ao 文獻通考 (Comprehensive Examination of Documents and Compositions). Enyclopedia of ca. 1319. By Ma Tuan-lin 馬端臨. 348 *chüan*. Shanghai: Commercial Press, 1936.

Wu pei chih 武備志 (Treatise on Armament Technology). 1621. By Mao Yüan-yi 茅元儀 (d. 1629). 240 *chüan*.

Wu-tai shih 五代史 (History of the Five Epochs). 973/974. By Hsüeh Chü-cheng 薛居正 (912–981). 150 chs. Covers period 907–960. Pal. ed. (1923).

Ya 雅, a division of the *Shih ching*.

Yang Hsiung (57 B.C.–A.D. 18). *T'ai hsüan ching* (Canon of Supreme Mystery).

Yang Hsüan-chih. See *Lo-yang ch'ieh-lan chi*

Yen Chih-t'ui 顏之推 (531–591). *Yen-shih chia-hsün* 顏氏家訓 (Family Instructions for the Yen Clan). 2 *chüan*. Tr. Teng Ssu-yü (1968).

Yen-ching sui-shih-chi 燕京歲時記 (Record of the Annual Seasons at Yen-ching [Peking]). 1900. By Tun Li-ch'en 敦禮臣 (1855–ca. 1924). Tr. Bodde (1965).

Yen Fo ku piao. *See* Han Yü

Yen-shih chia-hsün. *See* Yen Chih-t'ui

Yen t'ieh lun 鹽鐵論 (Discourses on Salt and Iron). Comp. soon after 81 B.C. By Huan K'uan 桓寬 (fl. ca. 73 B.C.). Report of an 81 B.C. conference of officials and scholars on governmental economic policies. 60 chs. *SPPY* ed. Tr. Gale (1967) of chs. 1–28.

Yi or *Yi ching* 易經 (Changes Classic). Anon. Early divination corpus, probably early Chou, containing 8 trigrams and 64 hexagrams; this text supplemented by 10 "wings" of varying late Chou or early Han date. Tr. Legge (1899b) and Wilhelm (1950).

Yi li 儀禮 (Observances and Ceremonies). Probably fourth or third centuries B.C. Anon. Tr. Couvreur (1916) and Steele (1917).

Ying Shao. See *Feng-su t'ung-yi*

Ying-tsao fa-shih 營造法式 (Treatise on Architectural Methods). 1097. By Li Chieh 李誡.

Yüan tao. *See* Han Yü

Yüeh ling 月令 (Monthly Ordinances). Third century B.C. In *Li chi* ch. 4. Tr. Legge (1885:27.249–310) and Couvreur (1913:1.330–410). Also (not bearing name *Yüeh ling*) in first 12 sects. of *Lü-shih ch'un-ch'iu* (240 B.C.).

Yung-lo ta-tien 永樂大典 (Yung-lo Great Encyclopedia). The world's largest encyclopedia, completed 1408 in 11,095 manuscript vols. Never printed because of great expense. Only 349 vols. extant.

B. Secondary Works

Journal Abbreviations

AHR	*American Historical Review* (Washington, D.C.)
AM	*Asia Major* (Leipzig then London)
BMFEA	*Bulletin of the Museum of Far Eastern Antiquities* (Stockholm)
BSOAS	*Bulletin of the School of Oriental and African Studies* (London)
Bull. Inst. Eth.	*Bulletin of the Institute of Ethnology, Academia Sinica* (*Chung-yang yen-chiu-yüan min-tsu hsüeh yen-chiu-so chi-k'an* 中央研究院民族學研究所集刊) (Taipei)
CS	*Chinese Science* (Philadelphia)
CSH	*Chinese Studies in History* (Armonk, N.Y.)
EC	*Early China* (Berkeley, Calif.)
HJAS	*Harvard Journal of Asiatic Studies* (Cambridge, Mass.)

HoR	*History of Religions* (Chicago)
JA	*Journal Asiatique* (Paris)
JAOS	*Journal of the American Oriental Society* (New Haven, Conn.)
JAS	*Journal of Asian Studies* (Ann Arbor, Mich.)
JCP	*Journal of Chinese Philosophy* (Honolulu)
JHI	*Journal of the History of Ideas* (Philadelphia)
JNChBRAS	*Journal of the North China Branch of the Royal Asiatic Society* (Shanghai)
MCB	*Mélanges Chinois et Bouddhiques* (Brussels)
MDAWB	*Mitteilungen, Deutsche (Preussische) Akademie der Wissenschaft zu Berlin (Institut für Orientforschung)* (Berlin)
PEW	*Philosophy East and West* (Honolulu)
RO	*Rocznik Orientalistyczny* (Warsaw)
TP	*T'oung Pao* (Leiden)

Cited Works

Abegg, Lily. 1952. *The Mind of East Asia*. New York and London: Thames and Hudson.

Acker, William B., tr. 1954, 1974. *Some T'ang and Pre-T'ang Texts on Chinese Painting*. 2 vols. Leiden: E. J. Brill.

Aijmer, Göran. 1964. *The Dragon Boat Festival in the Hupeh-Hunan Basin, Central China, a Study in the Ceremonial of the Transplantation of Rice*. Monograph Series publication no. 9. Stockholm: Ethnographical Museum of Sweden.

Altham, H. S. 1962. *A History of Cricket*. 2 vols. London: George Allen & Unwin (revised from 1st ed. of 1926).

Ames, Roger T. *See* Hall and Ames (1987)

Anderson, Perry. 1974. *Lineages of the Absolutist State*. London: NLB.

Anon. 1960. *Meng-tzu yi-chu* 孟子譯注 (*Mencius* Translated and Commented On). Prepared by the Mencius Translation and Commentary Section, Department of Chinese Literature, Lanchou, Kansu. Peking, 1960.

Anon. 1974. *Han T'ang pi-hua* 漢唐壁畫 (Murals from the Han to the T'ang Dynasty). Peking: Foreign Languages Press.

Aristotle (384–322). *Politics* and *History of Animals*. *See also* Thompson (1910).

Asās al-balāghah. Arabic dictionary by al-Zamakhsharī (d. 1143).

Aubin, Françoise. *See* Chan (1973)

Augustine, Saint (354–430). *Confessions*.

Ayscough, Florence. 1929–1934. *Tu Fu, the Autobiography of a Chinese Poet*. 2 vols. London: Jonathan Cape and Boston/New York: Houghton Mifflin.

Babbitt, Irving. 1919. *Rousseau and Romanticism*. Boston: Honghton Mifflin.

Bacon, Francis. 1605. *Advancement of Learning*. London: Henrie Tomes. Cited in Broad (1951).

Balazs, Etienne. 1964. *Chinese Civilization and Bureaucracy*. Tr. H. M. Wright. Ed. Arthur F. Wright. New Haven, Conn.: Yale University Press.

Bao Zhiming. 1985. Review of Hansen (1983) and "Reply to Professor Hansen." *PEW* 35:203–212, 425–429.

Barnard, Noel. 1973. "Records of Discoveries of Bronze Vessels in Literary Sources and Some Pertinent Remarks on Aspects of Chinese History." *Journal of the Institute of Chinese Studies, The Chinese University of Hong Kong* 6:455–544.

Barrett, Timothy. 1986. "Postscript to Chapter 16." In Twitchett and Loewe (1986:873–878).

Basham, A. L. 1959. *The Wonder That Was India*. New York: Evergreen (1st ed., London: Jackson, 1954).

Bauer, Wolfgang. 1964. "Icherleben und Autobiographie im älteren China." *Heidelberger Jarbücher* 8:12–40.

———. 1979. "The Problem of Individualism and Egoism in Chinese Thought." In *Studia Sino-Mongolica, Festschrift für Herbert Franke*. Ed. Wolfgang Bauer. Pp. 427–444.

Baynes, Cary F. *See* Wilhelm (1950)

Baxter, Glen W. *See* Yoshikawa Kojirō (1955)

Beaseley, W. G., and Pulleyblank, Edwin G., eds. 1961. *Historians of China and Japan*. London: Oxford University Press.

Bernal, J. D. 1954. *Science in History*. London: Watts.

Beurdeley, Michel. 1969. With Kristopher M. Schipper, Chang Fu-jui, Jacques Pimpaneau. *The Clouds and the Rain, the Art of Love in China*. Tr. from French by Diana Imber. Fribourg, Switzerland: Office du Livre and London: Hamond. Also published as *Chinese Erotic Art*. Rutland, Vt.: C. E. Tuttle.

Bielenstein, Hans. 1947. "The Census of China during the Period 2–742 A.D." *BMFEA* 19:125–163.

———. 1950. "An Interpretation of the Portents in the Ts'ien Han-shu." *BMFEA* 22:127–143.

———. 1953, 1959, 1967, 1979. *The Restoration of the Han Dynasty, with Prolegomena on the Historiography of the Hou Han Shu*. 4 vols. Stockholm: Museum of Far Eastern Antiquities.

———. 1976. "Lo-yang in Later Han Times." *BMFEA* 48:1–142.

Bieler, Ludwig. 1963. *The Irish Penitentials*. With an appendix by D. A. Binchy. Scriptores Latini Hibernidae V. Dublin: The Dublin Institure of Advanced Studies.

Biggerstaff, Knight. *See* Teng and Biggerstaff (1971)

Binchy, D. A. *See* Bieler (1963)

Biot, Edouard, tr. 1941. *Le Tcheou-li ou les rites des Tcheou*. 3 vols. Paris: Imprimerie Nationale (also anastatic reprint, Peking: Wen Tien Ko, 1940).

Birdwhistell, Anne D. 1989. *Transition to Neo-Confucianism: Shao Yung on Knowledge and Symbols of Reality*. Stanford, Calif.: Stanford University Press.

Blader, Susan. *See* Le Blanc and Blader (1987)

Bland, J. O. P., and Backhouse, Edmund. 1914. *Annals and Memoirs of the Court of Peking*. Boston: Houghton Mifflin.

Bloom, Alfred. 1981. *The Linguistic Shaping of Thought: A Study in the Impact of Language on Thinking in China and the West*. Hillsdale, N.J.: Lawrence Erlbaum Associates.

Boas, G. *See* Lovejoy and Boas (1935)

Bodde, Derk. 1933. "A Perplexing Passage in the Confucian Analects." *JAOS* 53:347–351. Reprinted in Bodde (1981:383–387).

———. 1939. "Types of Chinese Categorical Thinking." *JAOS* 59:200–219. Reprinted in Bodde (1981:141–160).

———. 1940. *Statesman, Patriot and General in Ancient China: Three Shih-chi Biographies*

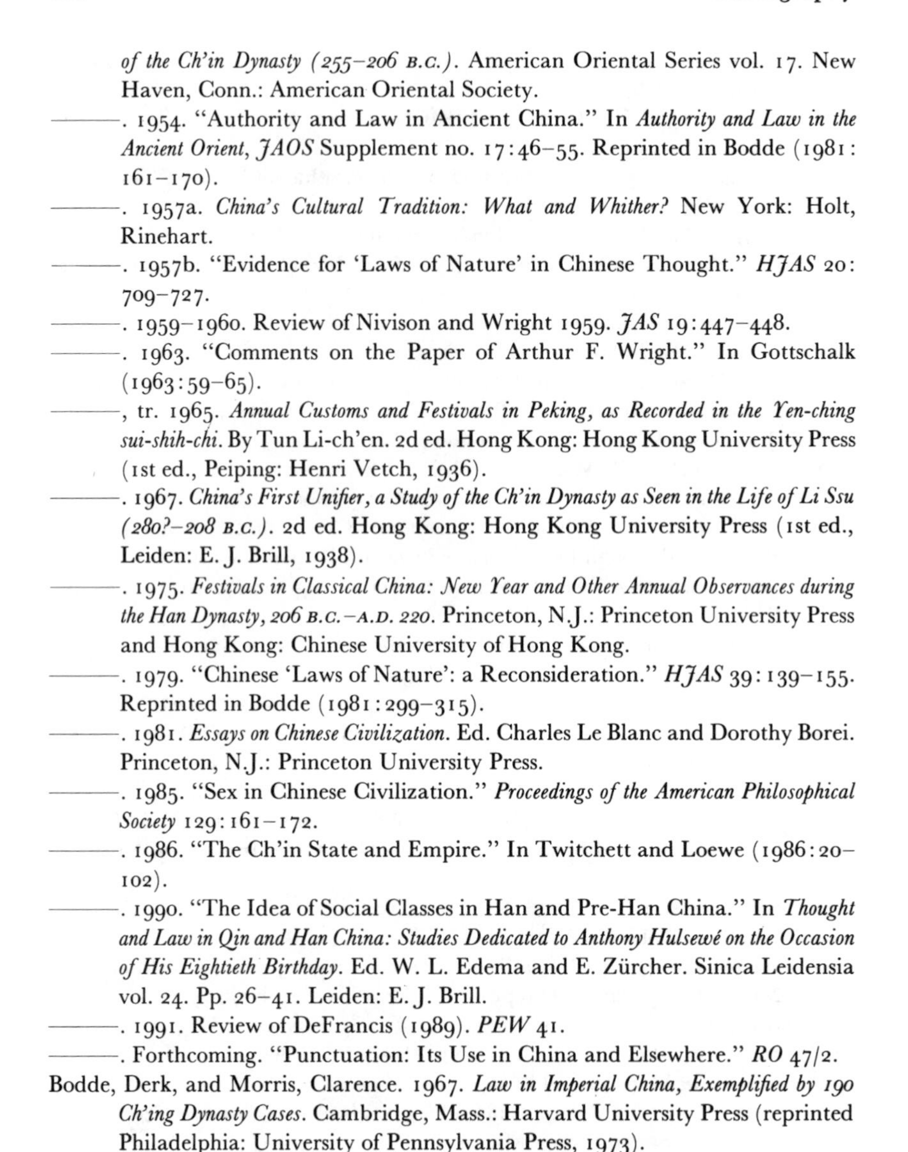

of the Ch'in Dynasty (255–206 B.C.). American Oriental Series vol. 17. New Haven, Conn.: American Oriental Society.

———. 1954. "Authority and Law in Ancient China." In *Authority and Law in the Ancient Orient*, *JAOS* Supplement no. 17:46–55. Reprinted in Bodde (1981: 161–170).

———. 1957a. *China's Cultural Tradition: What and Whither?* New York: Holt, Rinehart.

———. 1957b. "Evidence for 'Laws of Nature' in Chinese Thought." *HJAS* 20: 709–727.

———. 1959–1960. Review of Nivison and Wright 1959. *JAS* 19:447–448.

———. 1963. "Comments on the Paper of Arthur F. Wright." In Gottschalk (1963:59–65).

———, tr. 1965. *Annual Customs and Festivals in Peking, as Recorded in the Yen-ching sui-shih-chi*. By Tun Li-ch'en. 2d ed. Hong Kong: Hong Kong University Press (1st ed., Peiping: Henri Vetch, 1936).

———. 1967. *China's First Unifier, a Study of the Ch'in Dynasty as Seen in the Life of Li Ssu (280?–208 B.C.)*. 2d ed. Hong Kong: Hong Kong University Press (1st ed., Leiden: E. J. Brill, 1938).

———. 1975. *Festivals in Classical China: New Year and Other Annual Observances during the Han Dynasty, 206 B.C.–A.D. 220*. Princeton, N.J.: Princeton University Press and Hong Kong: Chinese University of Hong Kong.

———. 1979. "Chinese 'Laws of Nature': a Reconsideration." *HJAS* 39:139–155. Reprinted in Bodde (1981:299–315).

———. 1981. *Essays on Chinese Civilization*. Ed. Charles Le Blanc and Dorothy Borei. Princeton, N.J.: Princeton University Press.

———. 1985. "Sex in Chinese Civilization." *Proceedings of the American Philosophical Society* 129:161–172.

———. 1986. "The Ch'in State and Empire." In Twitchett and Loewe (1986:20–102).

———. 1990. "The Idea of Social Classes in Han and Pre-Han China." In *Thought and Law in Qin and Han China: Studies Dedicated to Anthony Hulsewé on the Occasion of His Eightieth Birthday*. Ed. W. L. Edema and E. Zürcher. Sinica Leidensia vol. 24. Pp. 26–41. Leiden: E. J. Brill.

———. 1991. Review of DeFrancis (1989). *PEW* 41.

———. Forthcoming. "Punctuation: Its Use in China and Elsewhere." *RO* 47/2.

Bodde, Derk, and Morris, Clarence. 1967. *Law in Imperial China, Exemplified by 190 Ch'ing Dynasty Cases*. Cambridge, Mass.: Harvard University Press (reprinted Philadelphia: University of Pennsylvania Press, 1973).

Boltz, William G. 1983. "Word and Word History in the Analects: The Exegesis of *Lun Yü* IX.l." *TP* 69:261–271.

Boodberg, Peter A. 1979. "On Chinese *Ts'ing*, 'Blue-Green.'" In *Selected Works of Peter A. Boodberg*. Ed. Alvin P. Cohen. Pp. 178–179. Berkeley and Los Angeles: University of California Press. *See also* Gale (1967)

Borei, Dorothy. *See* Bodde (1981)

Bourne, F. S. A. 1879–1880. "Essay of a Provincial Graduate with Translation." *China Review* 8:352–356.

Boxer, C. R. 1961. "Some Aspects of Western and Historical Writing on the Far East, 1500–1800." In Beaseley and Pulleyblank (1961:307–321).
Bray, Francesca. 1984. Pt. 2: *Agriculture*. In *Science and Civilisation in China*, vol. 6: *Biology and Biological Technology*. Ed. Joseph Needham. Cambridge, London, New York, etc.: Cambridge University Press.
Bridgman, R. F. 1952–1953. "La Médicine dans la Chine antique." *MCB* 10: 1–213.
Broad, C. D. 1951. "Bacon and the Experimental Method." In *A Short History of Science: Origins and Results of the Scientific Revolution, a Symposium*. Pp. 27–33. Glenco, Ill.: Free Press, 1951 (also Garden City, N.Y.: Anchor Books, 1959).
Brooks, R. *See* du Halde (1741)
Brown, T. Julian. 1974. "Punctuation." In *The New Encyclopaedia Britannica* (Chicago, 15th ed.). *Macropaedia* 15:274–277.
Burford, Alison. 1972. *Craftsmen in Greek and Roman Society*. London: Thames and Hudson.
Burke, Josph. *See* Hogarth (1955)
Bury, J. B. 1955. *The Idea of Progress, an Inquiry into Its Origin and Growth*. New York: Dover (1st ed., London: Macmillan, 1920).
Bury, R. G., tr. 1933–1949. *Sextus Empiricus* [early third century A.D.]. Loeb Classical Library. 4 vols. London: Heinemann.
Bush, Susan. 1971. *The Chinese Literati on Painting: Su Shih (1037–1101) to Tung Ch'i-ch'ang (1555–1636)*. Cambridge, Mass.: Harvard University Press.
Bushell, Stephen W., tr. 1910. *Description of Chinese Pottery and Porcelain, Being a Translation of the T'ao Shuo*. Oxford: Clarendon Press.
Butterfield, Herbert. 1969. "Reflections on Religion and Modern Individualism." *JHI* 22:33–46.
Cahill, James F. 1960. "Confucian Elements in the Theory of Painting." In Wright (1960a: 114–140).
Callus, D. A. *See* Crombie (1955)
Cammann, Schuyler. 1960. "The Evolution of Magic Squares in China." *JAOS* 80:116–124.
———. 1961. "The Magic Square of Three in Old Chinese Philosophy and Religion." *HoR* 1:37–80.
———. 1963. "Old Chinese Magic Squares." *Sinologica* 7: 14–53.
Campbell, Norman. 1953. *What Is Science?* New York: Dover (reprint of London: Methuen, 1921).
Chan, Hok-lam. 1979–1980. "Li Chih 1527–1602 in Contemporary Chinese Historiography: New Light on His Life and Works." *CSH* 12/1–2: 1–209.
———. 1980. "Li Chih (1527–1602): Additional Research Notes." *CSH* 13/3:81–84.
Chan, Wing-tsit. 1953. *Religious Trends in Modern China*. New York: Columbia University Press.
———. 1957. "Neo-Confucianism and Chinese Scientific Thought." *PEW* 6:309–332.
———. 1963. *A Source Book in Chinese Philosophy*. Princeton, N.J.: Princeton University Press.
———. tr. 1967. *Reflections on Things at Hand, the Neo-Confucian Anthology*. By Chu Hsi and Lü Tsu-ch'ien. New York and London: Columbia University Press.

———. 1973. "Chu Hsi's Completion of Neo-Confucianism." In *Études Song in Memorian Etienne Balazs*. Series 2, no. 1. Ed. Françoise Aubin. Pp. 59–90. Paris and The Hague: Mouton.

———. 1976a. "The Study of Chu Hsi in the West." *JAS* 35:555–577.

———. 1976b. "Wang Shou-jen." In Goodrich and Fang (1976:2.1408–1416).

Chang Kwang-chih. 1977a. *The Archaeology of Ancient China*, 3rd revised and enlarged ed. New Haven, Conn., and London: Yale University Press (1st ed., 1963; 4th rev. ed., 1987).

———. 1977b. "Ancient China." In Chang Kwang-chih (1977c:25–52).

———, ed. 1977c. *Food in Chinese Culture: Anthropological and Historical Perspectives*. New Haven, Conn., and London: Yale University Press.

Chavannes, C. G. *See* Groot (1886)

Chavannes, Edouard, tr. 1895–1905, 1969. *Les Mémoires historiques de Se-ma Ts'ien*. Vols. 1–5, Paris: Ernest Leroux. Vol. 6, Paris: Adrien Maisonneuve. Their coverage is as follows: 1895: vol. 1, Introduction and chs. 1–4; 1897: vol. 2, chs. 5–12; 1898–1899: vol. 3, pt. 1, chs. 13–22; vol. 3, pt. 2, chs. 23–30; 1901: vol. 4, chs. 31–42; 1905: vol. 5, chs. 43–47; 1969: vol. 6 (posthumous), chs. 48–52, appendices, general index.

———. 1909–1915. *Mission archéologigue dans la Chine septentrionale*. 2 vols. and portfolios of plates (vol. 1, 1913; vol. 2, 1915; portfolios of plates, 1909). Paris: Ernest Leroux.

Chavannes, Edouard, and Pelliot, Paul. 1911–1913. "Un Traité manichéen retrouvé en Chine." *JA* 10th series, 18:499–617; 11th series, 1:99–199, 261–395.

Chejne, Anwar G. 1969. *The Arabic Language: Its Role in History*. Minneapolis: University of Minnesota Press.

Ch'en Ch'i-yün, tr. 1980. *Hsün Yüeh and the Mind of Late Han China: a Translation of the Shen-chen with Introduction and Annotations*. Princeton, N.J.: Princeton University Press.

Ch'en Chih-mai. 1966. *Chinese Calligraphers and Their Art*. Melbourne: Melbourne University Press.

Ch'en, Kenneth. 1964. *Buddhism in China, a Historical Survey*. Princeton, N.J.: Princeton University Press.

Ch'en Shou-yi. 1961. *Chinese Literature, a Historical Introduction*. New York: Ronald Press.

Ch'en Teng-yüan 陳登原. 1958–1962. *Kuo-shih chiu-wen* 國史舊聞 (Reports on Chinese History). 2 vols. with continuous pagination. Peking (originally published 1938).

Cheng Pei-kai. 1983–1984. "Continuities in Chinese Political Culture: Interpretations of Li Zhi [Li Chih], Past and Present." *CSH* 17/2:4–29.

Chiang Yee. 1954. *Chinese Calligraphy, an Introduction to Its Aesthetic and Technique*. 2d ed. Cambridge, Mass.: Harvard University Press and London: Methuen (1st ed., 1938).

Ching, Julia. 1976a. *To Acquire Wisdom: the Way of Wang Yang-ming*. New York and London: Columbia University Press.

———. 1976b. "Truth and Ideology: the Confucian Way (Tao) and Its Transmission (Tao-t'ung)." *JHI* 35:371–388.

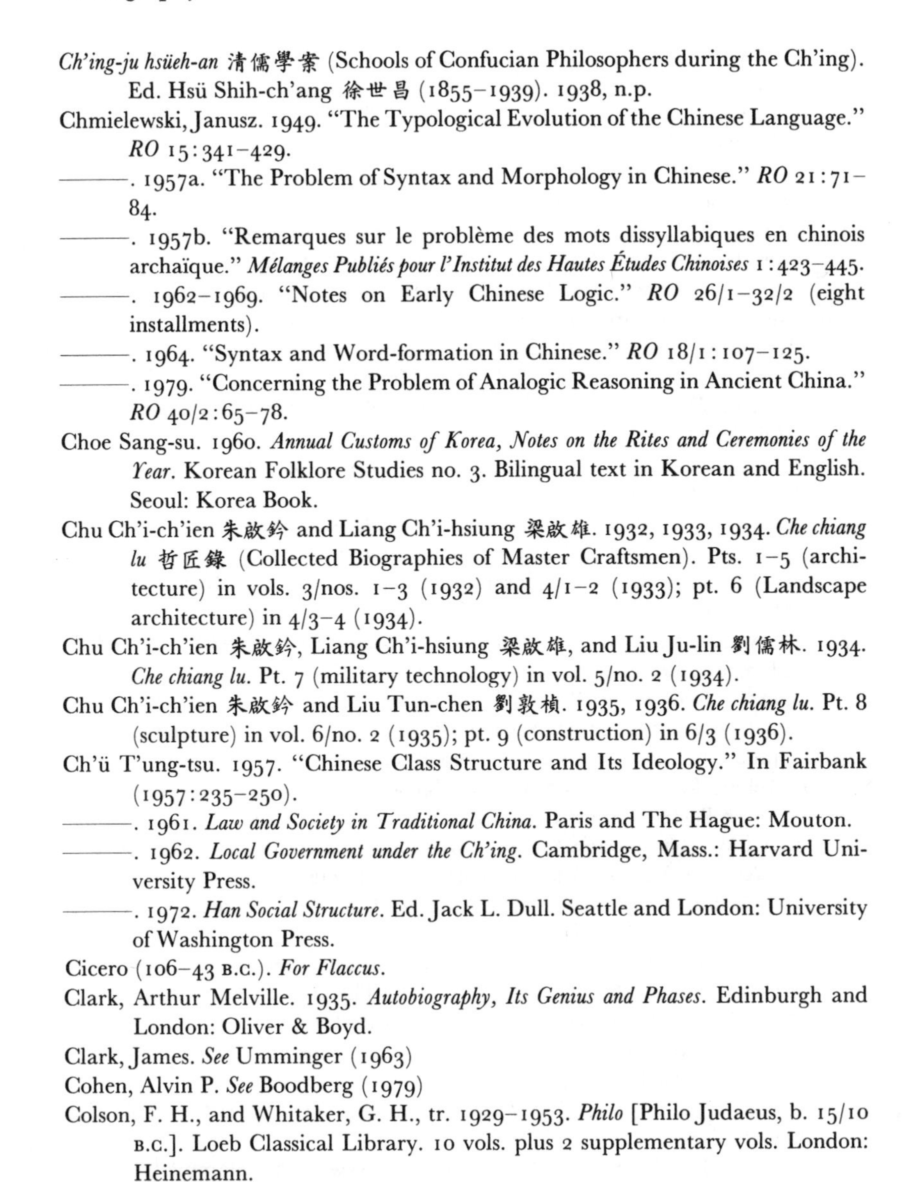

Ch'ing-ju hsüeh-an 清儒學案 (Schools of Confucian Philosophers during the Ch'ing). Ed. Hsü Shih-ch'ang 徐世昌 (1855–1939). 1938, n.p.

Chmielewski, Janusz. 1949. "The Typological Evolution of the Chinese Language." *RO* 15:341–429.

———. 1957a. "The Problem of Syntax and Morphology in Chinese." *RO* 21:71–84.

———. 1957b. "Remarques sur le problème des mots dissyllabiques en chinois archaïque." *Mélanges Publiés pour l'Institut des Hautes Études Chinoises* 1:423–445.

———. 1962–1969. "Notes on Early Chinese Logic." *RO* 26/1–32/2 (eight installments).

———. 1964. "Syntax and Word-formation in Chinese." *RO* 18/1:107–125.

———. 1979. "Concerning the Problem of Analogic Reasoning in Ancient China." *RO* 40/2:65–78.

Choe Sang-su. 1960. *Annual Customs of Korea, Notes on the Rites and Ceremonies of the Year*. Korean Folklore Studies no. 3. Bilingual text in Korean and English. Seoul: Korea Book.

Chu Ch'i-ch'ien 朱啟鈐 and Liang Ch'i-hsiung 梁啟雄. 1932, 1933, 1934. *Che chiang lu* 哲匠錄 (Collected Biographies of Master Craftsmen). Pts. 1–5 (architecture) in vols. 3/nos. 1–3 (1932) and 4/1–2 (1933); pt. 6 (Landscape architecture) in 4/3–4 (1934).

Chu Ch'i-ch'ien 朱啟鈐, Liang Ch'i-hsiung 梁啟雄, and Liu Ju-lin 劉儒林. 1934. *Che chiang lu*. Pt. 7 (military technology) in vol. 5/no. 2 (1934).

Chu Ch'i-ch'ien 朱啟鈐 and Liu Tun-chen 劉敦楨. 1935, 1936. *Che chiang lu*. Pt. 8 (sculpture) in vol. 6/no. 2 (1935); pt. 9 (construction) in 6/3 (1936).

Ch'ü T'ung-tsu. 1957. "Chinese Class Structure and Its Ideology." In Fairbank (1957:235–250).

———. 1961. *Law and Society in Traditional China*. Paris and The Hague: Mouton.

———. 1962. *Local Government under the Ch'ing*. Cambridge, Mass.: Harvard University Press.

———. 1972. *Han Social Structure*. Ed. Jack L. Dull. Seattle and London: University of Washington Press.

Cicero (106–43 B.C.). *For Flaccus*.

Clark, Arthur Melville. 1935. *Autobiography, Its Genius and Phases*. Edinburgh and London: Oliver & Boyd.

Clark, James. *See* Umminger (1963)

Cohen, Alvin P. *See* Boodberg (1979)

Colson, F. H., and Whitaker, G. H., tr. 1929–1953. *Philo* [Philo Judaeus, b. 15/10 B.C.]. Loeb Classical Library. 10 vols. plus 2 supplementary vols. London: Heinemann.

Cook, Daniel J., and Rosemont, Henry, Jr. 1981. "The Pre-established Harmony Between Leibniz and Chinese Thought." *JHI* 42:253–267. *See also* Rosemont and Cook (1977)

Cordier, Henri. 1904–1924. *Bibliotheca Sinica, dictionnarie bibliographique des ouvrages relatifs à l'empire chinois*. 4 vols. Paris: E. Guilmoto. Supplementary vol. 5., Paris: Paul Geuthner.

———. 1920. *Histoire générale de la Chine et de ses relations avec les pays étrangers depuis les temps les plus anciens jusqu'à la chute de la dynastie manchoue*. 4 vols. Paris: Paul Geuthner.

Couvreur, Séraphin, tr. 1895. *Les Quatre livres*. Ho Kien Fou: Imprimerie de la Mission Catholique.

———, tr. 1913. *Li Ki, ou mémoires sur les bienséances et les cérémonies*. 2 vols. 2d ed. Ho Kien Fou: Imprimerie de la Mission Catholique.

———, tr. 1914. *Tch'ouen Ts'iou et Tso Tchouan, la chronique de la principauté de Lou*. 3 vols. Ho Kien Fou: Imprimerie de la Mission Catholique.

———, tr. 1916. *I Li, le céremonial*. Sien Hsien: Imprimerie de la Mission Catholique.

Craig, Albert M. *See* Fairbank, Reischauer, and Craig (1965)

Creel, Herrlee G. 1954. *The Birth of China, a Study of the Formative Period of Chinese Civilization*. New York: Reynal & Hitchcock (1st ed., 1936).

———. 1970. *The Origins of Statecraft in China*. Vol. 1: *The Western Chou Empire*. Chicago and London: University of Chicago Press.

———. 1974a. "Shen Pu-hai: a Secular Philosopher of Administration." *JCP* 1:119–136.

———. 1974b. *Shen Pu-hai, a Chinese Political Philosopher of the Fourth Century* B.C. Chicago and London: University of Chicago Press.

———. 1987. "The Role of Compromise in Chinese Culture." In Le Blanc and Blader (1987:133–151).

Crombie, A. C. 1953. *Robert Grosseteste and the Origins of Experimental Science*. Oxford: Clarendon Press.

———. 1955. "Grosseteste's Position in the History of Science." In *Robert Grosseteste, Scholar and Bishop*. Ed. D. A. Callus. Pp. 98–120. Oxford: Clarendon Press.

Crump, James, tr. 1970. *Chan-kuo ts'e*. London: Oxford University Press.

Cutter, Robert Joe. 1989. "Brocade and Blood: The Cockfight in Chinese and English Poetry." *JAOS* 19:1–16.

Daly, Lloyd W. 1967. *Contributions to a History of Alphabetization in Antiquity and the Middle Ages*. Vol. 90. Brussels: Collection Latomus.

De Bary, William Theodore. 1959. "Some Common Tendencies in Neo-Confucianism." In Nivison and Wright (1959:25–49).

———. 1970. "Individualism and Humanitarianism in Late Ming Thought." In *Self and Society in Ming Thought*. Ed. William Theodore De Bary. Pp. 145–247. New York and London: Columbia University Press.

———. 1975a. "Neo-Confucian Cultivation and Seventeenth-century 'Enlightenment.'" In De Bary (1975b:141–216).

———, ed. 1975b. *The Unfolding of Neo-Confucianism*. New York: Columbia University Press.

———. et al., eds. 1960. *Sources of Chinese Tradition*. New York: Columbia University Press.

DeFrancis, John. 1950. *Nationalism and Language Reform in China*. Princeton, N.J.: Princeton University Press.

———. 1989. *Visible Speech: The Diverse Oneness of Writing Systems*. Honolulu: University of Hawaii Press.

de Groot, J. J. M. *See* Groot, J. J. M. de

Demiéville, Paul. 1973. "Gauche et droite en Chine." In Paul Demiéville, *Choix d'études sinologiques (1921–1970)*. Pp. 180–194. Leiden: E. J. Brill.

———. 1986. "Philosophy and Religion from Han to Sui." In Twitchett and Loewe (1986:808–872). *See also* Jao Tsung-yi and Demiéville (1971)

Des Rotours, Robert, tr. 1932. *Le Traité des examens*. Paris: Ernest Leroux.

———. 1933. "La Lutte à la corde (Pa-ho) en Chine." *JA* 222:341–350.

———, tr. 1947–1948. *Traité des fonctionnaires et traité de l'armée*. 2 vols. Leiden: E. J. Brill.

De Woskin, Kenneth J. 1982. *A Song for One or Two: Music and the Concept of Art in Early China*. Ann Arbor: Center for Chinese Studies, University of Michigan.

Diamadopoulos, P. 1961. Review of Sambursky 1959. *Philosophical Review* 70:257–259.

Diény, Jean-Pierre. 1968. *Aux Origines de la poésie classique en Chine: étude sur la poésie lyrique à l'époque des Han*. Leiden: E. J. Brill.

Diogenes Laertius (third century A.D.). *Lives of Eminent Philosophers*. *See* Hicks (1925), Saunders (1966)

Dobson, W. A. C. H. 1959. *Late Archaic Chinese*. Toronto: University of Toronto Press.

Dolby, William. 1983. "Early Chinese Plays and Theater." In *Chinese Theater, from Its Origins to the Present Day*. Ed. Colin Mackerras. Pp. 7–31. Honolulu: University of Hawaii Press.

Donnelly, P. J. 1967. *Blanc de Chine, the Porcelain of Tehua in Fukien*. New York and Washington: Praeger.

Draper, John W. 1874. *History of the Conflict Between Religion and Science*. New York: D. Appleton (also later eds.).

Driscoll, Lucy, and Toda, Kenji. 1935. *Chinese Calligraphy*. Chicago: University of Chicago Press.

Dubs, Homer H., tr. 1928. *The Works of Hsüntze*. London: Probsthain.

———. 1928–1929. "The Failure of the Chinese to Produce Philosophical Systems." *TP* 26:96–109.

———, tr. 1938–1955. *History of the Former Han Dynasty* (tr. of *Ch'ien Han shu* chs. 1–12, 99A–C). 3 vols. Baltimore, Md.: Waverly Press, 1938, 1944, 1955.

———. 1946. "Taoism." In *China*. Ed. Harley Farnsworth MacNair. Pp. 266–289. Berkeley and Los Angeles: University of California Press.

———. 1946–1947. "Han Yü and the Buddha's Relic, an Episode in Medieval Chinese Religion." *Review of Religion* 11:5–17.

———. 1960. "Comment on C. S. Goodrich's Review Article." *JAOS* 80:140–141.

Duyvendak, J. J. L. 1923. "A Literary Renaissance in China." *Acta Orientalia* 1:285–317.

———. 1928. *The Book of Lord Shang, a Classic of the Chinese School of Law*. London: Arthur Probsthain.

———. 1939. "The True Dates of the Chinese Maritime Expeditions in the Early Fifteenth Century." *TP* 34:341–412.

Eberhard, Wolfram. 1942. *Lokalkulturen im alten China*. Vol. 1: *Die Lokalkulturen des Nordens und Westens*. Leiden: E. J. Brill. Vol. 2: *Die Lokalkulturen des Südens und Ostens*. Peiping: Catholic University. *See also* Eberhard (1968) for English tr. of vol. 2.

———. 1957. "The Political Function of Astronomy and Astronomers in Han China." In Fairbank (1957:33–70).

———. 1960. *A History of China*. 2d rev. ed. Berkeley: University of California Press (1st ed., 1950).

———. 1967. *Guilt and Sin in Traditional China*. Berkeley and Los Angeles: University of California Press.

———. 1968. *The Local Cultures of South and East China*. Tr. Alide Eberhard of Eberhard (1942: vol. 2). Leiden: E. J. Brill.

Edwards, E. D. *See* Granet (1932)

Egbert, Virginia Wylie. 1967. *The Medieval Artist at Work*. Princeton, N.J.: Princeton University Press.

Eichhorn, Werner. 1954. "Description of the Rebellion of Sun En and Earlier Taoist Rebellions." *MDAWB* 2:325–352.

———. 1955. "Bemerkungen zum Aufstand des Chang Chio und zum Staate des Chang Lu." *MDAWB* 3:291–327.

Eliade, Mircea. 1954. *The Myth of the Eternal Return*. Tr. Willard R. Trask from 1949 French original. No. 46. New York: Bollingen.

Elman, Benjamin. 1982. "From Value to Fact: the Emergence of Phonology as a Precise Discipline in Late Imperial China." *JAOS* 102:493–500.

———. 1983. "Philosophy (*I-li*) Versus Philology (*K'ao-cheng*): the *Jen-hsin Tao-hsin* Debate." *TP* 69:175–222.

———. 1984. *From Philosophy to Philology, Intellectual and Social Aspects of Change in Late Imperial China*. Cambridge, Mass. and London: Harvard University Press.

Fairbank, John K., ed. 1957. *Chinese Thought and Institutions*. Chicago: University of Chicago Press.

———. 1974. "Introduction, Varieties of the Chinese Military Experience." In *Chinese Ways in Warfare*. Ed. Frank A. Kierman and John K. Fairbank. Pp. 1–26. Cambridge, Mass.: Harvard University Press. *See also* Teng and Fairbank (1954).

Fairbank, John K.; Reischauer, Edwin O.; Craig, Albert M. 1965. *East Asia: the Modern Transformation*. London: George Allen & Unwin.

Fang, Achilles, tr. 1952, 1965. *The Chronicle of the Three Kingdoms (220–265), Chapters 69–78 from the Tzu Chih T'ung Chien*. Ed. Glen W. Baxter. 2 vols. Cambridge, Mass.: Harvard University Press.

———. 1953. "Some Reflections on the Difficulty of Translation." In Wright (1953a:263–285).

Fang Chao-ying. 1943a. Biography of Hsieh Chi-shih (1689–1756). In *Eminent Chinese of the Ch'ing Period (1644–1912)*. Ed. Arthur W. Hummel. Pp. 306–307. 2 vols. with continuous pagination. Washington, D.C.: Government Printing Office.

———. 1943b. Biography of Yang Kuang-hsien (1597–1669) in Fang (1943a. Pp. 889–992). *See also* Goodrich and Fang (1976)

Fang Wan-chuan. 1984. "Chinese Language and Theoretical Thinking." *Journal of Oriental Studies* 22:25–32.

Fei Hsiao-t'ung. 1939. *Peasant Life in China, A Field Study of Country Life in the Yangtze Valley*. London: Routledge & Kegan Paul and New York: Dutton.

Feng-hsi Archaeological Team 灃西發掘隊. 1959. "1955–57 nien Shensi Ch'ang-an Feng-hsi fa-chüeh chien-pao" 1955-57年陕西長安灃西發掘簡報 (Brief Report on the Excavation at Feng-hsi near Sian, 1955–1957). *K'ao Ku* 考古 (Archaeology) 1959/10:516–530.

Feng Yu-lan. *See* Fung Yu-lan

Fenollosa, Ernest. 1934. *The Chinese Written Character as a Medium for Poetry*. Ed. Ezra Pound. London: S. Note and New York: Arrow Editions (reprinted San Franciso, 1964).

Feuer, Lewis S. 1963. *The Scientific Individual, the Psychological and Sociological Origins of Modern Science*. New York and London: Basic Books.

Filliozat, Jean. *See* Renou and Filliozat (1947–1953)

Finley, Moses I., and Pleket, H. W. 1976. *The Olympic Games: The First Thousand Years*. London: Chatto & Windus.

Fleming, Donald. 1965. Review of Feuer (1963). *Isis* 56:309.

Forke, Alfred. 1924. "Zu Lun-yü II.16." *AM* 1:112–118.

———, tr. 1962. *Lun Heng. By Wang Ch'ung (A.D. 27–ca. 96)*. 2 vols. New York: Paragon (1st ed., 1907, 1911).

Forrest, R. A. D. 1948. *The Chinese Language*. London: Faber & Faber.

Foster, J. R. *See* Gernet (1982)

Franke, Herbert. 1950. "Some Remarks on the Interpretation of the Chinese Dynastic Histories." *Oriens* 3:113–122.

Fried, Morton. 1952. "Military Status in Chinese Society." *American Journal of Sociology* 57:347–355.

Friedrich, Michael, and Lackner, Michael. 1983–1985. "Once Again: the Concept of *Wu-hsing*." *EC* 9–10:218–219.

Fung Yu-lan. 1935. "The Origins of Ju and Mo." *Chinese Social and Political Science Review* 19:151–163.

———. 1952. 1953. *A History of Chinese Philosophy*. Tr. Derk Bodde. Vol. 1 (1952): *The Period of the Philosophers*. Vol. 2 (1953): *The Period of Classical Learning*. Princeton, N.J.: Princeton University Press.

Gale, Esson M., tr. 1967. *Discourses on Salt and Iron, a Debate on Commerce and Industry in Ancient China*. Taipei: Ch'eng-wen (original ed. Leiden: E. J. Brill, 1931). Taipei ed. contains tr. of chs. 1–19 of original ed. plus tr. of chs. 20–28 published by Gale in collaboration with Peter A. Boodberg and T. C. Lin in *JNChBRAS* 65(1934):73–110.

Gálik, Marián. 1980. "The Concept of Creative Personality in Traditional Chinese Literary Criticism." *Oriens Extremus* 27:183–202.

Gallagher, Louis J., tr. 1953. *China in the Sixteenth Century: the Journals of Matthew Ricci, 1583–1610*. New York: Random House.

Gardner, Charles S. 1938. *Chinese Traditional Historiography*. Cambridge, Mass.: Harvard University Press (reprinted 1961).

Gardner, Daniel K. 1986. *Chu Hsi and the Ta-hsüeh, Neo-Confucian Reflection on the Confucian Canon*. Cambridge, Mass., and London: Council on East Asian Studies, Harvard University; distributed by Harvard University Press.

Garrett, M. 1983–1985. Review of Bloom (1981). *EC* 9–10:220–236.

Gawlikowski, Krzysztof. 1985. "The School of Strategy (*bing jia*) in the Context of Chinese Civilization." *East and West* n.s. 35:167–210.

Gernet, Jacques. 1956. *Les Aspects économiques du Bouddhisme, dans la société chinoise du Ve au Xe siècle*. Vol. 39. Saigon: Publications de l'École Française d'Extrême Orient.

———. 1960. "Les Suicides par le feu chez les Bouddhistes chinois du Ve au Xe siècle." *Mélanges publiées par l'Institut des Hautes Études Chinoises* 2:527–558.

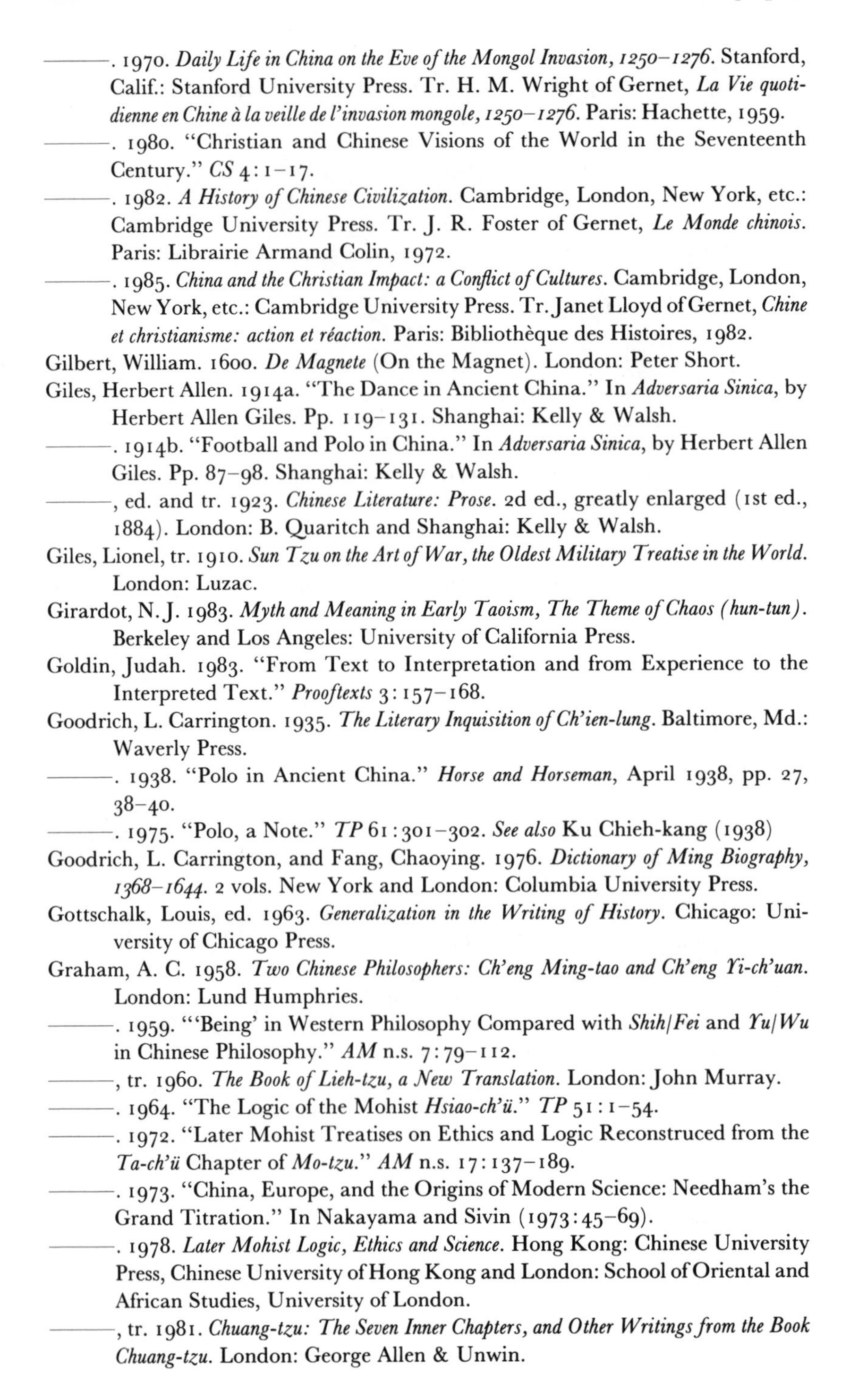

———. 1970. *Daily Life in China on the Eve of the Mongol Invasion, 1250–1276*. Stanford, Calif.: Stanford University Press. Tr. H. M. Wright of Gernet, *La Vie quotidienne en Chine à la veille de l'invasion mongole, 1250–1276*. Paris: Hachette, 1959.

———. 1980. "Christian and Chinese Visions of the World in the Seventeenth Century." *CS* 4: 1–17.

———. 1982. *A History of Chinese Civilization*. Cambridge, London, New York, etc.: Cambridge University Press. Tr. J. R. Foster of Gernet, *Le Monde chinois*. Paris: Librairie Armand Colin, 1972.

———. 1985. *China and the Christian Impact: a Conflict of Cultures*. Cambridge, London, New York, etc.: Cambridge University Press. Tr. Janet Lloyd of Gernet, *Chine et christianisme: action et réaction*. Paris: Bibliothèque des Histoires, 1982.

Gilbert, William. 1600. *De Magnete* (On the Magnet). London: Peter Short.

Giles, Herbert Allen. 1914a. "The Dance in Ancient China." In *Adversaria Sinica*, by Herbert Allen Giles. Pp. 119–131. Shanghai: Kelly & Walsh.

———. 1914b. "Football and Polo in China." In *Adversaria Sinica*, by Herbert Allen Giles. Pp. 87–98. Shanghai: Kelly & Walsh.

———, ed. and tr. 1923. *Chinese Literature: Prose*. 2d ed., greatly enlarged (1st ed., 1884). London: B. Quaritch and Shanghai: Kelly & Walsh.

Giles, Lionel, tr. 1910. *Sun Tzu on the Art of War, the Oldest Military Treatise in the World*. London: Luzac.

Girardot, N. J. 1983. *Myth and Meaning in Early Taoism, The Theme of Chaos (hun-tun)*. Berkeley and Los Angeles: University of California Press.

Goldin, Judah. 1983. "From Text to Interpretation and from Experience to the Interpreted Text." *Prooftexts* 3: 157–168.

Goodrich, L. Carrington. 1935. *The Literary Inquisition of Ch'ien-lung*. Baltimore, Md.: Waverly Press.

———. 1938. "Polo in Ancient China." *Horse and Horseman*, April 1938, pp. 27, 38–40.

———. 1975. "Polo, a Note." *TP* 61: 301–302. *See also* Ku Chieh-kang (1938)

Goodrich, L. Carrington, and Fang, Chaoying. 1976. *Dictionary of Ming Biography, 1368–1644*. 2 vols. New York and London: Columbia University Press.

Gottschalk, Louis, ed. 1963. *Generalization in the Writing of History*. Chicago: University of Chicago Press.

Graham, A. C. 1958. *Two Chinese Philosophers: Ch'eng Ming-tao and Ch'eng Yi-ch'uan*. London: Lund Humphries.

———. 1959. "'Being' in Western Philosophy Compared with *Shih/Fei* and *Yu/Wu* in Chinese Philosophy." *AM* n.s. 7: 79–112.

———, tr. 1960. *The Book of Lieh-tzu, a New Translation*. London: John Murray.

———. 1964. "The Logic of the Mohist *Hsiao-ch'ü*." *TP* 51: 1–54.

———. 1972. "Later Mohist Treatises on Ethics and Logic Reconstruced from the *Ta-ch'ü* Chapter of *Mo-tzu*." *AM* n.s. 17: 137–189.

———. 1973. "China, Europe, and the Origins of Modern Science: Needham's the Grand Titration." In Nakayama and Sivin (1973: 45–69).

———. 1978. *Later Mohist Logic, Ethics and Science*. Hong Kong: Chinese University Press, Chinese University of Hong Kong and London: School of Oriental and African Studies, University of London.

———, tr. 1981. *Chuang-tzu: The Seven Inner Chapters, and Other Writings from the Book Chuang-tzu*. London: George Allen & Unwin.

———. 1986. *Yin-Yang and the Nature of Correlative Thinking*. Singapore: Institute of East Asian Philosophies, University of Singapore.

———. 1989. *Disputers of the Tao: Philosophical Argument in Ancient China*. La Salle, Ill.: Open Court.

Granet, Marcel. 1932. *Festivals and Songs of Ancient China*. London: George Routledge and Sons. Tr. E. D. Edwards of Granet, *Fêtes et chansons anciennes de la China*. Paris: Ernest Leroux, 1919.

———. 1934. *La Pensée chinoise*. Paris: La Renaissance du Livre.

Green, Robert W., ed. 1959. *Protestantism and Capitalism: The Weber Thesis and Its Critics*. Boston: Heath.

Griffith, Samuel B., tr. 1963. *Sun Tzu, the Art of War*. London: Oxford University Press.

Groot, J. J. M. de. 1886. *Les Fêtes annuellement célébrées à Emoui (Amoy), étude concernant la religion populaire des chinois*. Tr. C. G. Chavannes from de Groot's Dutch work. 2 vols. Paris: Ernest Leroux.

———. 1903–1904. *Sectarianism and Religious Persecution in China*. 2 vols. Amsterdam: Johannes Müller.

Gulik, Robert H. van. 1940. *The Lore of the Chinese Lute*. Monumenta Nipponica Monographs no. 3. Tokyo: Sophia University.

———, tr. 1941. *Hsi K'ang and His Poetical Essay on the Lute*. Monumenta Nipponica Monographs no. 4. Tokyo: Sophia University.

———, ed. 1951. *Erotic Colour Prints of the Ming Period, with an Essay on Chinese Sex Life from the Han to the Ch'ing Dynasty, B.C. 206–A.D. 1644*. 3 vols. Tokyo. Privately published.

———. 1961. *Sexual Life in Ancient China: A Preliminary Survey of Chinese Sex and Society from ca. 1500 B.C. till 1644 A.D.* Leiden: E. J. Brill.

du Halde, Jean-Baptiste. 1741. *The General History of China. Containing a Geographical, Historical, Chronological, Political and Physical Description of the Empire of China, Chinese-Tartary, Corea and Thibet*... 3d ed., corrected. 4 vols. London: J. Watts. Tr. R. Brooks of du Halde, *Description géographique, historique, chronologique, politique de l'Empire de la Chine et de la Tartarie chinoise*. 4 vols. Paris: P. G. Lemercier, 1735.

Hall, David L., and Ames, Roger T. 1987. *Thinking Through Confucius*. Albany: State University of New York Press.

Hallpike Christopher R. 1986. *The Principles of Social Evolution*. Oxford: Clarendon Press.

Hamilton, Gary G. 1985. "Why No Capitalism in China? Negative Questions in Historical, Comparative Research." In *Max Weber in Asian Society*. Ed. Andreas E. Buss. Pp. 187–211. *Journal of Developing Societies*, vol. 1, no. 2. Leiden: E. J. Brill.

Hansen, Chad. 1983. *Language and Logic in Ancient China*. Ann Arbor: University of Michigan Press.

Harbsmeir, Christoph. 1981. *Aspects of Classical Chinese Syntax*. London and Malmo: Curzon Press.

———. 1983–1985. Review of Hansen (1983). *EC* 9–10:250–257.

Harding, Harry. 1981. *Organizing China, the Problem of Bureaucracy*. Stanford, Calif.: Stanford University Press.

Harper, Donald. 1987. "The Sexual Arts of Ancient China as Described in a Manuscript of the Second Century B.C." *HJAS* 47:539–593.

Harris, Harold Arthur. 1964. *Greek Athletes and Athletics*. London: Hutchinson.

———. 1972. *Sport in Greece and Rome*. London: Thames and Hudson.

Hartman, Charles. 1968. *Han Yü and the T'ang Search for Unity*. Princeton, N.J.: Princeton University Press.

Hartwell, Robert M. 1971. "Historical Analogism, Public Policy, and Social Science in Eleventh- and Twelfth-Century China." *AHR* 76:690–727.

Harvey, John. 1947. *Gothic England, a Survey of National Culture, 1300–1550*. London: B. T. Batsford.

Hawkes, David, tr. 1959. *Ch'u Tz'u: the Songs of the South, an Ancient Chinese Anthology*. Oxford: Clarendon Press.

Hedley, Geffrey. 1936–1937. "Yi-hsing Ware." *Transactions of the Oriental Ceramic Society* 14:70–86.

Henderson, John B. 1984. *The Development and Decline of Chinese Cosmology*. New York: Columbia University Press.

Hervouet, Yves, tr. 1972. *Le Chapitre 117 du Che-ki (biographie de Sseu-ma Siang-jou)*. Paris: Presses Universitaires de France.

———. 1976. "L'Autobiographie dans la Chine traditionelle." In *Études d'histoire et de littérature chinoises offertes au Professeur Jaroslav Prusek*. Ed. Yves Hervouet. Vol. 24. Pp. 107–141. Paris: Bibliothèque de l'Institut des Hautes Études Chinoises.

Hicks, R. P. tr. 1925. *Diogenes Laertius, Lives of Eminent Philosophers*. 2 vols. Loeb Classical Library. London: Heinemann.

Hightower, James R. 1961. "Individualism in Chinese Literature." *JHI* 22:159–168.

Hirsch, Rudolf. 1967. *Printing, Selling and Reading, 1450–1550*. Wiesbaden: Harrassowitz.

Hirth, Friedrich. 1887. "Ancient Porcelain: a Study in Chinese Mediaeval Industry and Trade." *JNChBRAS* n.s. 22:129–202.

Ho Ping-ti 1954. "The Salt Merchants of Yang-chou, a Study of Commercial Capitalism in XVIIIth-century China." *HJAS* 17:130–168.

———. 1968. "Salient Aspects of China's Heritage." In *China's Heritage and the Communist Political System*. Ed. Ho Ping-ti and Tsou Tang. 2 vols. Pp. 1–37. Chicago: University of Chicago Press.

———. 1975. *The Cradle of the East, an Inquiry into the Indigenous Origins of Techniques and Ideas of Neolithic and Early Historic China, 5000–1000* B.C. Hong Kong: Chinese University of Hong Kong and Chicago: University of Chicago Press.

Hoang, Pierre, 1902. "Exposé de l'origine et du développement du système de Tchou Hi et de son influence sur l'esprit des lettrés." In *Mélanges sur l'administration*, Pierre Hoang. Variétés Sinologiques no. 21. Pp. 147–164. Shanghai: Imprimerie Mission Catholique.

Hobbes, Thomas. 1651. *The Leviathan*. London.

Hodous, Lewis. *See* Soothill and Hodous (1937)

Hogarth, William. 1955. *The Analysis of Beauty, with the Rejected Passages from the Manuscript Drafts and Autobiographical Notes*. Ed. Joseph Burke (original Hogarth book 1753). Oxford: Clarendon Press.

Holzman, Donald. 1956. "The Conversational Tradition in Chinese Philosophy." *PEW* 6:223–230.

———. 1957. *La Vie et la pensée de Hi K'ang (223–262 ap. J.-C.)*. Leiden: E. J. Brill.

———. 1958. "Shen Kua and His Meng-ch'i pi-t'an." *TP* 46:260–292.

Homann, Rolf, tr. 1976. *Pai Wen P'ien or the Hundred Questions, a Dialogue Between Two Taoists on the Macrocosmic and Microcosmic System of Correspondences*. Leiden: E. J. Brill.

Hooykaas, Meijer. 1972. *Religion and the Rise of Modern Science*. Edinburgh and London: Scottish Academic Press.

Hosie, Lady. *See* Soothill (1951)

Howard, Richard C. 1962. "Modern Chinese Biographical Writing." *JAS* 21:465–475.

Hsia, C. T. 1978. "Yen Fu and Liang Ch'i-ch'ao as Advocates of New Fiction." In *Chinese Approaches to Literature from Confucius to Liang Ch'i-ch'ao*, ed. Adele Austin Rickett. Pp. 221–257. Princeton, N.J.: Princeton Unviersity Press.

Hsiao Kung-ch'üan. 1976. "Li Chih." In Goodrich and Fang (1976:807–818).

Hsü Cho-yun. 1965. *Ancient China in Transition: an Analysis of Social Mobility, 722–222 B.C.* Stanford, Calif.: Stanford University Press.

Hsü, Francis L. K. 1948. *Under the Ancestors' Shadow: Chinese Culture and Personality*. New York: Columbia University Press.

Hsü Fu-kuan 徐復觀. 1963. *Chung-kuo jen-hsing-lun shih: Hsien Ch'in p'ien* 中國人性論史先秦篇 (A History of the Chinese Philosophy of Human Nature: Pre-Ch'in Part). Taichung: Tung-hai University (also published Taipei, 1969).

Hsü Shih-ch'ang. See *Ch'ing-ju hsüeh-an*

Hu Hsien Chin. 1944. "The Chinese Concepts of 'Face'." *American Anthropologist* 46:45–64.

Hu Shih. 1922. *The Development of the Logical Method in Ancient China*. Shanghai: Oriental Book.

———. 1934. *The Chinese Renaissance*. Chicago: University of Chicago Press.

Huang Han 黃漢. 1936. *Kuan-tzu ching-chi ssu-hsiang* 管子經濟思想 (Economic Thought in the *Kuan-tzu*). Shanghai: Commercial Press. Tr. T'an and Wen (1954:213–403).

Huang Hsing-Tsung. 1986. "Plants and Insects in Man's Service." In *Science and Civilisation in China*, Joseph Needham et al., 6/1:471–553. Cambridge, London, New York, etc.: Cambridge University Press.

Huang, Ray. 1981. *1587, a Year of No Significance: The Ming Dynasty in Decline*. New Haven, Conn., and London: Yale University Press.

———. 1986. "The Merger of Chinese History with Western Civilization." *CSH* 20/1:51–122.

———. 1988. *China: A Macro-History*. Armonk, N.Y.: M. E. Sharpe.

Huang Siu-chi. 1944. *Lu Hsiang-shan, a Twelfth Century Chinese Idealist Philosopher*. American Oriental Series vol. 27. New Haven, Conn.: American Oriental Society.

Hucker, Charles O. 1959. "Confucianism and the Chinese Censorial System." In Nivison and Wright (1959:182–208).

Hudson, G. F. *See* Soothill (1951)

Hulsewé, A. F. P. 1985. *Remnants of Ch'in Law*. Leiden: E. J. Brill.

Hummel, Arthur W. *See* Fang Chao-ying (1943a)

Huxley, Thomas Henry. 1964. (originally written 1863). "The Method of Scientific Investigation." In *Science: Method and Meaning.* Ed. Samuel Rapport and Helen Wright. Pp. 2–10. New York: Washington Square Press.

Imber, Diana. *See* Beurdeley (1969)

Irwin, Richard Gregg. 1953. *The Evolution of a Chinese Novel: Shui-hu-chuan.* Cambridge, Mass.: Harvard University Press.

Ishihara, Ahira, and Levy, Howard S., tr. 1970. *The Tao of Sex, an Annotated Translation of the Twenty-eighth Section of the Essence of Medical Prescriptions (Ishimpō).* New York and London: Harper and Row.

Ivanhoe, Philip J. 1988. "Reflections on the *Chin-ssu lu.*" *JAOS* 108:269–275.

Jan Yün-hua. 1965. "Buddhist Self-Immolation in Medieval China." *HoR* 4:209–268.

Jao Tsong-yi and Demiéville, Paul. 1971. *Missions Paul Pelliot, Documents conservés à la Bibliothèque Nationale.* Vol. 2: *Airs de Touen-houang.* Paris: Editions du Centre National de la Recherche Scientifique.

Jenner, W. J. F., tr. 1981. *Memories of Loyang,* Yang Hsüan-chih. Oxford: Clarendon Press.

Jenyns, Soame. 1953. *Ming Pottery and Porcelain.* London: Faber & Faber.

Johnson, Wallace S., Jr. 1970. *A Concordance to the Kuan-tzu.* Research Aids Series no. 9. Taipei: Chinese Research Aids Center.

———. 1975. *A Concordancce to Han-fei-tzu.* Research Aids Series no. 13. San Francisco: Chinese Materials Center.

———, tr. 1979. *The T'ang Code.* Vol. 1: *General Principles.* Princeton, N.J.: Princeton University Press.

Jowett, B., tr. 1953. *The Dialogues of Plato.* 4th ed. Oxford: Clarendon Press.

Julien, Stanislas, tr. 1856. *Histoire et fabrication de la porcelaine chinoise.* Paris: Mallet-Bachelier.

———, tr. 1864. *Thsien-Tseu-Wen, le livre des milles mots.* Paris: Benjamin Duprat.

Jung, Carl G. 1928. *Die Beziehungen zwischen dem Ich und dem Unbewusten.* Darmstadt: Otto Reichl.

Kaltenmark, Max. 1963. "Les Danses sacrées en Chine." In *Les Danses sacrées: Egypte ancienne, Israël, Islam, Asie Centrale, Inde, Cambodge, Bali, Java, Chine, Japon.* Pp. 411–450. No named ed. Paris: Editions du Seuil.

———. 1979. "The Ideology of the T'ai-p'ing ching." In *Facets of Taoism: Essays in Chinese Religion.* Ed. Holmes Welch and Anna Seidel. Pp. 19–53. New Haven, Conn., and London: Yale University Press.

Karlgren, Bernhard. 1929. "The Authenticity of Ancient Chinese Texts." *BMFEA* 1:165–184.

———. 1930. "Some Fecundity Symbols in Ancient China." *BMFEA* 2:1–67.

———. 1942. "Some Ritual Objects of Prehistoric China." *BMFEA* 14:65–69.

———. 1944. "Glosses on the Siao ya Odes." *BMFEA* 16:25–169.

———. 1949. *The Chinese Language.* New York: Ronald Press.

———, tr. 1950a. *The Book of Odes.* Stockholm: Museum of Far Eastern Antiquities.

———, tr. 1950b. "The Book of Documents." *BMFEA* 22:1–81.

———. 1957. "Grammata Serica Recensa." *BMFEA* 29:1–332.

Keightley, David N. 1978. "The Religious Commitment: Shang Theology and the Genesis of Chinese Political Culture." *HoR* 17:211–225.

Kendall, Paul Murray. 1974. "Biographical Literature." In *The New Encyclopaedia Britannica* (Chicago, 15th ed.). *Macropaedia* 2:1006–10014.

Kennedy, George A. 1951. "The Monosyllabic Myth." *JAOS* 71:161–166.

———. 1953. *ZH Guide, an Introduction to Sinology*. New Haven, Conn.: Far Eastern Publications, Yale University.

Kierman, Frank A. *See* Fairbank (1974), Maspero (1981)

Kitāb al-'ayn. By al-Khalīl Ibn Ahmad (d. 786). Arabic dictionary.

Knechtges, David R., tr. 1982. *Wen Xuan or Selections of Refined Literature*. Vol. 1: *Rhapsodies on Metropolises and Capitals*, Xiao Tong [Hsiao T'ung] (501–533). Princeton, N.J.: Princeton University Press.

Knoblock, John. 1982–1983. "The Chronology of Xunzi's Works." *EC* 8:28–52.

Kracke, Edward A. 1953. *Civil Service in Early Sung China, 960–1067*. Cambridge, Mass.: Harvard University Press.

Kroll, J. L. 1985–1987. "Disputation in Ancient Chinese Culture." *EC* 11–12: 118–145.

Ku Chieh-kang. 1938. "A Study of Literary Persecution during the Ming." Tr. L. Carrington Goodrich. *HJAS* 3:254–311.

Kuchera, S. 1977. "Iz Istorii Kitayskogo Tantsevalnogo Iskusstva: *Tsi-pan u*" (From the History of Chinese Dancing: the *Ch'i-p'an wu*). In *Kitai: Istoriya, Kultura i Istoriografiya* (China: History, Culture and Historiography). Moscow: Nauka Publishing House.

Kuhn, Dieter. 1988. Pt. 9: *Textile Technology: Spinning and Reeling*. In *Science and Civilisation in China*. Vol. V: *Chemistry and Chemical Technology*. Ed. Joseph Needham. Cambridge, London, New York, etc.: Cambridge University Press.

Kunst, Richard A. 1977. "More on *Xiu* and *Wuxing*, with an Addendum on the Use of Archaic Construction." *EC* 3:67–69.

Kuo Po-kung 郭伯恭. 1938. *Yung-lo ta-tien k'ao* 永樂大典考 (A Study of the Yung-lo Great Encyclopedia). Shanghai: Commercial Press.

Lamont, H. G. 1973–1975. "An Early Ninth Century Debate on Heaven: Liu Tsung-yüan's *T'ien shuo* and Liu Yü-hsi's *T'ien lun*." *AM* n.s. 18:181–208; 19:37–85.

Lattimore, Owen. 1951. *Inner Asian Frontiers of China*. 2d ed. New York: American Geographical Society (1st ed., 1940).

Lau, D. C. 1967. "A Note on *Ke Wu*." *BSOAS* 30:353–357.

Laufer, Berthold. 1932. "The Early History of Polo." *Polo, the Magazine for Horsemen*, April 1932, pp. 13–14, 43–44.

Lawton, Thomas. 1973. *Chinese Figure Painting*. Washington, D.C.: Freer Gallery of Art, Smithsonian Institution.

Le Blanc, Charles. 1985. *Huai-nan Tzu, Philosophical Synthesis in Early Han Thought: The Idea of Resonance (Kan-ying) with a Translation and Analysis of Chapter Six*. Hong Kong: Hong Kong University Press.

———. 1985–1986. "A Re-examination of the Myth of Huang-ti." *Journal of Chinese Religions* 13–14:45–63.

———. 1987. "From Ontology to Cosmogony: Notes on *Chuang Tzu* and *Huai-nan Tzu*." In Le Blanc and Blader (1987: 117–129). *See also* Bodde (1981).

Le Blanc, Charles, and Blader, Susan, eds. 1987. *Chinese Ideas about Nature and Society: Studies in Honour of Derk Bodde*. Hong Kong: Hong Kong University Press.

Legge, James, tr. 1885. *The Li Ki, or Collection of Treatises on the Rules of Propriety or Ceremonial Usages*. In *The Sacred Books of the East*. Ed. F. Max Müller. Vols. 27–28. Oxford: Clarendon Press.

———, tr. 1899a. *The Sacred Books of China*. Part 1. *The Shu-king, the Religious Portions of the Shih king, the Hsiao king*. Constitutes *The Sacred Books of the East*. Ed. F. Max Müller. Vol. 3. 2d ed. Oxford: Clarendon Press.

———, tr. 1899b. *The Sacred Books of China*. Part 2. *The Yi King*. Constitutes *The Sacred Books of the East*. Ed. F. Max Muller. Vol 16. 2d ed. Oxford: Clarendon Press.

———, tr. 1960. *The Chinese Classics*. Hong Kong: Hong Kong University Press (reprint of the revised 2d ed.). Comprises the following vols.: 1. *The Confucian Analects, Great Learning, Doctrine of the Mean*; 2. *The Works of Mencius*; 3. *The Shoo-king* [*Shu ching*]; 4. *The She-king* [*Shih ching*]; 5. *The Ch'un Ts'ew* [*Ch'un ch'iu*], *with the Tso Chuen* (in two parts with continuous pagination).

Leibniz, Gottfried Wilhelm. 1686. *Discours de métaphysique*.

———. 1716. *Discours sur la théologie naturelle des chinois*. Tr. Rosemont and Cook (1977).

Leslie, Donald. 1954. *Man and Nature; Sources on Early Chinese Biological Ideas*. Inaugural Dissertation, Cambridge University.

Levenson, Joseph R. 1957. "The Amateur Ideal in Ming and Early Ch'ing Society: Evidence from Painting." In Fairbank (1957: 320–344).

Levy, Howard S. 1956. "Yellow Turban Religion and Rebellion at the End of Han." *JAOS* 76: 214–227.

———. 1966. *Chinese Footbinding, the History of a Curious Erotic Custom*. New York: W. Rawls. *See also* Ishihara and Levy (1970).

Li Chi. 1972. "Chu Hsi the Poet." *TP* 58: 55–119.

Liao, W. K., tr. 1939. 1959. *The Complete Works of Han Fei Tzu*. 2 vols. London: Arthur Probsthain.

Lin, T. C. *See* Gale (1967)

Lin Yutang. 1967. *The Chinese Theory of Art*. London: Heinemann.

Ling Shun-sheng 凌純聲. 1959a. "Chung-kuo tsu-miao ti ch'i-yüan" 中國祖廟的起源 (Origin of the Ancestral Temple in China). *Bull. Inst. Eth.* 7: 141–176; English summary: 177–184.

———. 1959b. "Shen-chu yü Yin-Yang hsing ch'i ch'ung-pai" 神主與陰陽性器崇拜 (Ancestral Tablet and Genital Symbolism in Ancient China). *Bull. Inst. Eth.* 8: 1–38; English summary: 39–46.

Liu, James J. Y. 1962. *The Art of Chinese Poetry*. London: Routledge and Kegan Paul.

———. 1967. *The Chinese Knight-errant*. London: Routledge and Kegan Paul.

———. 1975. *Chinese Theories of Literature*. Chicago and London: University of Chicago Press.

Liu, James T. C. 1973. "How Did a Neo-Confucian School Become the State Orthodoxy?" *PEW* 23: 483–505.

———. 1985. "Polo and Cultural Change from T'ang to Sung China." *HJAS* 45:203–223.

Lloyd, Janet. *See* Gernet (1985)

Loewe, Michael. 1967. *Records of Han Administration.* 2 vols. Cambridge: Cambridge University Press.

———. 1974. *Crisis and Conflict in Han China, 104 B.C. to A.D. 9.* London: George Allen & Unwin.

———. 1979. *Ways to Paradise: The Chinese Quest for Immortality.* London: George Allen & Unwin.

———. 1987. "The Cult of the Dragon and the Invocation for Rain." In Le Blanc and Blader (1987:195–213).

———. 1990. "The *Juedi* Games: A Re-enactment of the Battle Between Chiyou and Xianyuan?" In *Thought and Law in Qin and Han China: Studies Dedicated to Anthony Hulsewé on the Occasion of His Eightieth Birthday.* Ed. W. L. Edema and E. Zürcher. Sinica Leidensia vol. 24. Pp. 140–157. Leiden: E. J. Brill. *See also* Twitchett and Loewe (1986)

Loon, Piet van der. *See* Van der Loon, Piet

Lovejoy, A. O. 1948. "Nature as Aesthetic Norm." In *Essays in the History of Ideas,* A. O. Lovejoy. Pp. 69–77. Baltimore, Md.: Johns Hopkins University Press.

Lovejoy, A. O., and Boas, G. 1935. *A Documentary History of Primitivism and Related Ideas.* Vol. 1: *Primitivism and Related Ideas in Antiquity.* Baltimore, Md.: Johns Hopkins University Press.

Lukes, Steven. 1973. *Individualism.* Oxford: Basic Blackwell.

Lyall, Leonard A., tr. 1925. *The Sayings of Confucius.* 2d ed. London and New York: Longmans Green.

Mackerras, Colin. *See* Dolby (1983)

MacNair, Harley Farnsworth. *See* Dubs (1946)

de Mailla, J. A. M. de Moyriac, tr. 1777–1785. *Histoire générale de la Chine, ou annals de cet empire, traduite du Tong Kien Kang Mou.* 13 vols. Paris: Ph.-D. Pierres ... Clousier.

Major, John. 1976. "A Note on the Translation of Two Technical Terms in Chinese Science." *EC* 2:1–3.

———. 1977. "Reply to Richard Kunst's Comments on *Hsiu* and *Wu Hsing.*" *EC* 3:69–70.

———. 1978. "Myth, Cosmology, and the Origins of Chinese Science." *JCP* 5:1–20.

———. 1984. "The Five Phases, Magic Squares, and Schematic Cosmography." In *Explorations in Early Chinese Cosmology,* ed. Henry Rosemont, Jr. Pp. 133–166. Chico, Calif.: Scholars Press. This book constitutes *Journal of the American Academy of Religion Thematic Studies,* vol. 50, no. 2.

Margouliès, Georges, tr. 1926. *Le Kou-wen chinois, recueil de textes avec introduction et notes.* Paris: Paul Geuthner.

Martindale, Andrew. 1972. *The Rise of the Artist in the Middle Ages and Early Renaissance.* London: Thames and Hudson.

Maspero, Henri. 1948–1951. "Le Ming-t'ang et la crise religieuse chinoise avant les Han." *MCB* 9:1–71.

———. 1981. *Taoism and Chinese Religion.* Amherst: University of Massachusetts Press. Tr. Frank A. Kierman, Jr., of Maspero, *Le Taoïsme et les religions chinoises.* Paris: Gallimard, 1971. (The several quotations in the present book were made from Maspero's original French text before publication of the Kierman translation.)

Mather, Richard, tr. 1976. *Shih-shuo Hsin-yü: A New Account of Tales of the World.* Minneapolis: University of Minnesota Press.

Mathews, R. H. 1963. *Mathews' Chinese-English Dictionary.* Rev. Amer. ed. Cambridge, Mass.: Harvard University Press (1st ed., Shanghai: China Inland Mission and Presbyterian Mission Press, 1931).

Maverick, Lewis. *See* T'an and Wen (1954)

Mayers, William Frederick. 1924. *The Chinese Readers Manual.* Shanghai: Presbyterian Mission Press (1st ed., 1874).

McKeon, Richard. 1974. "Censorship." In *The New Encyclopaedia Britannica* (Chicago, 15th ed.). *Macropaedia* 3:1083–1090.

Mei, Yi-pao, tr. 1929. *The Ethical and Political Works of Motse.* London: Arthur Probsthain.

———. 1934. *Motse, the Neglected Rival of Confucius.* London: Arthur Probsthain.

Meisner, Maurice. 1977. *Mao's China: A History of the People's Republic.* New York: Free Press.

Merton, Robert K. 1970. *Science, Technology and Society in Seventeenth Century England.* New York: Howard Fertig. Reprint, with new Preface and Bibliography, of original ed. in *Osiris* (Brussels) 4/2(1938).

Michaud, Paul. 1958. "The Yellow Turbans." *Monumenta Serica* 17:47–127.

Milsky, Constantin. 1973. "New Developments in Language Reform." *China Quarterly* 53:98–133.

Miyakawa, Hisayuki. 1960. "The Confucianization of South China." In Wright (1960a:21–46).

Miyazaki, Ichisada. 1976. *China's Examination Hell: The Civil Service Examinations of Imperial China.* Tr. of Japanese original by Conrad Schirokauer. New York and Tokyo: Weatherhill.

Moorman, John R. H. 1945. *Church Life in England in the Thirteenth Century.* Cambridge: Cambridge University Press.

Morgan, Evan, tr. 1934. *Tao the Great Luminant, Essays from Huai Nan Tzu.* Shanghai: Kelly & Walsh. Tr. of *Huai-nan-tzu* chs. 1–2, 7–8, 12–13, 15, 19.

Morgan, M. H., tr. 1914. *Vitruvius, the Ten Books on Architecture* [ca. 30 B.C.]. Cambridge, Mass.: Harvard University Press.

Morris, Clarence. *See* Bodde and Morris (1967)

Mote, Frederick W. 1961. "The Growth of Chinese Despotism, a Critique of Wittfogel's Theory of Oriental Despotism as Applied to China." *Oriens Extremus* 8:1–41.

Mungello, David E. 1977. *Leibniz and Confucianism, the Search for Accord.* Honolulu: University Press of Hawaii.

Munro, Donald, ed. 1985. *Individualism and Holism: Studies in Confucian and Taoist Values.* Michigan Monographs in Chinese Studies no. 52. Ann Arbor: Center for Chinese Studies, University of Michigan.

Naish, G. P. B. 1957. "Ships and Shipbuilding." In *A History of Technology*. Ed. Charles Singer, E. J. Holmyard, A. R. Hall, and T. I. Williams. 5 vols. 3:471–500. London and New York: Oxford University Press, 1954–1958.

Nakayama, Shigeru, and Sivin, Nathan, eds. 1973. *Chinese Science: Explorations of an Ancient Tradition*. Cambridge, Mass., and London: MIT Press.

Naquin, Susan. 1976. *Millenarian Rebellion in China: The Eight Trigrams Uprising of 1813*. New Haven, Conn., and London: Yale University Press.

Needham, Joseph. 1954–. *Science and Civilisation in China* (abbr. *SCC*). Cambridge, London, New York, etc.: Cambridge University Press. Of the 15 vols. or parts of vols. published as of early 1990, vols. 1–4 have been written with the collaboration of Wang Ling; most of the vols. from vols. 4/pt. 3 onward with the collaboration of Lu Gwei-Djen; and vol. 5/pts. 3–4 with the collaboration of Ho Ping-yü (Ho Peng Yoke). Other parts, cited in this book and written by scholars other than Needham, include: Bray (1984), Huang Hsing-tsung (1986), Kuhn (1988), Sivin (1980), Tsien (1985). Those further vols. or parts of vols. that are cited in this book under Needham himself include:

———. 1954. *SCC* 1: *Introductory Orientations*.

———. 1956. *SCC* 2: *History of Scientific Thought*.

———. 1959. *SCC* 3: *Mathematics and the Sciences of the Heavens and the Earth*.

———. 1962. *SCC* 4/1: *Physics and Physical Technology: Physics*.

———. 1965a. *SCC* 4/2: *Physics and Physical Technology: Mechanical Engineering*.

———. 1971. *SCC* 4/3: *Physics and Physical Technology: Civil Engineering and Nautics*.

———. 1976. *SCC* 5/3: *Chemistry and Chemical Technology: Spagyrical Discovery and Invention: Historical Survey from Cinnabar Elixirs to Synthetic Insulin*.

———. 1983. *SCC* 5/5: *Chemistry and Chemical Technology: Physiological Alchemy*.

———. 1986. *SCC* 6/1: *Biology and Biological Technology: Botany*.

The following are further writings by Needham also cited in this book:

———. 1965b. *Time and Eastern Man*. Occasional Paper no. 21. London: Royal Anthropological Institute of Great Britain and Ireland. Reprinted in *The Voices of Time: A Cooperative Survey of Man's Views of Time as Expressed by the Sciences and the Humanities*. Ed. J. T. Fraser. Pp. 92–135. New York: George Braziller, 1966.

———. 1969. *The Grand Titration, Science and Society in East and West*. London: George Allen & Unwin.

———. 1970. "The Translation of Old Chinese and Technical Terms." In *Clerks and Craftsmen in China and the West*, Joseph Needham. Pp. 83–97. Cambridge: Cambridge University Press.

Nivison, David S. 1953. "The Problem of 'Knowledge' and 'Action' in Chinese Thought since Wang Yang-ming." In Wright (1953a:112–145).

———. 1959. "Introduction." In Nivison and Wright (1959:3–24).

———. 1960. "Protest against Conventions and Conventions of Protest." In Wright (1960a:177–201).

———. 1962. "Aspects of Chinese Traditional Biography." *JAS* 21:457–463.

Nivison, David S., and Wright, Arthur F., eds. 1959. *Confucianism in Action*. Stanford, Calif.: Stanford University Press.

Nylan, Michael, and Sivin, Nathan. 1987. "The First Neo-Confucian: An Introduction to Yang Hsiung's 'Canon of Supreme Mystery' (*T'ai hsuan ching*, c. 4 B.C.)." In Le Blanc and Blader (1987:41–99).
Paine, Robert Treat, Jr. 1955. "Chinese Ceramic Pillows from Collections in Boston and Vicinity." *Far East Ceramic Bulletin* 7:1–46.
Palmgren, Nils. 1934. *Kansu Mortuary Urns of the Pan Shan and Ma Chang Groups. Palaeontologia Sinica*, series D, vol. 3, fascicle 1. Peiping: Geological Survey of China.
Parsons, Talcott. *See* Weber (1930)
Pelliot, Paul. *See* Chavannes and Pelliot (1911–1913)
Peterson, Willard J. 1975. "Fang I-chih (1611–1671): Western Learning and the 'Investigation of Things.'" In De Bary (1975b:369–411).
———. 1979. *Bitter Gourd: Fang I-chih and the Impetus for Intellectual Change.* New Haven, Conn.: Yale University Press.
———. 1986. "Calendar Reform Prior to the Arrival of Missionaries at the Ming Court." *Ming Studies* 21:45–61.
Pfeiffer, Rudolf. 1968. *History of Classical Scholarship, from the Beginnings to the End of the Hellenistic Age.* Oxford: Clarendon Press.
Philo Judaeus (first century A.D.). *De Iosepho* and *De Aeternitate Mundi. See* Colson and Whitaker (1929–53); Saunders (1966)
Plato (ca. 428–348/347 B.C.). *Laws* and *Republic. See* Jowett (1953)
Pliny the Elder (A.D. 23–79). *Natural History.*
Plutarch (ca. A.D. 46–after 119). *Lives* or *Parallel Lives.*
Pokora, Timoteus. 1961. "On the Origins of the Notions *T'ai-p'ing* and *Ta-t'ung* in Chinese Philosophy." *Archiv Orientalni* 29:448–454.
Porkert, Manfred. 1974. *The Theoretical Fourndations of Chinese Medicine: Systems of Correspondence.* Cambridge, Mass. and London: MIT Press.
Pound, Ezra, tr. 1952. *Confucius, the Great Digest and Unwobbling Pivot.* London: P. Owen. *See also* Fenollossa (1934)
Price, Derek de Solla. 1974. "Gears from the Greeks, the Anikythera Mechanism—a Calendar Computer from ca. 80 B.C." *Transactions of the American Philosophical Society* n.s. 64/7:1–70.
Pulleyblank, Edwin G. 1960. "Neo-Confucianism and Neo-Legalism in T'ang Intellectual Life, 755–805." In Wright (1960a:77–114).
———. 1961. "Chinese Historical Criticism: Liu Chih-chi and Ssu-ma Kuang." In Beaseley and Pulleyblank (1961:135–166).
Purcell, Victor. 1936. *Problems of Chinese Education.* London: Kegan Paul.
Ramsey, S. Robert. 1987. *The Languages of China.* Princeton, N.J.: Princeton University Press.
Rapport, Samuel. *See* Huxley (1964)
Reding, Jean-Paul. 1986a. "Greek and Chinese Categories: A Reexamination of the Problem of Linguistic Relativism." *PEW* 36:349–374.
———. 1986b. "Analogical Reasoning in Early Chinese Philosophy." *Asiatische Studien* 40:40–56.
Reeve, Henry. *See* de Tocqueville (1835–1840)
Reischauer, Edwin O., tr. 1955a. *Ennin's Diary, The Record of a Pilgrimage to China in Search of the Law.* New York: Ronald Press.

———. 1955b. *Ennin's Travels in T'ang China.* New York: Ronald Press. *See also* Fairbank, Reischauer, and Craig (1965)

Renou, Louis, and Filliozat, Jean. 1947–1953. *L'Inde classique.* 2 vols. Paris: Payot.

Révah, I. S. 1960. *La Censure inquisitoriale portugaise au xvi*[e] *siècle.* Vol. 1. Lisbon: Instituto de Alta Cultura.

Ricci, Matteo. *See* Gallagher (1953)

Richards, I. A. 1932. *Mencius on the Mind, Experiments in Multiple Definition.* London: Kegan Paul.

Rickett, Adele Austin. *See* Hsia (1978)

Rickett, W. Allyn. 1965. *Kuan-tzu, a Repository of Early Chinese Thought.* Hong Kong: Hong Kong University Press.

———. 1985. *Guanzi: Political, Economic, and Philosophical Essays from Early China.* Vol. 1 (tr. of chs. 1–34). Princeton, N.J.: Princeton University Press.

Rideout, J. K. 1947. "The Context of the Yüan Tao and the Yüan Hsing." *BSOAS* 12:403–408.

Rosemont, Henry, Jr. *See* Major (1984)

Rosemont, Henry, Jr., and Cook, Daniel J., tr. 1977. *Discourse on the Natural Theology of the Chinese,* Gottfried Wilhelm Leibniz. Tr. with intro. and comm. of Leibniz, *Discours sur la théologie naturelle des chinois* (1716). Monographs of the Society for Asian and Comparative Philosophy, no. 4. Honolulu: University Press of Hawaii. *See also* Cook and Rosemont (1981)

Ross, W. D. *See* Thompson (1910)

Rotours, Robert Des. *See* Des Rotours, Robert.

Roy, Olivier. 1972. *Leibniz et la China.* Paris: J. Vrin.

Russell, Bertrand. 1945. *A History of Western Philosophy, and Its Connection with Political and Social Circumstances from the Earliest Times to the Present Day.* New York: Simon and Schuster.

Rygaloff, Alexis. 1958. "A propos de l'antonymie: l'example du chinois." *Journal de Psychologie Normale et Pathologique* 55:358–376.

Sambursky, S. 1959. *Physics of the Stoics.* London: Routledge and Kegan Paul.

Santillana, Giorgio de. 1961. *The Origins of Scientific Thought.* London: Weidenfeld and Nicolson.

Sargent, Galen Eugène. 1955. *Tchou Hi contre le Bouddhisme.* Paris: Imprimerie Nationale.

Sarton, George. 1927–1948. *Introduction to the History of Science.* 3 vols. Carnegie Institution of Washington, Publication No. 376. Baltimore, Md.: Williams and Wilkins.

Saunders, Jason L. 1966. *Greek and Roman Philosophy after Aristotle.* New York: Free Press and London: Collier-Macmillian.

Sayer, Geoffrey R., tr. 1951. *Ching-te-chen t'ao-lu or the Potteries of China.* London: Routledge and Kegan Paul.

Schafer, Edward H. 1951. "Ritual Exposure in Ancient China." *HJAS* 14:130–184.

———. 1958. "Falconry in T'ang Times." *TP* 46:293–338.

———. 1963. *The Golden Peaches of Samarkand: A Study of T'ang Exotics.* Berkeley and Los Angeles: University of California Press.

———. 1965. "The Idea of Created Nature in T'ang Literature." *PEW* 15:153–160.

———. 1968. "Hunting Parks and Animal Enclosures in Ancient China." *Journal of the Economic and Social History of the Orient* 11:318–341.

———. 1982. "Blue-Green Clouds." *JAOS* 102:91–92.

Schipper, Kristopher M. 1969. "Science, Magic and the Mystique of the Body: Notes on Taoism and Sexuality." In Beurdeley (1969:7–38).

Schirokauer, Conrad. *See* Miyazaki (1976)

Seidel, Anna. 1969. *La Divinisation de Lao Tseu dans le Taoïsme des Han.* Paris: Publications d'Ecole Française d'Extrême Orient, vol. 71.

———. 1969–1970. "The Image of the Perfect Ruler in Early Taoist Messianism: Lao Tzu and Li Hung." *HoR* 9:216–247. *See also* Kaltenmark (1979)

Serruys, Paul L.-M. 1959. *The Chinese Dialects of Han Time According to Fang Yen.* Berkeley and Los Angeles: University of California Press.

Sextus Empiricus (early third century A.D.). *Against the Physicists. See* Bury (1933–1949).

Shih Sheng-han. 1958. *A Preliminary Study of the Book Ch'i Min Yao Shu, an Agricultural Treatise of the Sixth Century.* Peking: Science Press.

———, tr. and comm. 1959. *On Fan Sheng-chih shu, an Agriculturalist Book of China Written by Fan Sheng-chih in the First Century B.C.* Peking: Science Press.

Shryock, John K. 1932. *The Origin and Development of the State Cult of Confucius.* New York and London: Century.

———, tr. 1937. *The Study of Human Abilities: The Jen wu chih of Liu Shao.* American Oriental Series vol. 13. New Haven, Conn.: American Oriental Society.

Sickman, Lawrence, and Soper, Alexander. 1956. *The Art and Architecture of China.* Baltimore, Md.: Penguin.

Sircar, D. C. 1965. *Indian Epigraphy.* Delhi: Motīlal Banarsidas.

Sivin, Nathan. 1966. "Chinese Conceptions of Time." *Earlham Review* 1:82–92.

———. 1968. *Chinese Alchemy, Preliminary Studies.* Cambridge, Mass.: Harvard University Press.

———. 1974. "Foreword." In Porkert (1974:xi–xvi).

———. 1975. "Shen Kua." In *Dictionary of Scientific Biography.* Ed. Charles C. Gillispie. 16 vols. 2:369–393. New York: Charles Scribner's Sons, 1970–1980.

———. 1976. "Chinese Alchemy and the Manipulation of Time." *Isis* 67:513–527. Reprinted in *Science and Technology in East Asia.* Ed. Nathan Sivin. Pp. 108–122. New York: History of Science, 1977.

———. 1978. "On the Word 'Taoist' as a Source of Perplexity with Special Reference to the Relations of Science and Religion in Traditional China." *HoR* 17:303–330.

———. 1980. "The Theoretical Background of Elixir Alchemy." In *Science and Civilisation in China.* Ed. Joseph Needham. 5/4:210–297. Cambridge, London, New York, etc.: Cambridge University Press.

———. 1982. "Why the Scientific Revolution Did Not Take Place in China—or Didn't It? *CS* 5:45–66.

———. 1988. "Science and Medicine in Imperial China—The State of the Field." *JAS* 47:41–90. *See also* Nakayama and Sivin (1973); Nylan and Sivin (1987)

Smith, J. A. *See* Thompson (1910)

Solomon, Bernard S. 1954. "'One Is No Number' in China and the West." *HJAS* 17:253–260.

Soothill, William Edward. 1951. *The Hall of Light, a Study of Early Chinese Kingship.* Ed. Lady Hosie and G. F. Hudson. London: Lutterworth Press.

Soothill, William Edward, and Hodous, Lewis. 1937. *A Dictionary of Chinese Buddhist Terms.* London: Kegan Paul, Trench, Trubner & Co.

Soper, Alexander C. 1948. "Hsiang-kuo Ssu, an Imperial Temple of the Northern Sung." *JAOS* 68:19–45.

———, tr. 1951. *Kuo Jo-hsü's Experiences in Painting (T'u-hua chien-wen chih).* Washington, D.C.: American Council of Learned Societies. *See also* Sickman and Soper (1956)

Sprat, Bishop Thomas. 1667. *History of the Royal Society of London.* London.

Sprenkel, O. B. Van der. *See* Van der Sprenkel, O. B.

Stannard, Jerry. 1961. Review of Sambursky 1959. *Philosophy of Science* 28:83–84.

Steele, John, tr. 1917. *The I-li or Book of Etiquette and Ceremonial.* 2 vols. London: Probsthain.

Stein, R. A. 1963. "Remarques sur les mouvements du Taoïsme politico-religieux au IIe siècle ap. J.-C." *TP* 50:1–78.

Steininger, Hans. 1967. "Ein Lexikographischer Beitrag zum Ersten Gesang des *Shih-Ching.*" In *Festschrift für Wilhelm Eilers.* Ed. Gernot Wiessner. Pp. 559–565. Wiesbaden: Harrassowitz.

von Strauss, Victor. 1879. "Bezeichnung der Farben Blau und Grün in Chinesischen Alterthum." *Zeitschrift der Deutschen Morgenländischen Gesellschaft* 33:502–508.

Sun Zen I-tu and Sun Shio-chuan, tr. 1966. *T'ien-kung K'ai-wu: Chinese Technology in the Seventeenth Century,* Sung Ying-hsing. University Park: Pennsylvania State University Press.

Swann, Nancy Lee, tr. 1950. *Food and Money in Ancient China: The Earliest Economic History of China to A.D. 25, Han Shu 24 with Related Texts, Han Shu 91 and Shih-chi 129.* Princeton, N.J.: Princeton University Press.

T'an Po-fu and Wen Kung-wen, tr. 1954. In *Economic Dialogues in Ancient China: Selections from the Kuan-tzu.* Ed. Lewis A. Maverick. Tr. by T'an and Wen of *Kuan-tzu*'s economic chs. and of a study on the *Kuan-tzu* by Huang Han 1936. Carbondale, Ill.: Lewis A. Maverick (privately printed).

T'ang Yung-t'ung 湯用彤. 1938. *Han Wei Liang-Chin Nan-pei Ch'ao Fo-chiao shih* 漢魏兩晉南北朝佛教史 (A History of Buddhism during the Han, Wei, Two Chin, and Northern and Southern Dynasties). 2 vols. with continuous pagination. Changsha: Commercial Press. A slightly earlier 1938 Shanghai edition is identical save that its 2 vols. are separately paginated.

Tawney, Richard H. 1926. *Religion and the Rise of Capitalism.* London: John Murray.

Tchang, Mathias. 1905. *Synchronismes chinois, chronologie complète et concordance avec l'ère chrétienne de toutes les dates concernant l'histoire de l'Extrême-Orient.* Variétés Sinologiques no. 24. Shanghai: Imprimerie de la Mission Catholique.

Teng Ssu-yü, tr. 1968. *Family Instructions for the Yen Clan, Yen-shih chia-hsün,* Yen Chih-t'ui. Leiden: E. J. Brill.

Teng Ssu-yü and Biggerstaff, Knight. 1971. *An Annotated Bibliography of Selected Chinese Reference Works.* 3d ed. (1st ed., 1936). Cambridge, Mass.: Harvard University Press.

Teng Ssu-yü and Fairbank, John K. 1954. *China's Response to the West: A Documentary Survey, 1839–1923*. Cambridge, Mass.: Harvard University Press.

Thompson, d'Arcy Wentworth, tr. 1910. *Historia Animalum* in *The Works of Aristotle*. Ed. J. A. Smith and W. D. Ross. Vol. 4. Oxford: Clarendon Press.

Thompson, Laurence G., tr. 1958. *Ta T'ung Shu, the One-world Philosophy of K'ang Yu-wei*. London: George Allen & Unwin.

Ting Fu-pao 丁福保. 1970. *Fo-hsüeh ta tz'u-tien* 佛學大辭典 (Great Buddhist Dictionary). 3 vols. preceded by 1 intro. vol. Taipei reprint: Hua-yen Lien-she (Avataṁsa Lotus Society) (original ed., Shanghai, 1921).

Tjan, Tjoe Som, tr. 1949–1952. *Po Hu T'ung, the Comprehensive Discussions in the White Tiger Hall*. 2 vols. with continuous pagination. Leiden: E. J. Brill.

de Tocqueville, Alexis. 1835–1840. *Democracy in America*. Tr. Henry Reeve of French original. London.

Toynbee, Arnold. *See* Twitchett (1973)

Trask, Willard R. *See* Eliade (1954)

Tsien, Tsuen-hsuin. 1962. *Written on Bamboo and Silk, the Beginnings of Chinese Books and Inscriptions*. Chicago: University of Chicago Press.

———. 1985. Pt. 1: *Paper and Printing*. In *Science and Civilisation in China*, vol. 5: *Chemistry and Chemical Technology*. Ed. Joseph Needham. Cambridge, London, New York, etc.: Cambridge University Press.

Tsou Tang. *See* Ho (1968)

Tu, Ching-yi. 1974–1975. "The Chinese Examination Essay: Some Literary Considerations." *Monumenta Serica* 31:393–406.

Tun Li-ch'en. *See* Bodde (1965)

Turban, Helga, tr. 1971. *Das Ching-Ch'u sui-shih chi, ein chinesischer Festkalender*. Augsberg: Dissertationsdruck W. Blasaditsch.

Twitchett, Denis C. 1961. "Chinese Biographical Writing." In Beaseley and Pulleyblank (1961:95–114).

———. 1962. "Problems of Chinese Biography." In Wright and Twitchett (1962: 24–42).

———. 1963. *Financial Administration under the T'ang Dynasty*. Cambridge: Cambridge University Press (reissued 1970).

———. 1968. "Merchant, Trade and Government in Late T'ang." *AM* n.s. 14:63–95.

———. 1973. "Chinese Politics and Society from the Bronze Age to the Manchus." In *Half the World*. Ed. Arnold Toynbee. Pp. 31–78. London: Thames and Hudson.

Twitchett, Denis C., and Loewe, Michael, eds. 1986. *The Cambridge History of China*, vol. 1: *The Ch'in and Han Empires, 221 B.C.–A.D. 220*. Cambridge, London, New York, etc.: Cambridge University Press.

Tz'u hai 辭海 (Sea of Phrases). 1939. Dictionary in 2 vols. Shanghai: Chung-hua (China) Book Co.

Tz'u yüan 辭源 (Origin of Phrases). 1915. Dictionary in 2 vols. Shanghai: Commercial Press.

Umminger, Walter. 1963. *Supermen, Heroes and Gods, the Story of Sport through the Ages*. Tr. from German and adapted by James Clark. London: Thames and Hudson.

Unwin, Joseph Daniel. 1934. *Sex and Culture*. Oxford: Oxford University Press.

Van der Loon, Piet. 1961. "The Ancient Chinese Chronicles and the Growth of Historical Ideas." In Beaseley and Pulleyblank (1961:24–30).

Vandermeersch, Léon. 1965. *La Formation du légisme, recherches sur la constitution d'une philosophie politique caractéristique de la Chine ancienne*. Paris: Publications Ecole Française d'Extrême Orient, vol. 56.

Van der Sprenkel, O. B. 1960. "Chronographie et historiographie chinoises." In *Bibliothèque de l'Institut des Hautes Études Chinoises* 14:407–421 (equivalent to *Mélanges Publiées par l'Institut des Hautes Études Chinoises* 2:407–421). Paris: Institut des Hautes Études Chinoises.

van Gulik, Robert H. *See* Gulik, Robert H. van

Vitruvius (first century B.C.). *On Architecture*. *See* Morgan (1914)

Waley, Arthur, tr. 1918. *One Hundred and Seventy Chinese Poems*. London: Constable.

———, tr. 1934. *The Way and Its Power, a Study of the Tao Te Ching and Its Place in Chinese Thought*. London: George Allen & Unwin.

———, tr. 1937. *The Book of Songs*. Boston and New York: Houghton Mifflin.

———, tr. 1938. *The Analects of Confucius*. London: George Allen & Unwin.

Wang Hsiao-ch'uan 王曉傳. 1958. *Yüan Ming Ch'ing San-tai chin-hui hsiao-shuo hsi-ch'ü shih-liao* 元明清三代禁毀小說戲曲史料 (Historical Materials on the Prohibition of Fiction and Drama during the Yüan, Ming, and Ch'ing Dynasties). Peking.

Wang, John Ching-yu. 1972. *Chin Sheng-t'an*. New York: Twayne.

Wang Li-ch'i 王利器. 1958. *Yen t'ieh lun chiao chu* 鹽鐵論校注 (The *Yen t'ieh lun* Collated and Commented on). Shanghai.

Wang Yi-t'ung, tr. 1984. *A Record of Buddhist Monasteries in Lo-yang*, Yang Hsüan-chih. Princeton, N.J.: Princeton University Press.

Ware, James R., tr. 1966. *Alchemy, Medicine and Religion in the China of A.D. 320, the Nei p'ien of Ko Hung (Pao-p'u tzu)*. Cambridge, Mass.: MIT Press.

Watson, Burton. 1958. *Ssu-ma Ch'ien, Grand Historian of China*. New York and London: Columbia University Press.

———, tr. 1961. *Records of the Grand Historian of China, Translated from the Shih Chi of Ssu-ma Ch'ien*. 2 vols. New York and London: Columbia University Press.

———, tr. 1963. *Hsün Tzu: Basic Writings*. New York and London: Columbia University Press.

———, tr. 1968. *The Complete Works of Chuang Tzu*. New York and London: Columbia University Press.

Weber, Max. 1930. *The Protestant Ethic and the Spirit of Capitalism*. Tr. Talcott Parsons. London: Unwin.

Welch, Holmes. 1967. *The Practice of Chinese Buddhism, 1900–1950*. Cambridge, Mass.: Harvard University Press. *See also* Kaltenmark (1979)

Wen Kung-wen. *See* T'an and Wen (1954)

White, Andrew Dickson. 1896. *A History of the Warfare of Science with Theology*. 2 vols. London: Macmillan (also later eds.).

Wiessner, Gernot. *See* Steininger (1967)

Wilbur, C. Martin. 1943. *Slavery in China during the Former Han Dynasty, 206 B.C.–A.D. 25*. Chicago: Field Museum of Natural History.

Wilhelm, Richard, tr. 1928. *Frühling und Herbst des Lü Bu We*. Tr. of *Lü-shih ch'un-ch'iu*. Jena: E. Diederichs.

———, tr. 1930. *Li Gi, das Buch der Sitte des älteren und jüngeren Dai.* Jena: E. Diederichs.

———, tr. 1950. *The I Ching or Book of Changes.* Tr. from Wilhelm's German *I Ging* (Jena: E. Diederichs, 1924) by Cary F. Baynes. 2 vols. New York: Pantheon.

Williams, Edmund T. 1913. "The State Religion of China during the Manchu Dynasty." *JNChBRAS* 44:11–45.

Wilson, H. S. 1941. "Some Meanings of 'Nature' in Renaissance Literary Theory." *JHI* 2:430–448.

Wright, Arthur F. 1951. "Fu I and the Rejection of Buddhism." *JHI* 12:33–47.

———, ed. 1953a. *Studies in Chinese Thought.* Chicago: University of Chicago Press.

———. 1953b. "The Chinese Language and Foreign Ideas." In Wright (1953a: 286–303).

———. 1959. *Buddhism in Chinese History.* Stanford, Calif.: Stanford University Press.

———, ed. 1960a. *The Confucian Persuasion.* Stanford, Calif.: Stanford University Press.

———. 1960b. "Sui Yang-ti: Personality and Stereotype." In Wright (1960a:47–76).

———. 1963. "On the Uses of Generalization in the Study of Chinese History." In Gottschalk (1963:36–58).

———. 1965. "Symbolism and Function, Reflections on Changan and Other Great Cities." *JAS* 24:667–679. *See also* Balazs (1964); Nivison and Wright (1959).

Wright, Arthur F., and Twitchett, Denis C., eds. 1962. *Confucian Personalities.* Stanford, Calif.: Stanford University Press.

Wright, H. M. *See* Balazs (1964) and Gernet (1970)

Wu Kuang-ming. 1987. "Counterfactuals, Universals and Chinese Thinking." *PEW* 37:84–94.

Wu, Nelson I. 1962. "Tung Ch'i-ch'ang (1555–1636): Apathy in Government and Fervor in Art." In Wright and Twitchett (1962:260–293).

Wu Pei-yi. 1979. "Self-examination and Confession of Sins in Traditional China." *HJAS* 39:5–38.

Wylie, Alexander. 1867. *Notes on Chinese Literature.* Shanghai: American Presbyterian Mission Press (reprinted 1922).

Yabuuti, Kiyosi. 1954. "Indian and Arabian Astronomy in China." Pp. 385–603 in Silver Jubilee Volume of the *Zinbun-Kagaku-Kenkyyusyo.* Kyoto: Kyoto University.

Yang, C. K. 1959. "Some Characteristics of Chinese Bureaucratic Behavior." In Nivison and Wright (1959:134–164).

———. 1961. *Religion in Chinese Society.* Berkeley and Los Angeles: University of California Press.

Yang Lien-sheng. 1949. "Numbers and Units in Chinese History." *HJAS* 12:216–225; reprinted in Yang (1961b:75–84).

———. 1954. "Toward a Study of Dynastic Configurations." *HJAS* 17:329–345; reprinted in Yang (1961b:1–17).

———. 1957. "Economic Justification for Spending—an Uncommon Idea in Traditional China." *HJAS* 20:36–52; reprinted in Yang (1961b:58–74).

———. 1961a. "The Organization of Chinese Official Historiography: Principles and Methods of the Standard Histories from the T'ang through the Ming Dynasty." In Beaseley and Pulleyblank (1961:44–59); reprinted in Yang

Lien-sheng, *Excursions in Sinology*. Pp. 96–111. Cambridge, Mass.: Harvard University Press, 1969.
———. 1961b. *Studies in Chinese Institutional History*. Harvard-Yenching Institute Series 20. Cambridge, Mass.: Harvard University Press.
———. 1970. "Government Control of Urban Merchants in Traditional China." *Tsing Hua Journal of Chinese Studies* n.s. 8: 186–206.
Yang, Martin C. 1945. *A Chinese Village, Taitou, Shantung Province*. New York: Columbia University Press.
Yao Shan-yu. 1948. "The Cosmological and Anthropological Philosphy of Tung Chung-shu." *JNChBRAS* 73: 40–68.
Yeh Te-hui 葉德輝. 1973. *Shu-lin ch'ing-hua* 書林清話 (Chats about Books). Taipei reprint of 1st ed. of 1920.
Yi Shui. 1980. "Wrestling in Ancient China Recalled." *China Reconstructs* 29/7 (July 1980): 68–69.
Yoshikawa Kojirō. 1955. "The *Shih-shuo hsin-yü* and Six Dynasties Prose Style." Tr. Glen W. Baxter from Japanese. *HJAS* 18: 124–141.
Young, Percy M. 1968. *A History of British Football*. London: Stanley Paul.
Yü Ying-shih. 1977. "Han China." In Chang Kwang-chih (1977c: 55–83).
Zachariae, Theodor. 1897. *Die Indischen Wörterbücher (Kośa)*. Strassburg: Trübner.
Zi, Etienne. 1894. *Pratique des examens littéraires en Chine*. Variétés Sinologiques no. 5. Shanghai: Imprimerie de la Mission Catholique.
Zilsel, Edgar. 1942. "The Genesis of the Concept of Physical Law." *Philosophical Review* 51: 245–279.
Zürcher, Erik. 1959. *The Buddhist Conquest of China, the Spread and Adaptation of Buddhism in Early Medieval China*. Leiden: E. J. Brill.
———. 1980. "Buddhist Influence on Early Taoism." *TP* 66: 84–117.
———. 1981. "Eschatology and Messianism in Early Chinese Buddhism." In *Leyden Studies in Sinology*. Ed. W. L. Edema. Pp. 34–56. Leiden: E. J. Brill.

INDEX

This index includes entries as cited in the text, not necessarily as they are listed in the Glossary and Bibliography, and omits persons, places, and terms the author deems to be of less general importance. "Chinese Original Sources" also found in the Bibliography are indexed usually either by author or by title. "Secondary Sources" are indexed only by author. Because the main focus of this book is China, citations such as "Buddhism" or "Organicism" should be understood as referring to China unless further qualifying words are used, for example: "Buddhism: Tibet, Mongolia, Japan," or "Organicism: Aristotelian." Terms and names such as "Confucianism," "Taoism," or "Chu Hsi" occur so ubiquitously that they have been indexed selectively, not exhaustively.

ABOUT THE AUTHOR

Derk Bodde is professor emeritus of Chinese Studies at the University of Pennsylvania. He holds degrees from Harvard College and the University of Leiden, Holland. During his long and distinguished career he has been a Harvard-Yenching Travelling Fellow, Guggenheim Fellow, National Endowment for the Humanities Senior Fellow at Cambridge University, and he was awarded the Fulbright Senior Research Fellowship that inaugurated the Fulbright Program in 1948.

He is a member of the American Philosophical Society and the American Academy of Arts and Sciences and a past president of the American Oriental Society. Dr. Bodde has received the Distinguished Scholar Award from the Association for Asian Studies.

Among his many publications are the two-volume translation from the Chinese of *A History of Chinese Philosophy,* by Fung Yu-lan, *Peking Diary: A Year of Revolution, China's First Unifier,* and *Festivals in Classical China.* He coauthored *Tolstoy and China,* and *Law in Imperial China.*